A TEXT BOOK OF GENETICS

A TEXTBOOK OF GENETICS

By

H. S. Bhamrah
C. M. Chaturvedi

ANMOL PUBLICATIONS PVT LTD
NEW DELHI-110 002 (INDIA)

ANMOL PUBLICATIONS PVT LTD
4374/4B, Ansari Road, Daryaganj
New Delhi-110 002

A Textbook of Genetics

Second Revised Edition 1997

ISBN 81-7488-481-5
Reprint, 2001

PRINTED IN INDIA

Published by J. L. Kumar for Anmol Publications Pvt Ltd, New Delhi and Printed at Mehra Offset Press, Delhi.

Contents

Contents

Preface

The science of genetics has a short history of development as compared to other fundamental sciences. It has expanded so rapidly in its few decades of existence that there is a wealth of material to draw on. For the sake of simple and direct presentation of the basic principles, examples have been carefully chosen. The title covers all areas of genetics, emphasizing current trends of thoughts and techniques. The subject matter has been treated in the sequence in which discoveries in the science of genetics were made, commencing with Mendelian genetics, covering modern and molecular genetics, and culminating with applied genetics.

As a comprehensive introduction to the molecular biology of the gene, this book assumes no prior knowledge and is upto date with current research. We hope that this overview will make the subject more accessible and readily allow readers to proceed to more advanced work.

The authors are grateful to all their friends and colleagues, whose continuous inspiration have initiated them to bring out this book.

Constructive suggestions from the readers will always be welcomed and entertained whole heartedly.

Authors

1

Mendel's Law of Heredity

Gregor Johann Mendel was the first man to conduct decisive experiments in heredity and to formulate the basic laws of genetics. He is appropriately called the *Father of Genetics* and his conclusion constitute the foundation of this branch of Science. *Mendel* was born on July 22, 1822 in Heinzendort, a village in the Austro Hungarian empire (now Czechoslovakia). Second son of a poor farmer and fruit grower, he entered the Augustinian monastery at Brunn in Austria in 1843 after graduation. He became a monk in 1947 on his twenty-fifth birthday. In 1851 he joined the University of Vienna to study Natural Science after completion of which he was appointed as a substitute teacher in Physics and Natural history in the preparatory school at Brunn in 1854. He continued at this post for about 14 years. He utilized his spare time and vacation for conducting experiments on plant hybridization. He begin his experiments in a small monastry garden, on the garden pea, *Pisum sativum* in 1856 and continued till 1864. The results of these classic experiments and Mendel's conclusion were published in the Annual proceedings of the *Natural History History Society of Brunn,* 1865, which appeared in 1866. The proceedings were distributed to more than 120 other Societies, Universities and Academies in different countries. Unfortunately, no attention was paid to the Mendelian work, and therefore, remained unknown for a considerable period probably because the biologists of that time were busy in removing the confusions and controversies which urgent have been arisen by the publication of Darwin's *"Origin of Species."* In 1900 it was rediscovered independently by three botanists: *Hugo de Vries* of Holland, *Von Tschemark* of Austria, and *Correns* of Germany. They realized that because Mendel's results were published nearly 34 years before their own, he had priority claims to recognition for the discovery and so they named these principles of heredity after him. The work of these botanists and later on of several others showed that Mendel's principles applied not only to the garden peas but to other plants and to animals as well. This led to the establishment of genetics as a science.

Mendel's Experiments

Mendel grew in his garden 34 different varieties of garden peas *(Pisum sativum)* the seeds of which he had obtained from the local seedsman. He first made sure that every one of these varieties bred true to its type. For his experiments, he selected those varieties which contrasted sharply in certain well-defined and easily recognisable traits. For example, he selected a tall variety (6 to 7 ft) and a short one (9 to 18 inches), or a red flowering variety and a white flowering variety. Mendel adopted the following procedure in all his experiments. He crossed two plants which differed constantly in one or several characters and planted the seeds so obtained to produce offspring. The plants thus obtained from the crossing of two original pure strains were called *hybrids* by him. In genetical language a hybrid may be defined as an individual which results form the crossing of two individuals differing in at least one pair of characters. The process of crossing which results in the production of genetical hybrids is called *hybridization*. The hybrids, on attaining maturity, were allowed to self-fertilize to produce the second generation. Before proceeding further, the meaning of certain terms commonly used in the explanation of such experiments must be clearly understood. The plants of unlike character with which the hybridisation is first made constitute the *parental* or *P* generation and the progeny of hybrids obtained from the seeds of this cross is called the *first filial* or F_1 generation. When the F_1 hybrids are *self-fertilized* or *inhred,* i.e., bred amongst themselves, the progeny resulting therefrom is called the *Second filial* of F_2 generation. The subsequent generations resulting from self-fertilization are called F_3, F_4 and so on. When an F_1 hybrid is crossed with one of the *P* parents, the operation is called *back-crossing*.

Selection of Material

Mendel selected garden pea as his experimental material because it has the following characters:

1. *Variation:* The organisms which are to be chosen for the genetical experiments, should have a number of detectable differences and at a time only single detectable character should be considered.

2. *Reproduction:* The chosen organisms should be sexually reproducing (i.e., by fusion of male and female sex cells) because only then the offsprings will be able to receive different characters from both the male and female parents.

3. *Controlled Mating:* The chosen organisms should be able to mate in controlled or well-planned conditions. Because in genetical experiments sometimes we have to rear genetically pure parents by methods of controlled mating. One should maintain careful records of the offsprings of many generations.

4. *Short life cycle:* The chosen organisms should have very short life cycles.

5. *Large number of offsprings:* The organisms which have been chosen for the genetical experiments should produce large number of offsprings after each successive mating because it will help in deducing the correct conclusions.

6. *Convenience in handling.* The experimental species should be of a type that can be raised and maintained conveniently and inexpensively in the laboratory. For instance, the elephant will prove entirely useless material for genetical experiments than the *Drosophila,* pea plants, tomato, rats, guinea pigs, etc., which have been generally used and are still used in hybridization experiments.

Selection of Characters

Mendel selected seven pairs of contrasting characters for his experiments. He made artificial crosses between two plants showing at least a pair of contrasting characters. The members of each such pair of characters are called allelomorphs.

Crossing Techniques

Since garden pea is self-pollinating, a great care was taken during the experiments; (i) To prevent the possibility of self-fertilization, the anothers were removed from the plant taken as seed parents, even before the stigma was fully mature. (ii) The pollen from the pollen parent was transferred at the dehiscence stage and was dusted on the feathery stigma of those flowers from which anthers were removed *(emasculated)*. (iii) These cross pollinated flowers were enclosed in separate bags to avoid further deposition of pollen from other sources. (iv) Seeds were collected separately in marked bottles. In case of seed character, seeds were studied immediately but for other plant character, seeds were sown to raise the plants and the pod, flower and stem character were noted. (v) Even the reciprocal crosses were conducted but no change was observed in the expected ratio of offsprings.

TABLE 1.1

List of Seven Pairs of Contrasting Characters in Pea Plant

	Character	*Dominant*	*Recessive*
1.	Stem length	Tall	Dwarf
2.	Flower position	Axial	Terminal
3.	Pod shape	Inflated	Constricted
4.	Pod colour	Green	Yellow
5.	Seed shape	Round	Wrinkled
6.	Seed colour	Yellow	Green
7.	Seed coat colour	Grey	White

Symbol and Terminology

Symabolism is not the important part of genetics and no universal notation has been established. But symbols are useful for providing a language to describe important aspects of genetics. *Mendel* used letters of the alphabet as symbols for genes. A capital letter signified a dominant and a lowercase letter a recessive member of a pair of alleles. *Mendel* considered factors as abstract units, any one of which could be symbolized by A or B or some other letters. Consider that many genes are known and the several may be involved in a single series of experiments. To avoid confusion as to which gene is indicated, appropriate letter symbols are chosen to represent particular genes. The *mutant trait* that deviates from the "wild type" is usually chosen as the basis for the symbol. This trait usually is produced by the recessive allele, because most mutations occur as recessives. The dominant allele usually produces a functional product. A mutational change removes all or part of this product and leaves the recessive to come to expression only in the absence of a dominant allele. For example, the mutant vermilion is a recessive for eye colour in fruit flies. As compared with the dominant wild type, the colour production is diminished and the gene symbol is *v*. A few mutant alleles, such as the one of wrinkled wings in the fruit fly, however, are dominant and are therefore represented by a capital letter (e.g., W).

If the mutant is to provide the symbol, the history of the orgaism under investigation must be kown well enough tot suggest which member of the contrasting pair (e.g., tall or dwarf in peas) is the mutant trait. World collections of pea species show no dwarfs in natural populations. Dwarf peas occur only in certain cultivated stocks that have been developed. Dwarf is probably the recessive mutant *(d)*, and tall is the allele for "wild type" *(D)*. With the aid of these symbols, Mendel's experiment may be reconstructed in steps. Parents *(P)*, with either of the two alleles *(DD* or *dd)*, are represented as follows:

Tall parent		Dwarf parent
DD	X	*dd*

1. Phenotype and Genotype

As has already been seen, the F_2 dominants of any cross consist of two types of individuals, i.e., heterozygotes and homozygotes. A cross between a true breeding tall and a true breeding dwarf pea plant produces 3/4 tall and 1/4 dwarf plants in F_2 progeny. The tall plants are of two types, 1/4 of the total progeny of F_2 breed true and are homozygous tall, i.e. contain only the tall alleles *TT*, whereas 2/4 of the total F_2 progeny do not breed true and are heterozygous *(Tt)*, i.e. contain both the alleles, for tallness as well as dearfness The external appearance of all the tall plants is same, i.e. they have the same *phenotype*. Their genetic make-up or genotype is different, i.e. one group has a *TT* pair of alleles whereas the other group has *Tt* pair. F_2 dwarf pants have the genotype *tt*. All the genes possessed by an individual constitute its *genotype*.

The phenotype may be visible to the eye as in the case of tall and dwarf pants, the colour of the flowers, the colour of the skin and eyes in humans, etc. or it can be detected by special tests of identification, as the determination of blood types by serological tests. It may also be the expression of the genotype and environmental interaction. For example, Himalayan breed rabbits normally have black pigment at the tips of their feet, tail, nose and ears. When the rabbits are raised at very high temperatures, the enzyme (which is temperature-sensitive) produced by the gene for Himalayan colour pattern is inactivated. This results in loss of pigmentation and all white rabbits are produced. Here the genotype is the same but it produces different phenotypes in different environments (temperature differences).

The F_2 Mendelian ratio is represented as 3:1 phenotypically and 1 : 2 : 1 genotypically. The F_2 progency consists of two phenotypes but of three genotypes.

2. Allelomorphs or Alleles, Homozygous and Heterozygous

Alleles, the abbreviated form of the term "Allelomorphs" (meaning one form or the other) indicates alternative form of the same gene. For instance, in the above example *"T"* and *"t"* are two allelomorphs of the gene for plant height. In pure tall or pure dwarf plants, same allele is duplicated (*TT* or *tt*), while in hybrid tall both the alleles will be present *(Tt)*. An individual, having only one allele or in other words, two identical alleles, is known as homozygous *(TT* or *tt)*. Similarly, an individual, having two different alleles will be called heterozygous or hybrid *(Tt)*. In the present usage the term gene and the allele are interchangeable, but while gene can be used for any factor, allele is used with reference to another allele. For instance, while *"T"* and *"t"* are alleles to each other, they cannot be be allelic to any other gene.

3. Backcross and Testcross

The F_1 individuals obtained in a cross are usually selfed to get the F_2 progeny. They can also be crossed with one of the two parents from which they were derived. Such a cross of F_1 individual with either of the two parents is known as a "backcross." In such backcrosses, when F_1 is backrossed to the parent with dominant phenotype, no recessive individuals are obtained in the progeny. On the other hand, when it is crossed with recessive parent, both phenotypes appear in the progeny. While both these crosses are "backcrosses" only the cross with the recessive parent is known as a "testcross." It is called a testcross, because it is used to test whether an individual is homozygous (pure) or heterozygous (hybrid).

This can be illustrated with the help of the example of tall *(TT)* and dwarf (*tt*) plants. In the F_1 generation only tall plants *(Tt)* appear. These plants can be backcrossed with either of the two parents.

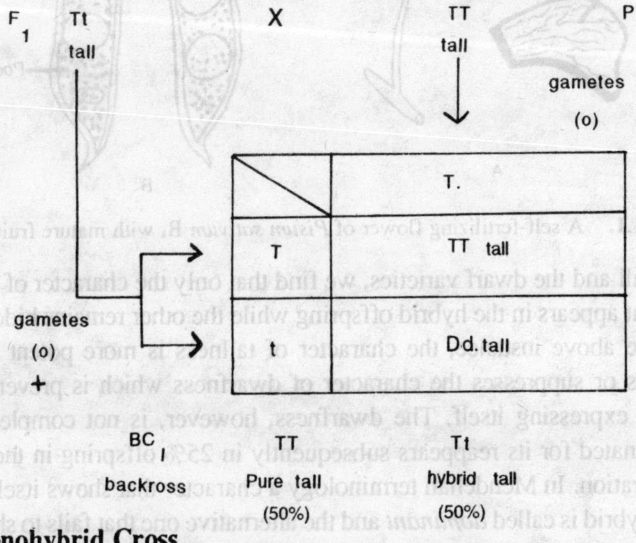

Monohybrid Cross

Monohybrid crosses are basic of Mendelian genetics. Pertinent information about genetic segregation as it occurs in monohybrid combination and discussed in the following pages. Such crosses may occur in all major groups of sexually reproducing organisms. Dominance is the major form of interaction between allele because the dominant alleles is usually the one that produces a functional product while the recessive allele does not. Therefore, the normal phenotype is produced if the dominant alleie is present.

Mendel gave some very good explanations for the observed results of his hybridization experiments. Reverting to the cross between

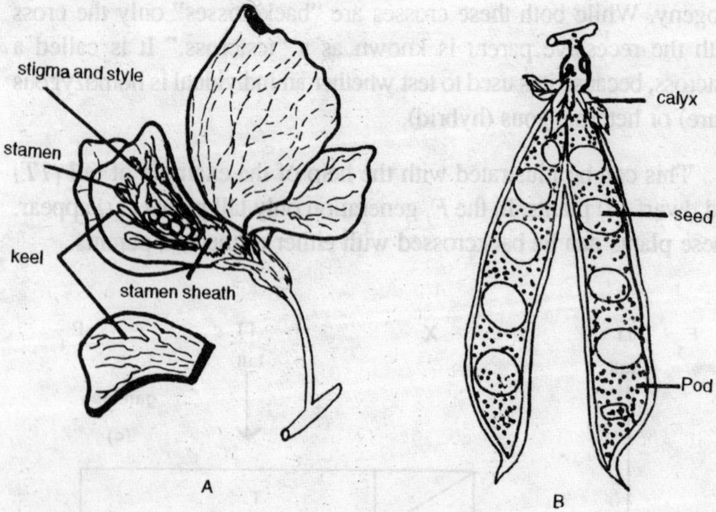

Fig. 1.1. A self-fertilizing flower of *Pisum sativum* B. with mature fruit.

the tall and the dwarf varieties, we find that only the character of one parent appears in the hybrid offspring while the other remains hidden. In the above instance, the character of tallness is more potent and marks or suppresses the character of dwarfness which is prevented from expressing itself. The dwarfness, however, is not completely eliminated for its reappears subsequently in 25% offspring in the F_2 generation. In Mendelian terminology a character that shows itself in the hybrid is called *dominant* and the alternative one that fails to show itself is called *recessive*. Thus, tallness is dominant and dwarfness is recessive. Similarly, in the cross between the red-flowering and white-flowering varieties, the F_1 hybrids are red-flowering because red flower colour is dominant over white flower colour.

Examples of Monohybrid Cross in Plants

Example 1. Seed Colour in Pea: When plants with yellow seeds were crossed with plants having green seeds, only yellow seeds were formed in F_1 generation. It means yellow seed colour is dominant over green. When F_1 plants were self-pollinated, the yellow and green seeds were obtained in the ratio of 3 : 1. Of the yellow seeds of F_2 generation, only one-third (1/3rd) gave rise to yellow seeds in the next generation (F_3 generation), while the remaining 2/3 produced yellow and green in the ratio of 3 : 1. The plants raised from F_2 green seeds were expected to produce green seeds only.

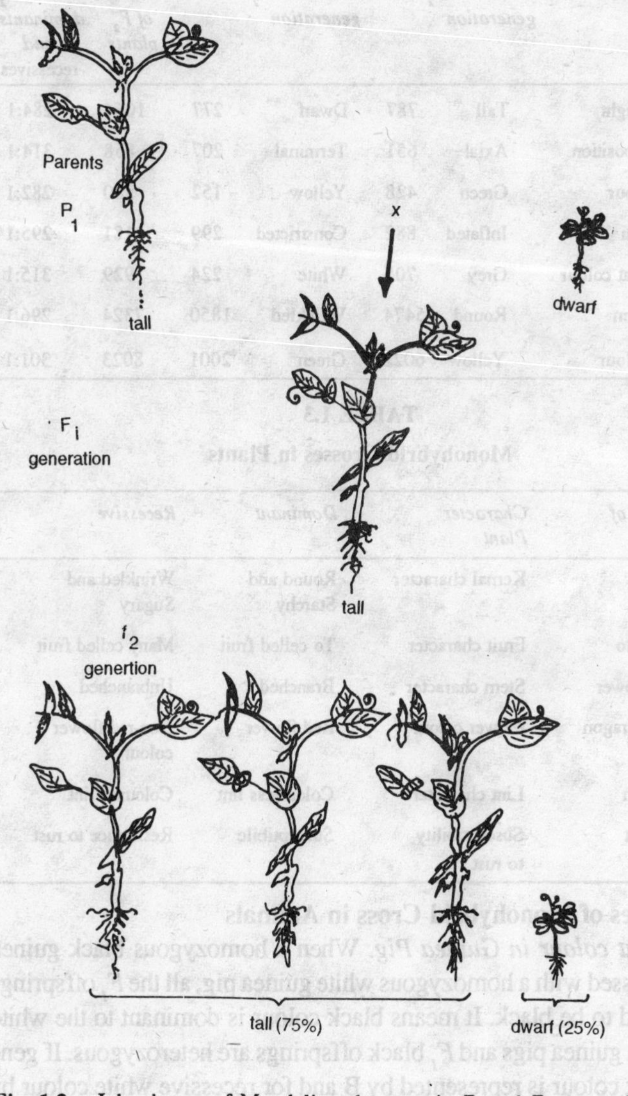

Fig. 1.2. Inhenitance of Mendelian character in F_1 and F_2 generations.

TABLE 1.2

List of Actual Results of Monohybrid Cross Obtained by Mendel for Different Characters in Garden Pea Plant

Character	*Dominant in F_2 generation*		*Recessive in F_2 generation*		*Total of F_2 plants*	*Ratio of F_2 dominants and* recessives
1. Plant height	Tall	787	Dwarf	277	1064	284:1
2. Flower position	Axial	651	Terminal	207	858	314:1
3. Pod colour	Green	428	Yellow	152	580	282:1
4. Pod form	Inflated	882	Constricted	299	1181	295:1
5. Seed coat colour	Grey	705	White	224	929	315:1
6. Seed form	Round	5474	Wrinkled	1850	7324	296:1
7. Seed colour	Yellow	6022	Green	2001	8023	301:1

TABLE 1.3

Monohybrid Crosses in Plants

S.No.	*Name of*	*Character Plant*	*Dominant*	*Recessive*
1.	Maize	Kernal character	Round and Starchy	Wrinkled and Sugary
2.	Tomato	Fruit character	To celled fruit	Many celled fruit
3.	Sunflower	Stem character	Branched	Unbranched
4.	Snapdragon	Flower colour	Red flower	Non-red flower colour
5.	Cotton	Lint character	Colourless lint	Coloured lint
6.	Wheat	Susceptibility to rust	Susceptibile	Resistance to rust

Examples of Monohybrid Cross in Animals

Coat colour in Guinea Pig. When a homozygous black guinea pit is crossed with a homozygous white guinea pig, all the F_1 offsprings are found to be black. It means black colour is dominant to the white colour in guinea pigs and F_1 black offsprings are heterozygous. If gene for black colour is represented by B and for recessive white colour by b the black guinea pigs of F_1 will be BB and white bb. When crossed, the heterozygous hybrid of F_1 hybrids produced ova and sperms, 50% of ova will carry gene B for blackness and 50% will carry the gene b for white. The same will be true for the sperms. Their union will result in 25% homozygous black (BB), 50% heterozygous black (Bb) and 25% homozygous white (bb) i.e. black and white in the ratio of 3 : 1.

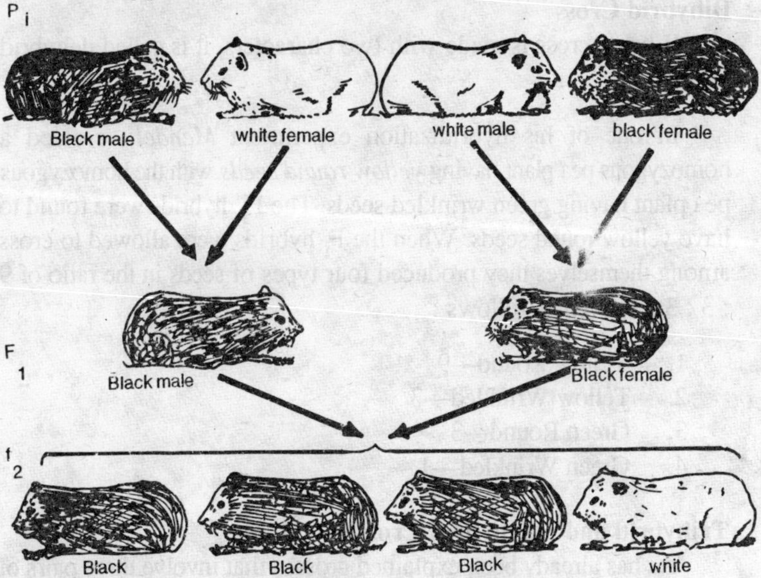

Fig. 1.3. A cross between black and white guinea pigs.

TABLE 1.4

Monohybrid Crosses in Animals

S.No.	Name of Animal	Character	Dominant	Recessive
1.	*Drosophila*	*(i) Eye colour*	*Red*	White
		(ii) Body colour	Grey	Black
		(iii) Wings	Normal	Vestigial
2.	Cat	(i) Skin colour	Tabby	Black or dark blue
		(ii) Length of body hair	Short	Long (angora hairs)

3.	Cattle	(i) Face colour	White	Coloured
		(ii) Horn	Hornless	Horned
4.	Dog	(i) Coat colour	Grey	Black
		(ii) Tail	Stumy	Normal

Dihybrid Cross

When a cross is made with two characters, it is called dehybrid cross.

In one of his hybridization experiment *Mendel* crossed a homozygous pea plant having *yellow round seeds* with the homozygous pea plant having green wrinkled seeds. The F_1 hybrids were found to have yellow round seeds. When the F_1 hybrids were allowed to cross among themselves they produced four types of seeds in the ratio of 9 : 3 : 3 : 1 given as follows :

1. Yellow Round—9
2. Tellow Wrinkled—3
3. Green Round—3
4. Green Wrinkled—1

Trihybrid and Polyhybrid Crosses

As has already been explained crosses that involve three pairs of contrasting characters are known as *trihybrid crosses.* If the characters involved are more than three, they are known as *polyhybrid crosses.* The F1 hybrids resulting from these crosses are therefore known as *trihybrids* and *polyhybrids* respectively. In a typical experiment *Mendel crossed* two varieties of garden pea plants differing in three pairs of contrasting characters viz., a tall pea plant with yellow round seeds (TT YY RR) with a dwarf possessing green wrinkled seeds (tt yy rr). The F1 offspring of such a cross will be a *trihybrid* and phenotypically it will be a tall plant with yellow round seeds (Tt Yy Rr). When these F1 hybrids are crossed among themselves each will form eight different kinds of gametes in equal numbers. Therefore, the possible combinations in the F2 will be 64. The nature of the gametes produced and he possible random combinations of these can be best illustrated by what is known as the *Forked-Line method.* In this method the entire trihybrid cross is first visualised as three monohybrid crosses i.e. Tt x Tt; Yy X Yy; Rr X Rr operating together. Since one member of each pair is dominant from each monohybrid cross a 3: 1 ratio will result Since the three pairs are independent, each monohybrid segregant may occur with any combination possible from any other pair of alleles. The combinations can be systematically arranged together. The 3 : 1 ratio from Tt X Tt may be combined with 3 : 1 ratios from each of the other monohybrid crosses namely Yy X Yy and Rr X Rr as indicated below. For the complete solution of such a combination usually the genotypes and phenotypes are necessary. The same forked-line system can be utililed to represent and combine genotypes that are expected from monohybrid crosses. From each monohybrid cross in the given example a genotypic frequency of 1 : 2 : 1 can be predicted and the three monohybrid units can be combined as shown in the diagram. This method is an alternative to the checker board method.

Tt Yy Rr x Tt Yy Rr

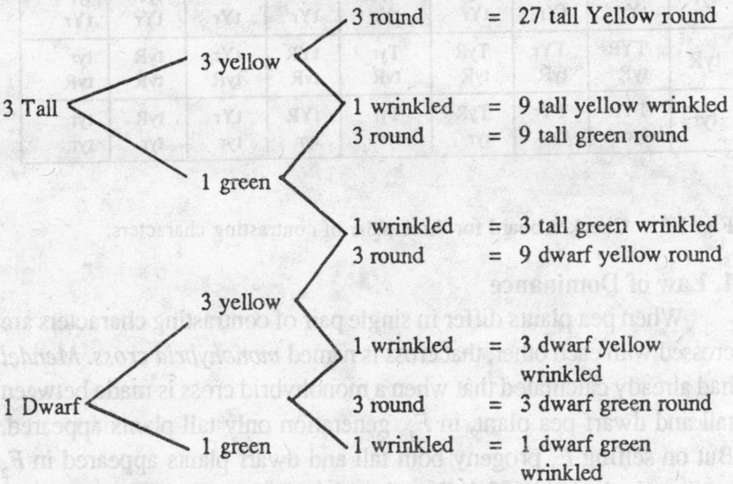

Fig. 1.4. Forked-line method showing the phenotypes of the progeny. For further details on Forked-Line method refer figures 10 and 11.

P_1 Generation	Tall Stem, Yellow and Round Seed TTYYRR	X	Short Stem, Green and Wrinkled Seed ttyyrr
F_1 Generation	Tall Stem, Yellow seed colour and Round Seeds Tt Yy Rr		

gametes

	TYR	TYr	TyR	Tyr	tYR	tYr	tyR	tyr
TRF	TYR TYR	TYr TYR	TyR TYR	Tyr TYR	tYR TYR	tYr TYR	tyR TYR	tyr TYR
Tyr	TYR TYr	TYr TYı	TyR TYr	Tyr TYr	tYR TYr	tYr TYr	tyR TY	tyr TYr
TyR	TYR TyR	TYr TyR	TyR TyR	Tyr TyR	tYR TyR	tYr TyR	tyR TyR	tyr TyR
TyR	TYR Tyr	TYr Tyr	TyR Tyr	Tyr Tyr	tYR Tyr	tYr Tyr	tyR Tyr	tyr Ty
tYr	TYR tYR	TYr tYR	TyR tYR	Tyr tYR	tYR tYR	tYr tYR	tyr tYR	tyr tYR
tyr	TYR tYr	TYr tYr	TyR tYr	Tyr tYr	tYR tYr	tYr tYr	tyR tYr	tyr tYr
tyR	TYR tyR	TYr tvR	TyR tyR	Tyr tvR	tYR tvR	tYr tvR	tyR tvR	tyr tvR
tyr	TYR tyr	TYr tyr	TyR tyr	Tyr tyr	tYR tyr	tYr tyr	tyR tyr	tyr tyr

Fig. 1.5. Checker board for three pairs of contrasting characters.

1. Law of Dominance

When pea plants differ in single pair of contrasting characters are crossed with each other, that cross is named *monohybrid cross. Mendel* had already calculated that when a monohybrid cross is made between tall and dwarf pea plant, in F_1 generation only tall plants appeared. But on selfing F_1 progeny both tall and dwarf plants appeared in F_2 progeny in the ratio of 3:1. From here it is evident that in F_1 generation, tallness conceals the dwarfness character so that only tallness appears. Thus that character which appears in F_1 generation is called as *dominant character*, and the character which remains unexpressed is called as *recessive charcer*. Thus according to Mendel, the law of dominance states that *on crossing homozygous organisms for a single pair of contrasting characters, only one character makes its appearance in F1 generation and is named as dominant character.*

Critical appreciation of Law of Dominance : A large number of workers conducted cross-breeding experiments to find out the applicability of law of dominance. The experiments were conducted by Correns on peas and maize, by Tschermak on peas, by Devries on maize etc., by Bateson and his collaborators on a variety of organisms, by Davenport on poultry, by Hurst on rabbits, by Toyama on silkmoth and by many others. These scientists observed that a large number of characters in various organisms are related as dominant and recessive. Seven pairs of characters in pea have already been mentioned. Some more characters are listed in the table given on the next page.

Practical importance of dominance. The fact that many hereditary traits show dominance and recessiveness is of practical importance. Let us illustrate this by an example from the inheritance of human beings. There is a form of idiocy due to a recessive gene. Let the dominant gene for normal be designated as *N* and its recessive allete for idiocy as *n*. A person who was hybrid for idiocy (Nn) would appear perfect'y normal. If he married a normal woman (NN), all his children would be apparently normal because each would get a dominant gene for normal from the mother. But half of the children would on an average be hybrids (Nn) as they would get the gene for idiocy from their father. If these hybrids in turn married pure normals, all the children again would be normal in appearance, though on an average half would be hybrids. Thus, the idiot gene might exist for several generations in hybrids without showing itself and the hybrid might look back over his family tree for many generations without finding an idiot ancestor. But if he married a hybrid woman of similar family history, he might have some idiot children, for both parents would now be hybrids (Nn X Nn). They would produce children in the ratio of 3 normals: 1 idiot (nn). Thus, hybrid parents, neither of whom suspected idiocy in himself (herself) or his (her) ancestors, might have an idiot child of the hereditary type under discussion.

Low of Dominance in Plants : Mendel and several other geneticists made many crosses to prove the validity of the law of dominance in plants some important characters are mentioned in the following table

TABLE 1.5

The Dominant and Recessive Characters in Plants

Name of plant	*Dominant*	*Recessive*
1. Nettle	Serrated leaves	Smooth margined leaves
2. Sunflower	Branched habit	Unbranched habit
3. Cotton	Coloured lint	White lint
4. Maize	Round starchy kernel	Wrinkle, sugary kernel
5. Snapdragon	Red flower	Non-red flower
6. Barley	Beardlessness	Beardness
7. Wheat	Susceptibility to rust	Immunity to rust
8. Tomato	Two celled fruit	Many celled fruit

Law of Dominance in Animals : The Mendelian law of dominance is equally applicable to the animals. It is well-studied in guinea pigs. When a cross is made between a homozygous block coated guinea pig and a homozygous brown/gray coated guinea, all black coated guinea pigs are produced in F_1. The black colour is due to a dominant gene and brown or gray is due to a recessing gene. On selfing F_1 black and brown coated offsprings are produced in a ratio of 3:1 as already shown in garden pea.

TABLE 1.6

The Dominant and Eecessive Characters in Animals

Name of animal	*Body character*	*Dominant*	*Recessive*
1. Cat	Skin colour	Tabby	Black or blue
	Length of hair	Short hairs	Long hairs (Angora)
2. Dog	Skin colour	Grey	Black
	Tall	Stumpy	Normal tall

3. Cattle	Colour of face Horn	White Polled or hornless	Coloured Horned
4. Horse	Skin colour Movement	Black Trotting	Red Pacing
5. Sheep	Hair or Wool or fleece	White	Black
6. Swine or pig	Skin colour Hoof	Black Uncleft	Red Normal
7. Salamander	Body colour	Dark	Light
8. Fruit fly (*Drosophila*)	Eye colour Wings Body colour	Red Flat and yellow Grey	White Curled and White Black
9. Land snail	Shape of shell	Unbanded shell	Banded shell

Mechanism of Dominance : Mendel carefully studied the results of his experiments and concluded that there is *something* present in sex cells (sperm or ova) which is the only connecting bridge between the parents and their offspring and this *something* is character controlling material which was named as *unit* or *factor* by *Mendel* himself. These Mendelian factors are now called as *genes*. So gene is the unit of heredity. Recent investigations have confirmed that genes are nothing but deoxyribonucleic acid (DNA) molecules which constitute the thread-like *chromosome*, alongwith RNA and other nucleoprotein.

In adult individual, the cells are diploid i.e. having two sets of chromosomes, coming from two different parents (male and female). The identical chromosomes are called the *homologous chromosomes*. Each contrasting character of the pair is represented by a single gene. If we represent tallness character by a gene T, then in a homozygous tall pea plant the genotype will be TT. Similarly on the other hand for homozygous dwarf pea plant, the genotype will be tt.

During the process of gametogenesis, the two chromosomes of a homozygous pair go to two different gametes. Thus each gamete will carry one gene i.e. either T or T. These gametes after fertilization produce a individual which will have both the genes of T and t for tall and dwarfness respectively. Thus in F_1 generation, the plants have two different gene, T and t and is named as heterozygous or hybrid individual. In this plant only one character which suppresses the character of dwarfism is *dominant* and other whose expression is suppressed is—the *recessive*.

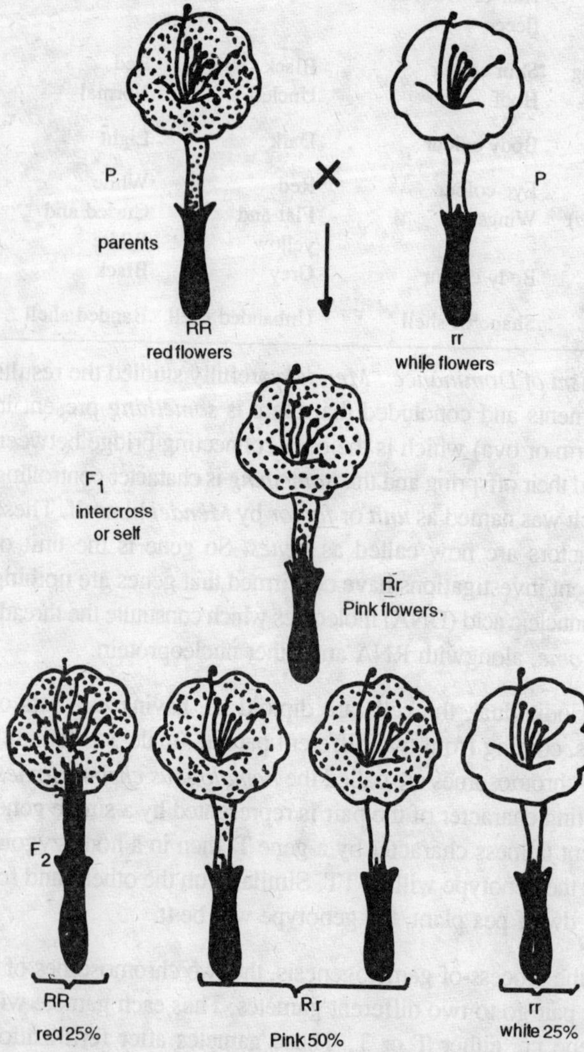

Fig. 1.6. Incomplete dominance in *Mirabilis jalapa*.

Incomplete Dominance : From the experiments conducted by *Mendel* it is found that when t‥ๅ alleles are brought together in gametic union one of them completely dominates over the other, manifesting itself in the hybrid. However, in some cases it is seen that the dominant gene fails to suppress the effect completely. An intermediate effect is developed. This is called the *incomplete dominance* or *parital dominance* or *intermediate inhintance* or *blending* inheritance.

Incomplete Dominance in Plants

This type of inheritance has been studied in the 4'O clock plant *Mirabilis jalapa* by *Correns*. In this plant there are two distinctive types of flowers colours namely the Red and the White. Both types breed true to the character when allowed to self-fertilize. When a pure red flowered (RR) 4'O clock plant is crossed with a pure white flowered (rr) one, the F_1 hybrids produce pink flower (Rr) a character which is intermediate between red and white coloured flowers of the parental generation. Thus an incomplete dominance was observed in F_1 hybrids. On self-fertilization the hybrid pink flowered plants produced in F_2 generation consisting of red flowered (RR), pink flowered (Rr) and white flowered (rr) individuals in a simple ratio of 1:2:1. Thus, the parental types segregated without any blending in the F_2 generation. The red varieties bred true to type and so too the white varieties. But the pink ones on self-fertilization again produced red flowered, pink flowered and white flowered plants in the ratio of 1:2:1 just like the F_1 hybrids. This proves the purity of the gametes formed in the hybrids and the observed ratio 1:2:1 agrees with Mendel's principle of random fertilization.

Incomplete Dominance in Animals

Incomplete dominance has also been demonstrated in breeding experiments on animals. When a pure breeding black *Andalusian fowl* is mated with a pure breeding splashed white one, the F_1 hybrids are all bluish in colour. When the bluish hybrids are crossed the resulting F_2 progeny consisted of one black, two blues and one splashed white, whereas according to the Mendelian law of dominance the offspring should consists of black and whites in the ratio of 3:1. The black individual is homozygous and hence breeds true to type. The blue ones are heterozygous and when mated among themselves again give rise to

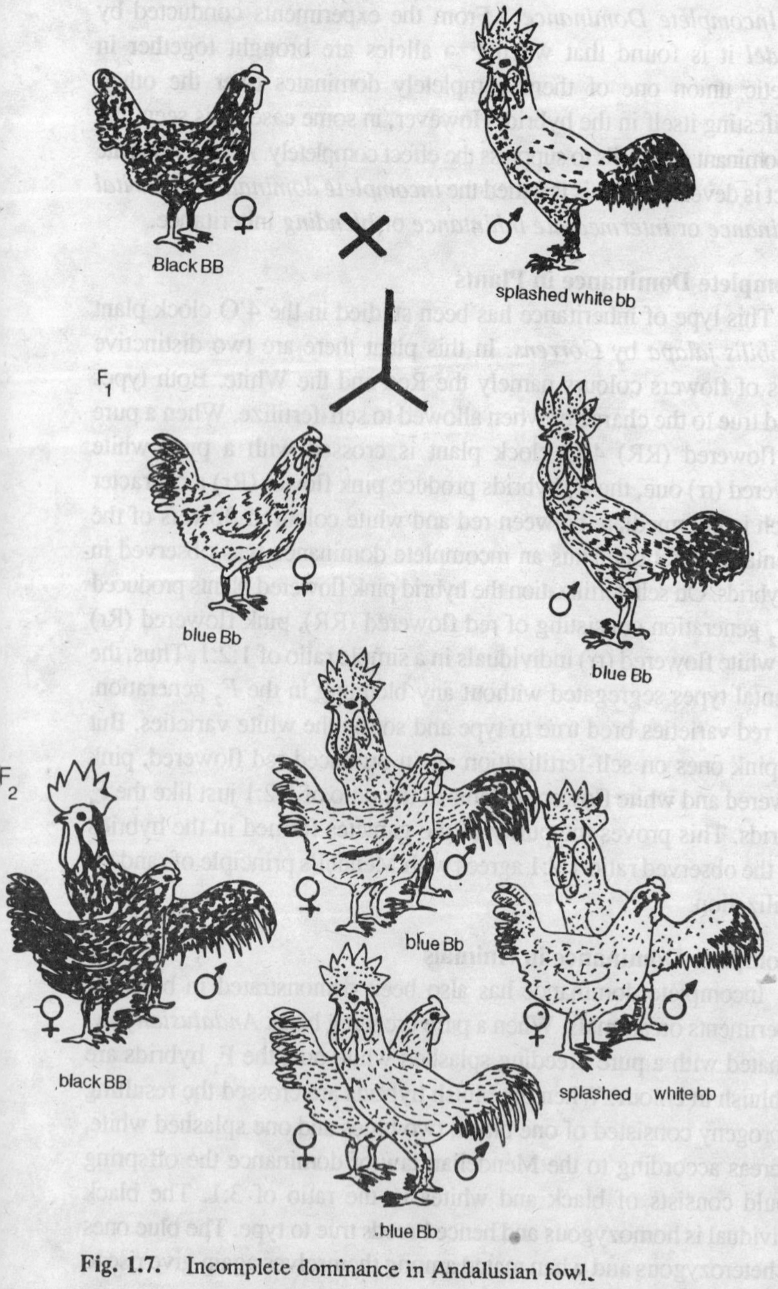

Fig. 1.7. Incomplete dominance in Andalusian fowl.

blacks, blues and splashed whites in the ratio of 1:2:1. The splashed whites which are homozygous and recessive breed true to type.

Codominance

When both alleles of a pair are fully expressed in a heterozygote, they are called *codominants*. Such alleles act in distinctive ways. In general, the two products are the same with respect to function but different in exact amino acid sequence. In humans, the ABO blood group antigens are a good example. Allele I^A for type blood is codominant with its allele I^2 for B-type blood. The heterozygote (I^AI^B) expresses the characteristics of both A and B antigens (AB-type blood). Since the two alleles control different protein products, a mating between a homozygous A-type person (I^AI^A) and a homozygous B-type person (I^BI^B) would result in all heterozygous (I^BI^B) offspring. Mating between heterzygotes (I^AI^B): X I^AI^B) would result in a ratio of 1 A-type (I^AI^A) : 2 AB type (I^AI^B) : I^B type (I^BI^B). A phenotypic ratio of 1:2:1 has thus replaced the 3:1 ratio because the alleles are codominant. In the ABO blood group, a third allele (i) occurs, and the homoygous arrangement (ii) produces O-type blood. This a good example of triple allelism in humans.

2. Law of Segregation (Purity of Gametes)

It is the first law of Mendel. This law states that the hybrids a heterozygotes of F_1 generation have two contrasting characters a allelomorphis of dominant and recessive nature. These characters remain together in F_1 for a long but do not mix with each other and segregate during gametogenesis and thus enter the different gametes. So each gamete receives only one character either dominant or recessive. It indicates the purity of gemetes.

To understand the idea of the law of segregation a monohybrid cross is taken. For example, we take a cross between a plea plant bearing axial flowers (AA) and a pea plant bearing terminal flowers (aa). If two pure varieties are crossed to each other, in F_1 generation, a heterozygous or hybrid plant is produced bearing the genotype as 'Aa'. On selfing the F_1 generation, in F_2 generation, the two varieties appear in the ratio of 3:1 i.e., 3 plants bearing axial flowers and one plant with terminal flowers. The appearance of terminal flower in F_2 generation indicates towards the process of generation.

Critical appreciation of law of segregation : The law of segregation of genes without their admixture in the hybrid was conception in original but purely theoretical, propagated by *Mendel*. It has since been confirmed by cytological studies. Dominance or no dominance, the law of segregation holds good to all cases. Its far reaching applicability has made it a rare biological generalization.

Mechanism of Segregation : The mechanism of segregation in above mentioned monohybrid cross between red and white flowered pea plant can be understood by assuming that the homozygous red flowered pea plants has the allele RR for redness and white flowered pea plant has the alleles rr for whiteness. The pea plant with RR alleles produces the gametes with single allele R and pea plant with rr alleles produces the gametes with the allele r. The gametes of both united to form a hybrid or heterozygote having the alleles Rr both for redness and whiteness. Due to the phenomenon of incomplete dominance the allele R for red colour partially expresses itself in hybrids of F_1, while the allele r for white colour remains latent or recessive. Both the allele R and r remain together for long time but they do not effect each other. Neither they mix nor they contaminate each other. Because a gametes can contain only one chromosome of a homologous pair, therefore, each gamete can carry single allele R or r. At the time of gametogenesis two types of gametes are produced by F_1 hybrids in equal numbers. Half of the gametes carry the allele R and other half carry the allele r. These gametes during the process of fertilization can unite in three possible combinations, viz., RR, Rr and rr, to produce three types of individuals in F_2 generation. Thus in F_2 75% individuals have coloured flowers and 25% white flowers. The appearance of white colour in F_2 generation indicates that in the hybrid the allete (r) for white colour remains alongwith allele (R) for red colour but do not mix with it or contaminated by it and it separates or segregates during gametogenesis.

Law of Segregation in Drosophila: T.H. Morgan described law of segregation in *Drosophila*. He crossed a homozygous long winged fly with homozygous vestigeal winged fly, in F_1 generation a heterozygous long winged fly is produced. On selfing F_1 two varieties of fly i.e. long winged and vestigeal winged flies are obtained in a ratio of 3:1 in F_2 generation.

3. Law of Independent Assortment

Mendel studied the law of independent assortment in dihybrid cases. In this he tried to find out how different characters would behave in relation to each other in their inheritance from generation to generation. For this purpose Mendel crossed two varieties of pea plants which were differing in two pairs of contrasting characters. The cross was made between plants having yellow and round cotyledons and plants having green and wrinkled colyledons. The F_1 hybrids all had yellow and round seeds. On selfing F_1 four types of plants are produced in the ratio of 9:3:3:1 as given below:

Yellow and round = 9

Yellow and wrinkled = 3

Green and round = 3

Green and wrinkled = 1

Critical appreciation of law of Independent Assortment. The law of independent assortment fails to have a universal applicability. Cytological studies have revealed that only those allelomorphs assort independently during meiosis, which are located in different homologous pairs. But, if the allelomorphs for different characters are present in the same homologous pair of chromosomes, these are passed on to the same gamete. Law of independent assortment does not apply to such cases.

Mechanism of Independent Assortment: To understand the mechanism involved in the independent assortment of characters assume that homozygous pea plant bearing yellow and round seeds is represented by RR YY, (RR for roundness and YY for yellow colour of the seeds) alleles, while homozygous plant with green and wrinkled seeds by rryy. During gametogenesis, RRYY plant will produce gamete with RY alleles and rryy plant will produce gamete with ry alleles with after fertilization will result into a hybrid plant with RrYy, alleles. This plant will produce yellow and round seeds because R dominates

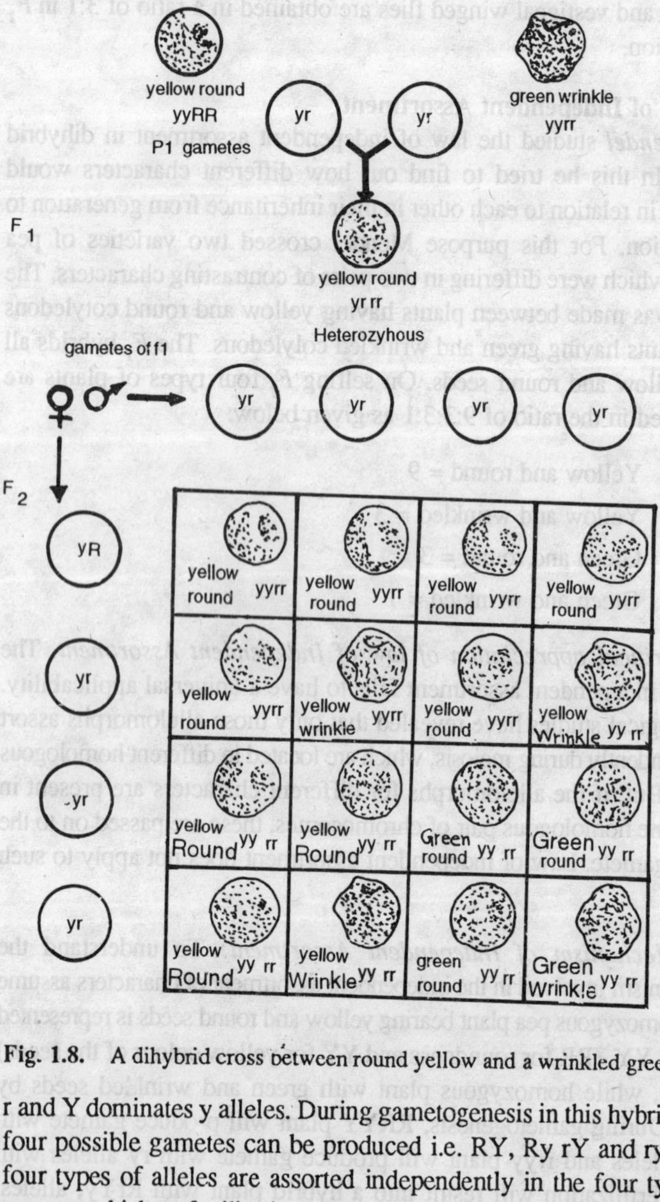

Fig. 1.8. A dihybrid cross between round yellow and a wrinkled green plant. r and Y dominates y alleles. During gametogenesis in this hybrid plant four possible gametes can be produced i.e. RY, Ry rY and ry. Thus four types of alleles are assorted independently in the four types of gametes which on fertilization can produce sixteen plants in F_2 generation as shown in the figure.

Out of 16 plants, 9 plants bear yellow round seeds, 3 yellow wrinkled seeds, 3 green round seeds, and one green wrinkled seeds in a ratio of 9:3:3:1. Thus it has proved the law of independent assortment and showed that each pair of contrasting characters behave independantly and bears permanent association or relation with particular character.

Deviation from Mendelian dominancy in dihybrid crosses: In the absence of complete dominance every genotype has a distinguishable phenotype. In *Snapdragons,* for example, a cross between plants with red flowers and broad leaves, (RRBB) and those with white flowers and narrow leaves (rrbb), results into F_1 progeny with pink flowers and intermediate leaves. F_2 progeny give a phenotypic or genotypic ratio of 1:2:1:2:4:2:1:2:1 which is the product [(1:2:1) X (1:2:1)] of two monohybrid ratio obtained in the case of incomplete dominance.

TABLE 1.7

Genotypes, Phenotypes and their Ratios in F_2 Progeny of a Cross Between Red-flowered, Bbroad-Leaved (RRBB) and White-flowered, Narrow-leaves (*rrbb*) Plants of *Antirrhinum* (snapdragon)

Genotypes	*Genotypic ratio*	*Phenotypes*	*Phenotypic ratio*
RRBB	1	Red-flowered, broad-leaved	1
RRBb	2	Red-flowered, intermediate-leaved	2
RRbb	1	Red-flowered, narrow-leaved	1
RrBB	2	Pink-flowered, broad-leaved	2
RrBb	4	Pink-flowered, intermediate-leaved	4
Rrbb	2	Pink-flowered, narrow-leaved	2
rrBB	1	White-flowered, broad-leaved	1
rrBb	2	White-flowered, intermediate-leaved	2
rrbb	1	White-flowered, narrow-leaved	1

Practical Application of Independent Segregation : A reference to the dihybrid ratio will show that the progeny of the F_1 dihybrid consists of four phenotypes. Two of these have the same combination of characters as the parents but two are of new types. The new types are produced as a result of recombination of the characters of the parents. The F_2 of a trihybrid cross shows six new combinations. Knowledge of the basic Mendelian principles tells us in advance the new combinations that will appear in the progeny of hybrids and enables us to predict their frequency. This is of great importance to the plant and animal breeder. New types of plants having new combinations of useful characters that do not occur in nature can be produced by hybridisation and further breeding. In this way resistance to disease and pests, higher yield and improved quality in crop and ornamental plants, and fruit trees can be combined in one plant. In fact, on the basis of Mendel's two laws, the plant breeder can set before himself a goal and proceed to produce plants to order by recombinations.

Why Mendal Succeeded

Having considered at some length *Mendel*'s methods and his results, it is time to say a word as to the cause of his success in a field in which many people before him had ventured but worked. In evaluating the work of *Mendel* it should be remembered that he was not a biologist but a monk who had some training in mathematics and physics.

In their hybridisation experiments, Mendel's predecessors had considered the organism as a whole and studied the inheritance of all the characters at once. This caused confusion and made matters complicated. *Mendel* simplified the problem and studied the inheritance of easily recognisable single characters at a time. Having done that, he studied the behaviour of two characters together at a time. In all his experiments. *Mendel* carefully counted the numbers of each type of offspring and kept statistical records of successive generations with the accuracy of a mathematician. He thus knew the pedigree of his plants. As a result of this technique, the phenomenona of inheritance were reduced to a mathematical problem. His clear and critical mind was a great asset in solving this problem. In this, perhaps, his training in mathematics was also helpful. More important than these, factors was that dame luck was smiling upon *Mendel*. It was fortuitos stroke of good luck that he selected a self-fertilizing annual plant garden pea as a material for his experiments. On account of repeated self-fertilization several distinct and true breeding cultivated varieties (or pure lines) had developed which differed from one another in a few characters only. The selection of the characters with which he worked was equally fortunate as they showed complete dominance. The flowers of pea are normally self-pollinated. Self-pollination could be prevented by opening the covering petals of the flowers and removing the stamens before the anthers had ripened and burst. The stigma of this emasculated flower could then be pollinated by pollen from another plant which was desired to be used as the male parent in a cross. Mendel adopted this very procedure in his experiments on hybridisation. He also took care to cover the flowers thus pollinated to prevent contamination by pollen of unknown origin by insects. To obtain the F_2 generation all he had to do was to allow the flowers to self-pollinate.

SUMMARY OF MENDEL'S IDEAS

Mendel's work and his ideas on heredity may be summarised as follows :

(1) Traits appear in two alternative forms such as tall stem and dwarf stem, coloured flowers or white flowers, round seeds or wrinkled seeds, etc. The unlike or contrasting forms of a characters are now called *allelomorphs* or *alleles*.

(2) Every trait in an organism is determined by a single pair of hereditary units, which are now called *genes*.

(3) One member of each contrasting pair, the *dominant*, could completely mask the other or *recessive* trait in a cross.

(4) When an organism which contains both members of an allelic pair forms gametes, the two members of a pair do not fuse or mix but separate or *segregate* from each other, so that each gamete carries only one of them. In other words, a gamete is *pure* for a character.

(5) Every character is inherited independently of every other character. In other words, the different maternal and paternal characters (or their genes) do not always stay together in the

offspring of the hybrid but enter into new combination in the gametes. The result is that in each generation a random *recombination* of characters or their genes takes place. The recombination is a matter of pure chance, so that when there are many offsprings every possible combination of characters is produced.

The principles of *segregation* and *recombination* were *Mendel's* original contribution to understanding the phenomena of heredity and have been repeatedly confirmed and extended by the work of several investigators on a great variety of plants and animals. The principles of segregation and independent recombination are the foundations of Mendelian heredity.

REVIEW QUESTIONS

1. In certain breeds of chickens, the factor (B) is responsible for the production of black feathers and its allele (b) produces feathers that are basically white (except for flecks of black.) The heterozygote is blue-feathered. Another factor (F) produces straight feathers, whereas its allele (f) results in the frizzled condition, giving brittle feathers. The heterozygote is mildly frizzled. What is to be expected from the following crosses? Give the phenotypic ratio.

A. A black, frizzled hen x a white, straight rooster.

B. A white, frizzled hen x a blue, straight rooster.

C. A blue, mildly frizzled hen x a white frizzled rooster.

D. A blue, mildly frizzled hen x a blue, mildly frizzled rooster.

2. In guinea pigs, short hair (L) is dominant to long (I). Black fur (A) is dominant to albino (a). A female from a strain which is pure breeding for black fur and short hair is mated to a male from a strain pure breeding for the albino condition and long hair.

A. What will be the phenotypes of the F_1?

B. If members of the F_1 are mated among themselves, what

percentage of offsprings can be expected to be homozygous for both traits? Give the genotypes and phenotypes of these homozygotes.

3. Write as much as possible of the following genotypes:

   A. A horned white bull.
   B. A hornless roan cow.
   C. A hornless red cow whose male parent was horned and roan.

4. In cattle, the alleles for red coat (R) and white coat (r) behave as codominants. Both red and white hair are produced so that the heterozygote is intermediate or roan coloured.

   A. Give the phenotypic and genotypic ratios to be expected among the offspring from a cross of two roan animals.
   B. What are the expected genotypic and phenotypic ratios from a cross of a roan animal and a white one.

5. In cattle, the hornless condition (H) is dominant to the horned condition (h).

   A. A horned bull is mated to a hornless cow that is heterozygous for the condition. What kinds of offspring are to be expected and in what ratio?
   B. If the cow is next mated to a hornless bull that is a heterozygote, what is the chance that the first calf will be horned?
   C. Assuming that the first calf born has horns, what is the chance that the second calf will be hornless?

6. Two short-haired female cats are mated to the same long-haired male. Several litters are produced. Female No. 1 produced eight short-haired and six long-haired kittens. Female No. 2 produced 24 short-haired ones and no long-haired. From these observations, what deductions can be made concerning hair length in heritance in these animals? Assuming the allelic pair S and s, give the likely genotypes of the two females cats and the male.

7. Using information from the above question, give the genotypic and phenotypic ratios to be expected in regard to fruit shape

from each of the following crosses:

A. Homozygous round x heterozygous round
B. Heterozygous round x heterozygous round
C. Oblong x oblong
D. Heterozygous round x oblong

8. In tomatoes, round fruit (O) is dominant to oblong fruit (o). Write as much as possible of the genotypes of:

A. A plant from a homozygous round-fruited stock.
B. A plant from an oblong strain.
C. A round-fruited plant which resulted from a testcross.
D. A oblong-fruited plant which from the cross of two round-fruited ones.

9. In the human, the allele for normal skin pigmentation (A) is dominant to (a) for the absence of pigment, the albino trait. A certain type of migraine headache (M) behaves as a dominant to the normal condition (m), absence of headache. Write as much as possible of the genotypes of the following three persons:

A. A phenotypically normal individual whose mother was an albino with migraine.
B. A person suffering from migraine who has normal skin pigmentation. Both parents have normal skin pigmentation, and one of them suffers from migraine.
C. An albino person who does not have headaches, whose normally pigmented parents both suffer from migraine.

10. A person who is dihybrid with respect to both the skin pigmentation and headache characteristics marries a person who is heterozygous at the skin pigmentation locus and who does not have headaches.

A. Give the genotype of each and the kinds of gametes each will produce.
B. What kinds of offspring are to be expected and in what phenotypic ratio?

11. The factors for normal hemoglobin (S) and for sickle-cell hemoglobin (s) may be considered codominants. A blood examination can detect the heterozygote, since a small percentage of the cells will show sickling. A man whose cells show no sickling upon examination marries a healthy woman who is found to have a certain percentage of sickle-cells.

A. What is the chance of their having a baby with the severe anemia?

B. The woman later marries a healthy man whose blood examination reveals sickle-cells. They have three children free of sickle-cell anemia. What is the chance that the next child will have the severe disorder?

C. What is the chance that any one of the healthy children is a carrier of the sickle-cell allele?

12. Assume that medical science finds a treatment for sickle-cell anemia. When instituted in infancy, it permits an otherwise doomed person to live a normal life span without any serious effects of the anemia. If two such persons, saved from the fatal effects of the disease, become parents, what will be the genotypes of their offspring? Would they require treatment? Explain. Also give an explanation which would be expected from a person who supports the Lamarckian concept.

13. In the human, free ear lobes (F) are dominant to attached lobes (f). Using this information and that on sickle-cell hemoglobin in Question 11, answer the following:

A. Give the genotypes of the following persons:

(1) A healthy woman with attached ear lobes whose red blood cells show some sickling.

(2 A man with free ear lobes whose mother also had free lobes but whose father had attached ones. This man's red blood cells show some sickling.

B. What kinds of offspring are to be expected if the man and woman marry, and in what ratio?

14. The fatal disorder, Huntington's disease, results in progressive nervous deterioration and death. It usually strikes well after the age of puberty so that afflicted persons may have produced offspring. The genetic factor responsible for the disorder is a dominant one (H), which is very rare in the population and does not seem to be on the increase.

A. Offer an explanation for the fact that more and more persons do not seem to be suffering from the disorder resulting from this dominant.

B. If a young person had one parent who died of the disorder and another who is apparently free of it, what does he or she know about chances of developing the disorder?

15. Give the number of phenotypic classes expected from each of the following:

A. Testcross of AaBbCC.
B. Testcross of AaBbCcDd.
C. Testcross of AaBbCcDdEE.

16. Give the number of phenotypic classes to be expected when the following crosses are made and when there is dominance:

A. Cross of two individuals who are AaBbCcDd.
B. Cross of two individuals who are AaBbCcDDEE.
C. Cross of two individuals who are AaBbCcDdEe.

17. Give the number of genotypic classes to be expected in A,B,and C of the above question.

In the pea plant, tallness (T) is dominant to shortness (t). Yellow seed colour is dominant (G) to green (g), and round seed shape (W) is dominant to wrinkled (w).

18. Show the different kinds of gametes which can be formed by individuals of the following genotypes: (1) TtGG. (2) TtGGWw, (3) TtGgWw.

19. What would be the expected phenotypic ratio if each of the above three plants were testcrossed?

20. Suppose plants (2) and (3) in Question 18 were crossed. What would be expected phenotypic ratio among the offspring?

21. Usually, though not invariably, human eye colour is inherited as if brown eyes were due to a dominant factor and blue to the corresponding recessive. Assume that this is correct in this and the following problems. A man with blue eyes marries a brown eyed woman whose mother had blue eyes. What proportion of the children would be expected to have blue eyes?

22. A brown eyed man marries a blue eyed woman. The first child is blue eyed. What is the man's genotype?
23. Which is more likely, two brown eyed parents producing a blue eyed child or two blue eyed parents producing a brown eyed child?
24. A brown eyed man marries a blue eyed woman and they have eight children, all with brown eyes. Can you be sure whether the man is homozygous or heterozygous? Which is more likely? If the ninth child has blue eyes will that settle the question?
25. What offspring would be expected from mating a roan cow with a roan bull? A roan with a white? A red with a white?
26. What would be the easiest way to determine whether a rose combed rooster is homozygous or heterozygous?
27. Which would be easier to go rid of in a flock of chickens, a recessive factor for single comb or a dominant factor for feathered legs?
28. In mice the genotype y/y is gray, Y/y is yellow, and Y/Y dies as small embryo. What offspring would be expected from a cross between a yellow and a gray? Between two yellows? In which cross would you expect the largest size of litter?
29. In man the gene for ability to taste phenyl-thio-carbamide (A) is dominant to the gene for inability (a). The brown eye gene (B) is dominant to the blue (b). What proportion of the offspring of two parents each of genotype A/a B/b would be blue eyed tasters? Blue eyed non-tasters?
30. What offspring would be expected from two parents of genotypes A/a B/b and a/a b/b?
31. What proportion of the children of two parents of genotypes A/a B/B and a/a b/b would be expected to be brown eyed tasters? Blue eyed tasters?

32. In Cocker Spaniels, the genotype A/-B/- is black, a/a B/- is liver, A/- b/b is red, and a/a b/b is lemon. A black Cocker is mated to a lemon one and produces a lemon pup. If this black dog is mated to another of his own genotype what proportions would be expected among the offspring?

33. In cattle polled is dominant over horned, and roan is the result of the heterozygous condition of the genes for red and white. What proportion of the offspring of a roan heterozygous polled bull and a roan, horned cow would be expected to be roan and horned?

34. How many kinds of gametes can be produced by an individual of genotype A/a B/b C/C d/d E/e F/F?

35. What proportion of the offspring of two parents, each of genotype A/a B/b C/c will be a/a b/b c/c ?

36. What proportion of the offspring of the cross A/a B/b C/c x A/a B/b C/c would be expected to be homozygous?

ANSWERS TO REVIEW QUESTIONS

1. A. All blue and mildly frizzled.

B. I blue mildly frizzled: 1 white, mildly frizzled.

C. 1 blue, mildly frizzled: 1 blue frizzled: 1 white, mildly frizzled: 1 white frizzled.

D. 1 black frizzled: 2 blue frizzled: 1 white frizzled: 2 black, mildly frizzled: 4 blue, mildly frizzled: 2 white, mildly frizzled: 1 black straight: 2 blue straight: 1 white straight.

2. A. All with black fur and short hair.

B. One out of four. BBLL—black short. BBll—black long. bbLL—albino short. bbll—albino long.

3. A. hhrr

B. H__Rr

C. HhRR

4. A. Red (RR), roan (Rr), white (rr) in a ratio of 1:2:1

B. One roan (Rr); one white (rr)

5. A. Horned and hornless in a ratio of 1:1
B. One-fourth
C. Three-fourths

6. Short hair is the dominant trait, Genotypes; female No. 1 is Ss; female No. 2 is SS. The male is ss.

7. A. 1 OO:1 Oo. All round
B. 1 OO: 2 Oo: 1 oo. 3 round: 1 oblong
C. oo. All oblong
D. 1 Oo: 1 oo. 1 round: 1 oblong

8. A. OO
B. oo
C. Oo
D. oo

9. A. Aamm
B. A—Mm
C. aamm

10. A. Parental genotype: AaMm. Gametes: AM, Am, aM, am
Parental genotype: Aamm. Gametes: Am, am.
B. Offspring: non-albino with migraine, non-albino without migraine, albino with migraine, albino without migraine in a ratio of 3:3:1:1

11. A. No chance
B. One-fourth
C. Two-thirds

12. All the children will be genotype "ss", homozygous for the defective allele, just as the parents are. The children would require treatment, since the environment has not altered the genetic material which is transmitted through the gametes. A Lamarckian would claim that the treatment has produced the normal condition and this trait can now be transmitted so that the children would be normal eventually regardless of treatment.

13. A. (1) ffSs (2) FfSs

B. Free, no sickling; free, some sickling; free, with anemia; attached, no sickling; attached, some sickling; attached with anemia. Ratio 1:2:1:2:1.

14. A. An allele does not increase in frequency simply because it is dominant. The frequency is the result of the interaction of many factors. Dominance by itself does not make an allele increase. Dominance means that a gene form will express itself in the presence of its allele.

B. The person has a 50:50 chance of developing the disorder. The afflicted parent is most likely genotype Hh, since the allele is so rare.

15. A. 4
B. 16
C. 16

16. A. 16
B. 8
C. 32

17. A. 81
B. 27
C. 243

18. (1) TG, tG
(2) TGW, TGw,tGW, tGw
(3) TGW, TgW, tGW, tgW,TGw, Tgw, tGw, tgw.

19. (1) 1 tall yellow: 1 dwarf yellow
(2) 1 tall yellow round: 1 tall yellow wrinkled: 1 dwarf yellow round: 1 dwarf yellow wrinkled.
(3) 1 tall yellow round: 1 tall green round: 1 dwarf yellow round: 1 dwarf green round: 1 tall yellow wrinkled: 1 tall green wrinkled: 1 dwarf yellow wrinkled: 1 dwarf green wrinkled.

20. 9 tall yellow round: 3 dwarf yellow round: 3 tall yellow wrinkled: 1 dwarf yellow wrinkled.

21. 1/2.

22. B/b.
23. Brown eyed parents with blue eyed child
24. No. More likely homozygous. Yes; in this case he would be heterozygous.
25. 1/2 roan, 1/4 red, 1/4 white, 1/2 roan, 1/2 white, All roan.
26. Cross with a single combed hen, and see if any single combed offspring appear. If so, he is heterozygous.
27. The dominant factor.
28. 1/2 yellow, 1/2 gray, 2/3 yellow, 1/3 gray. The first, since 1/4 of the embryos in the second cross would die.
29. 3/16. 1/16.
30. 1/4 blue eyed taster, 1/4 brown eyed taster, 1/4 blue non-taster, 1/4 brown eyed non-taster.
31. 1/2, None.
32. 9/16 black, 3/16 liver, 3/16 red, 1/16 lemon
33. 1/4.
34. 8.
35. 1/64.
36. 1/8.

2

Chromosomes

Chromosomes were first seen by *Hofmeister* (1848) in the pollen mother cells of *Tradescantia* in the form of darkly stained bodies. The term chromosome (Gr: *chrom* =colour; *soma*= body) was used by *Waldeyer* (1888) to designate their great affinity to basic dyes. Their functional significance was described by *W.S. Sutton* (1900) when he traced parallelism between segregation of chromosomes during meiosis and transmission of hereditary factors during gametogenesis. General reviews on the morphology of chromosomes have been published by *Heitz* (1935), *Kuwada* (1939), *Geitler* (1940) and *Kaufmann* (1948).

Chromosomes are the most significant components of the cell, particularly they are apparent during mitosis and meiosis. Their presence was demonstrated long before they were named "chromosomes" by *Waldeyer* in 1888. A chromosome can be considered as a nuclear componant having special organisation, individuality and function. It is capable of self-reproduction while maintaining its morphologic and physiologic properties through successive cell divisions.

Morphology

The morphology of chromosome can be best studied at the metaphase or anaphase of mitosis when they are present as definite organelles, being most condensed or coiled.

Number. The number of chromosomes in a given species is usually constant containing diploid number (2n) of chromosomes in their somatic cells and haploid (gametic or reduced) number (n) of chromosomes in their sex cell (sperms and ova). The number of chromosomes is variable from one to several hundreds among diferent species. For example, in *Ascaris megalocephala* it is 2, while in certain protozonans (*Aggregta*), there are more than 300 chromosomes; in *Paramecium* 30 to 40, in radiolarians as many as 1600, in *Hydra vulgaris* 32, *Musca domestica* 12, *Rana esculenta* 26, *Columba livia* 80, *Oryctolagus cuniculus* 44, *Gorilla gorilla* 48 and *Homo sapiens* (man) 46.

The chromosome numbers are also helpful for taxonomy. In the angiosperms the most frequent haploid number is 12 and members of this group have a range from 3 to 16. Similarly, in fungi, haploid number ranges from 3 to 8. In primates this haploid number is from 16 to 30. This haploid set of chromosomes present in the nucleus of gametes in generally called *genome* while in a diploid cell there will be two genomes. The diploid cells are the somatic cells in body. The diploid cells get the diploid set of chomosomes by the union of the haploid male and female gametes in the sexual reproduction.

TABLE 2.1

Showing the Diploid (2n) Numbers of Chromosomes in Some Important animals

Common name	*Zoological name*	*Chromosomes*
Man	*Homo sapiens*	48 or 46(?)
Rhesus monkey	*Maccaca mulatta*	48
Horse	*Equus caballus*	66 or 64 (?)
Wild horse	*Equus przewalshii*	66
Pig	*Sus scrofa*	40
Ox	*Boa taurus*	60
Sheep	*Ovis aries*	54
Cat	*Felis maniculata*	38
Dog	*Canis familiaris*	78
Goat	*Capra hircus*	60
Ass	*Equus asinus*	66
Rat	*Rattus rattus*	42
Domestic mouse	*Mus musculatus*	40
Guinea pig	*Cavia cobaya*	64
Rabbit	*Oryctolagus cuniculus*	44
Opossum	*Didelyphys virginiana*	22
Frog (Indian)	*Rana tigrina*	24

Frog (British)	*Rana esculanta*	26
Hyla (Tree frog)	*Hyla sp.*	24
Toad	*Bufo vulgaris*	22
Platyfish	*Platypoecilus maculatus*	48
Gold fish	*Carassius auratus*	94
Pigeon	*Columba livia*	79, 80
Turkey	*Meleagris gallopavo*	81, 82
Chicken	*Gallus domesticus*	77, 78
Duck	*Anas indicus*	80
Aulocantha (Protozoan)	*Aulocantha*	1600
Hydra	*Hydra vulgaris*	32
Ascaris	*Ascaris megalocephala*	2
House fly	*Musca domestica*	12
Honey bee	*Apis mellifica*	32, 16
Cricket	*Gryllus*	23, 24
Mosquito	*Culex sp.*	6
Lac insect	*Bombyx mori*	32
Hermit crab	*Eupagurus*	254

TABLE 2.2

Showing the Diploid Numbers of Chromosomes in Some Plants

Common name	*Botanical name*	*Chromosomes*
Yellow pine	*Pinus ponderosa*	24
Cabbage	*Brassica oleracea*	18
Radish	*Raphanus sativus*	18
Cucumber	*Cucumin sativus*	14
Cotton	*Gossipium hirsutum*	52
Plum	*Prunus domestica*	48
Cherry	*Prunus carasus*	32
Apple	*Malús sylvestris*	34, 51, 68
Pear	Pyrus comming	34, 51, 68
Pea (Garden)	*Pisum sativum*	14
Sweet pea	*Lathyrus odoratus*	14
Bean	*Phascolus vulgaris*	22
Scarlet	*Quercus coccinea*	24

Orange	*Citrus sinensis*	18, 27, 36
Sun flower	*Helanthus annuus*	34
Tobacco	*Nicotiana tobaccum*	48
Potato	*Solanum tuberosum*	18
Banana	*Musca paradisiace*	22, 44, 55, 77, 78
Cron	*Zea mays*	20
Onion	*Allium cepa*	16
Berley	*Hordeum vulgare*	14
Wheat (Summer)	*Triticum dicoccum*	28
Rye	*Secale cereale*	14
Rice	*Oryza sativa*	24

Size. Chromosomes range, on an average from 0.5 to about 30μ in length and from 0.2 to 3μ in diameter. The relative number of chromosomes generally differ in the nucleus but at a time all chromosomes of a cell may be of the same size. Plant cell normally posses larger chromosomes then animal cells. *Trillium* has chromosomes which may reach upto the length of 32μ at metaphase. Monocotyledon plants usually have larger chromosomes, than the dicotyledon which contain greater number of chromosomes. Among the animals, grasshoppers, crickets, mantids, newts and salamanders have large chromosomes.

Variation in size of the chromosomes can be induced by a number of environmental agents:

(1) Cells dividing at low temperature have shorter, more compact chromosomes than those dividing at high temperature.

(2) Colchicine is an alkaloid which interferes with spindle formation and cell division. It tends to shorten the chromosomes.

(3) Rapid and repeated division tend to result in smaller chromosomes. It appears that rate of cell division proceeds more rapidly than the formation of chromatin material as usual.

(4) In plants, the amount of phosphate in the nutritory medium has marked effect on the size of the chromosomes; high concentration gives larger chromosomes than those plants deficient of phosphates. Since the phosphate is an integral part of nucleic acid molecule, it

would seem that the amount of nucleic acid in the chromosome can be varied to give alterations in size.

Shape. The shape of the chromosomes is changeable from phase to phase in the continuous process of the cell growth and cell division. In the resting phase or interphase stage of the cell, the chromosomes occur in the form of thin, coiled, elastic and contractile, thread-like stainable structures, the *chromatin threads.* In the metaphase and the anaphase, the chromosomes become thick and filamentous. Each chromosome contains a clear zone, known as *centromere* or *kinetocore,* along their length. The centromere divides the chromosomes into two parts, each part is called *chromosome arm.* The position of centromere vary from chromosome to chromosome and it provides different shapes to the later which are following:

1. Telocentric. The rod-like chromosomes which have the centromere on the proxmal end are known as the telocentric chromosomes.

2. *Acrocentric.* The acrocentric chromosomes are also J-like in shape but these have the centromere at one end and thus giving a very short arm and an exceptionally long arm. The locusts (Acrididae) have the acrocentric chromosomes.

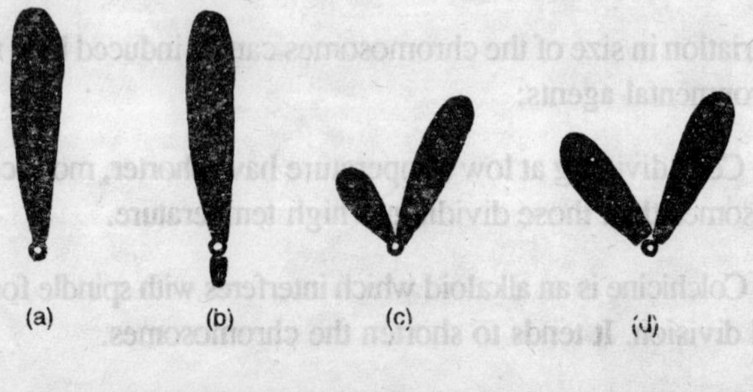

Fig. 2.1. Four morphologic types of chromosomes (a) Telocentric (b) Acrocentric (c) Submetacentric (d) Metacentric.

3. *Submetacentric.* The submetacentric chromosomes are L-shaped. In these, the centromere occurs near the centre or at medium portion of the chromosome and thus forming two unequal arms.

4. *Metacentric.* The metacentric chromosomes are V-shaped and in these chromosomes the centromere occurs in the centre and forming

two equal arms. The amphibians have metacentric chromosomes.

Structure of the Chromosome

In earlier light microscopic descriptions the chromosome, or chromatid, was though to consist of a coiled thread called the chromonema lying in a matrix. The chromosome was supposed to be covered by a membranous cellicle. Electron microscopic studies later showed that there is no definite membranous pellicle surrounding the chromosome. Other structures present in the chromosome include the chromatids, centromere, secondary constrictions, nucleolar organizers, telomeres and satellites as given under the following heads:

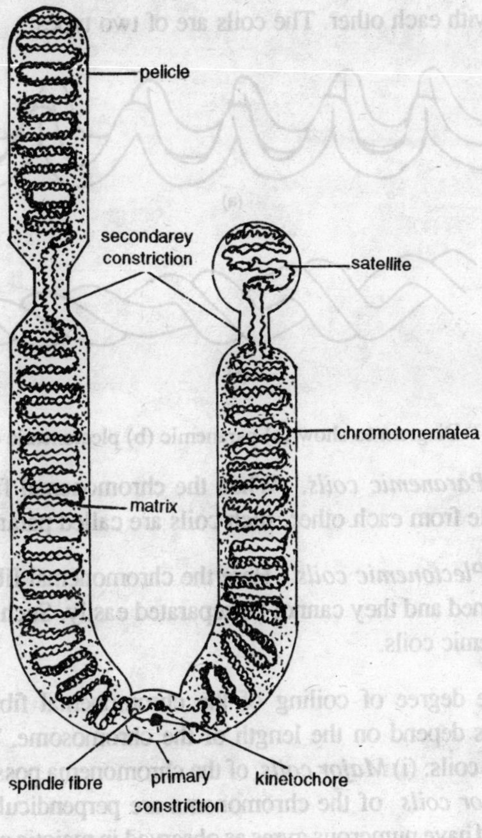

Fig. 2.2. Structure of a typical somatic chromosome at anaphase.

Chromatids. During metaphase a chromosome appears to possess two threads called *chromatids,* which become intertwined in the matrix of chromosome. These two chromatids are held together at a point along their length in the region of constriction of the chromosome. These chromatids are really spirally coiled *chromonemata* (sing., chromonema) at metaphase. The coiled filament was first of all observed by *Baranetzky*, in 1880, in the pollen mother cells of *Tradescantia* and was called chromonema by *Vejdovsky* in 1912. The chromonema may be composed of 2, 4 or more fibrils depending upon the species. This number of fibrils in the chromonema may depend on the diferent phases since at one phase it may contain one fibril and at other phase it may contain two or four fibrils. These fibrils of the chromonema are coiled with each other. The coils are of two types:

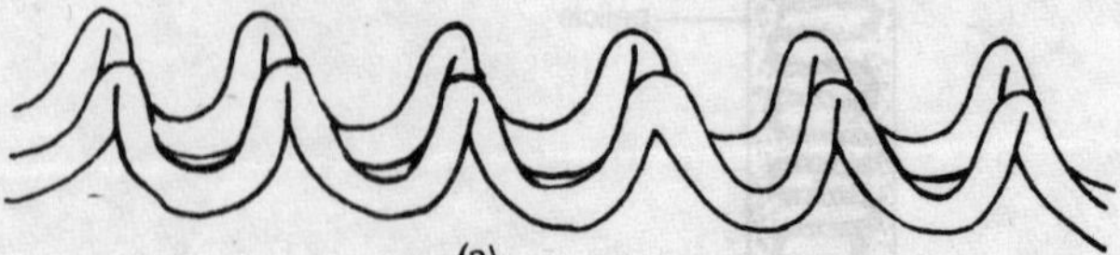

(a)

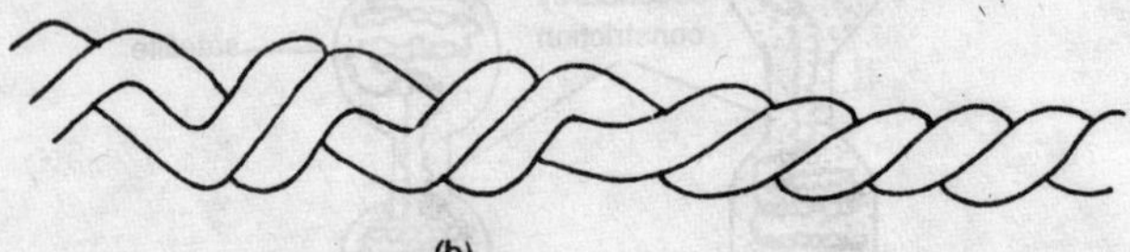

(b)

Fig. 2.3. Diagram to show (a) paranemic (b) plectonemic coils.

1. Paranemic coils. When the chromonemal fibrils are easily separable from each other, such coils are called paranemic coils.

2. Plectonemic coils. Here the chromonemal fibrils are closely intertwined and they cannot be separated easily. Such coils are called plectonemic coils.

The degree of coiling of the chromonemal fibrils during cell divisions depend on the length of the chromosome. There are three types of coils: (i) *Major coils* of the chromonema posses 10-30 gyres. (ii) *Minor coils* of the chromonema are perpendicular to the major coils and have numerous gyres as observed in meiotic chromosomes. If splitting has not yet occurred at this stage, there will be a single chromonema, if it has already taken place there would be two chromonemata. (iii) *Standard* or *somatic coils* are found in the chromonema of mitosis where chromonemata possess helical structures, resembling the major coils of the meiotic chromosome.

Chromomeres. The chromonema of thin chromosomes of mitotic and meitoc prophases have been found to contain alternating thick and thin regions and thus giving the appearance of a necklace in which several beads occur on a string. The thick or bead-like structures of the chromonema are known as the *chromomeres* and the thin region in between the chromomeres is termed as the *inter-chromomeres.* The position of the chromomeres in the chromonema is found to be constant for a given chromosome.

The cytologists have given various interpretations about the chromomeres. Some consider chromomeres as condensed nucleo-protein material, while other postulated that the chromomeres are regions of the super-imposed coils. The later view has been confirmed by the election microscopic observations. For long time most geneticists considered these chromomeres as *genes,* i.e., the units of heredity.

Centromere. It is an indispensable part of the chromosome and forms the *primary constriction* of metaphase. Without centromeres chromosomes are unable to orient themselves properly on the metaphase plate; come, later on, at anaphase and ultimately, are eleminated. As centromeres occupy a constant position, therefore, the centromeres are responsible for the shape of the chromosomes. Thus, the shape of the chromosomes is determined by the primary constriction, situated at the meeting point of the arms of the chromosomes. Within the primary constriction, there is a clear zone, having a small granule or spherule. This clear region is known as centromere (*Gr. meros,* part) or *kinetochore* or *kinetomere.* Its function is in terms of movement. They are responsible for the formation of chromosomal fibers in the spindle. The structure of the centromere is ovoid, non-stainable, with a large diameter, as in maize or may be like small granules or spherules, as in *Tadescantia.* In the centromere, there may be one or more small granules or spherules, called *chromomeres* and the spindle fibres.

Usually, each chromosome has only one centromere. In such cases, chromosome is called *monocentric*. There may be two i.e., *dicentric* or more *polycentric* or with a *diffused centromere,* as found in *Ascaris megalocephala* and Hemiptera.

After recent studies, it has been known that the centromere is made up of three zones present in duplicate. The middle zone maintain the relation of the chromosomes to the spindle. The diagram given below slows two sister-chromatids forming each metaphase-chromosome held by a region having a special division cycle.

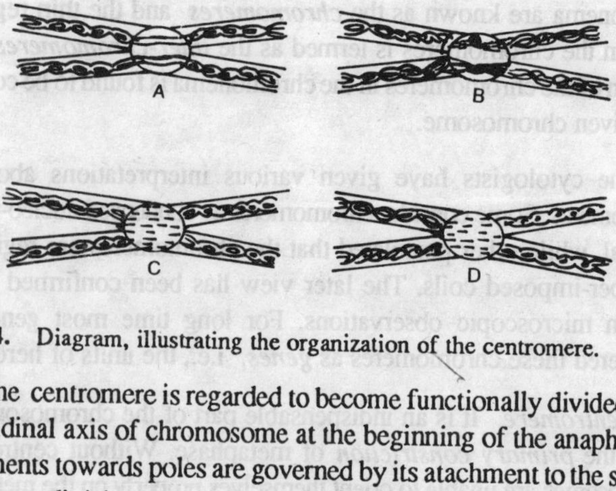

Fig. 2.4. Diagram, illustrating the organization of the centromere.

The centromere is regarded to become functionally divided to the longitudinal axis of chromosome at the beginning of the anaphase. Its movements towards poles are governed by its atachment to the spindle. Sometime, divisions occur also at right angles to the longitudinal axis forming two segment-centromeres, to which are attached the two chromatids of each arm. This structure, made up of two arms, is known a *isochromosome;* the name was suggested by *Darlingtion* in 1939. *Mc Clintock* (1932) has reported that such breakage is also possible by *x*-rays. In such cases, each fragment-part of the centromere is functional. It is also known that the centromere is a compound structure, whose parts are co-ordinated in division and movement. *Sachrader* (1936) and *Darlington* (1939) have suggested that the centromere can be considered to be homologous with the centrioles on structural as well as on theoretical basis.

Secondary Constriction. In addition to the primary constriction or centromere the arms of the chomosome may show one or more secondary constrictions (called *secondary consiriction II)*. These are different from nucleolar organizers (called secondary constriction I), although some cytologists also refer to the nucleolar organiser as the secondary constriction. The location of the secondary constriction II is constant for a particular chromosome, and is, therefore, useful for identification of chromosomes. It has been suggested that the secondary constrictions represent sites of breakage and subsequent fusion. In man secondary constrictions II are found on the long arms of chromosoms 1,10,13,16 and Y.

Nucleolar Organiser (secondary constriction I). Normally in each diploid set of chromosomes, two homologous chromosomes have additional 'constrictions' called nucleolar organisers. These are so-called because they are necessary for the formation of the nucleolus. The nucleolus is formed in the post-mitotic reconstruction phase. Under the light microscope the nucleolar organiser appears as a 'constriction' near one end of chromosome. The part of the chromosome beyond the nucleolar organiser is very short and appears like a sphere (*satellite*). In man chromosomes 13, 14, 15, 21, 22 and Y have nucleolar organisers and satellites. Chromosomes bearing satellites are called *SAT-chromosomes*. The prefix Sat-stands for 'Sine Acid Thymonucleionico' (without thymonucleic acid or DNA), since the chromosome on staining shows relative deficiency of DNA in the nucleolar organiser region. There are at least two SAT-chromosomes in each diploid nucleus.

Telomeres. The ends of a chromosome act differently from the interstitial portions. If an end or telomere is broken off either spontaneously or by induction, it is usually lost from the nucleus in subsequent cell division because it lacks centromere. The broken end of the remaining free chromosome is unstable and may unite with another broken end of the chromosome in the vicinity. However, the broken end will not unit with normal end. In meiotic prophase the telomeres are sometimes attracted to the centriole and seen to migrate to the nuclear membrane near the centriole. This behaviour results in what has been described as *Bouquet stage*.

Marix of Chromosome. As Supposed by some cytologists that

chromonemata are embedded in the chromatic matrix which is bounded by a pellicle. However, recent observation from electron microscopic studies revealed the absence of pellicle. Matrix is not defined as the main mass of the chromosome, which is characteristically Feulgen-positive. It can be removed by enzymatic means leaving behind a Feulgen-negative residual chromosome.

Heteropycnosis

It has generally been observed during various stages of mitosis that certain chromosomes or parts of the chromosomes are not as such, but are more condensed than rest of the karyotype. This refers to the *heteropycnosis.* This phenomenon results in clumping of the chromosomes during cell division. In animals pycnotic sets, on elimination, culminate in a male which was cytologically haploid. Heteropycnosis may be positive, followed by over condensation or negative, showing under or no condensation. It has been also observed that a particular part of chromosome or entire chromosome may not exhibit the condensation or heteropycnosis in all the phases. Since heteropycnosis is peculiarity of the heterochromatin, it helps in demarcating it from the euchromatin. It is much more prevalent in sex chromosome, though others too exhibit it.

Euchromatin and Heterochromatin

Although, during interphase, the chromatin of the chromosomes spreads out in the form of the fine threads of linin but at certain regions, the cromatin remains condensed in the form of darkley stained chromatin mass. These regions are known as the heterochromatin regions or heterochromatin.

Heterochromatin is of two types:

1. Facultative heterochromatin, and
2. Constitutive heterochromatin.

1. Facultative heterochromatin. This represents a temporary state of inactivation of chromatin in which one chromosome of the pair becomes partially or totally heterochromatic. For example, in mammals one of the two X-chromosomes in female somatic cells becomes heterochromatic and forms the sex-chromatin of *Barr body (Bar* and

Bertram, 1944). In male somatic cell, there is only one X chromosome and remains euchromatic (no barr body).

2. *Constitutive heterochromatin.* This type of heterochromatin presents a more permanent feature and is found in both the chromosomes of a pair. It is very often found in the centromeric regions, telomeres, in the regions of nucleolar organizers or as bands in other regions of the chromosomes. It is closely asociated with nucleoli in both plants and animals.

Chromosome Banding

T.C. Hsu and others (1969) introduced new methods for staining chromosomes by which distinct patterns of stained bands and lightly stained interbands became evident. These staining methods were very important since they permitted each chromosome to be identified uniquely, even if the over all morphology was identical. Distinctions can now be made among the relatively similar A-group chromosomes. For example, we may now say 1 or 2 or 3 chromosome instead of mearly a A-group chromosome of *Denver system.*

Followings are the methods for chromosome banding:

1. G-Banding. The most useful chromosome banding mathod is G-banding. This technique was developed by *Hsu* and *Arrighi.* It is observed that when the chromosomes are incubated in salive are stained with *Giemsa stain* or treated with urea or detergents. G-bands appear in the areas which are S-rich proteins. Giemsa stained preparations are more permanent and require ordinary microscope optics and illumination.

2. Q-Banding. This technique was developed by *Casperson.* It is observed, when the chromosomes are stained with *quinacrine mustrard* and observed through fluorescence - microscope, the regions of chromosomes rich in adanine and thymine get stained intensely. The guanine-cytosine regions remain unstained. These regions are called Q-bands. The defect of this staining is that, the stains fade after a short-time, moreover, special microscopic optics plus ultraviolet illumination are needed to see these bands.

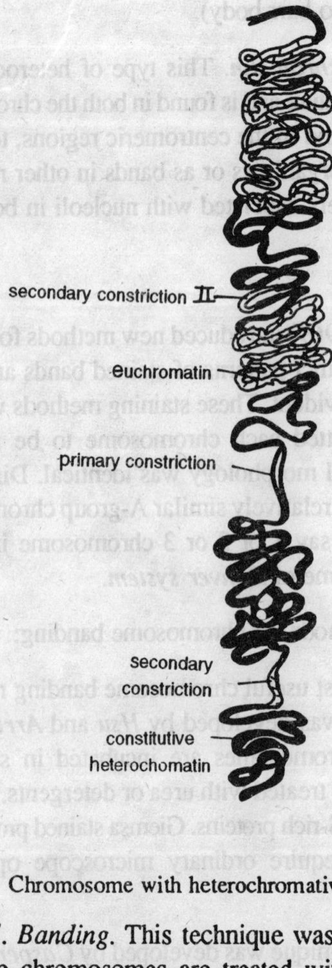

Fig. 2.5. Chromosome with heterochromative and enchromatin.

3. *C. Banding*. This technique was developed by *Pardue* and *Gall*. The chromosomes are treated with strong sodium hydroxide followed by warm saline and than stained with Giemsa stain. C-bands are especially evident around the centromere and in other chromosomes that contain substential amounts of highly repetitive constitutive hetero-chromatin.

4. R. Banding. These bands appear when chromosomes are incubated in a buffer at high temperature and stained with Giemsa stain. R-bands correspond to the regions on charomosomes having proteins lacking sulphur. These are receprocal of G-bands.

Bending technique of chromosome staining is highly useful in knowing various types of chromosomal aberrations such as deletion, duplication, inversion or translocation. The greater certainty of identifying whole chromosomes or parts of chromosomes by G-bands often allows the investigator to know exactly which chromosomes are present and which chromosome parts have undergone structural rearrangement.

Banding also provides a means to compare karyotypes of related species and to discribe differences that apparently have an evolutionary basis.

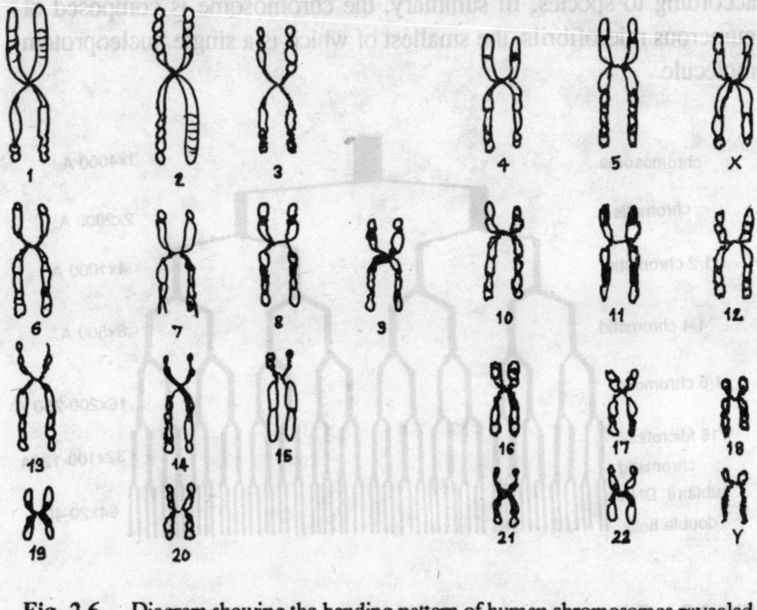

Fig. 2.6. Diagram showing the banding pattern of human chromosomes revealed with Giema stain.

ULTRASTRUCTURE OF CHROMOSOME

Two views have been proposed for ultra-structure of chromosomes:

(a) Multistranded view: This was proposed by *Ris* (1966). By

electron microscope the smallest visible unit of the chromosome is the fibril which is 100 A° in thickness. This fibril contains two DNA double helix molecules separated by a space of 25A° across and associated protein. Next largest unit is the half chromatid. The half chromatid consists of four 100 A° fibrils so that it is 400 A° in thickness and contains eight double helixes of DNA and the associated protein. Two half-chromatids from a complete chromatid consisting of 16 double DNA helix molecules. As the chromosome consists of two chromatids, thus total number of helixes will be 32 and dimeter 1600 A° thick before duplication or synthesis. After duplication chromosome has 64 double helixes of DNA with coresponding diameter of 3200 A°. The number of DNA helixes in each unit above fibril level varies according to species. In summary, the chromosome is composed of numerous microfibrils, the smallest of which is a single nucleoprotein molecule.

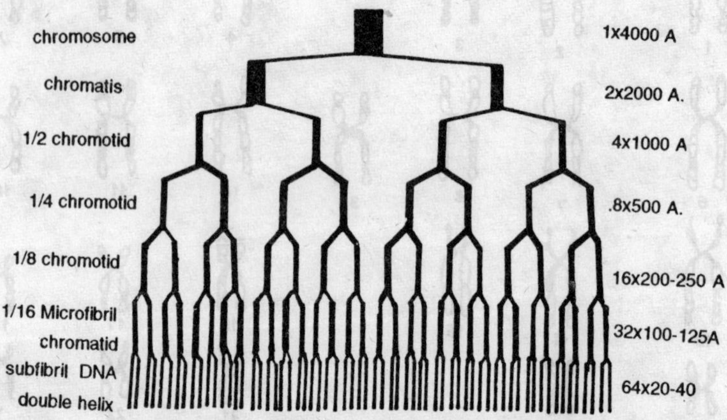

Fig. 2.7. Scheme of chromosome according to multiple strand hypothesis.

(b) Folded-Fibril Model. DuPraw (1965) presented this model for the fine structure of chromosme. According to this model, a chromosome consists of single long chain of DNA and protein forming what is called as fibril. The fibril is folded many times and irregularly entwinend to form the chromatid. This measure 250-300 A° in thickness.

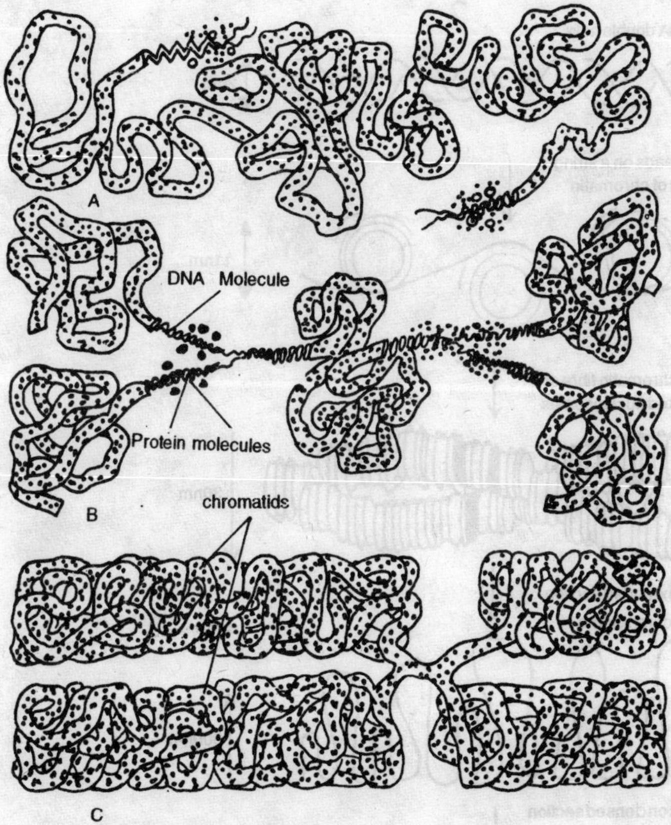

Fig. 2.8. DuPraw's folded fibre model of chromatin in interphase (A&B) and in Metaphase (C).

The Nucleosome sub-unit of chromatin. The chromatin is formed of repeating units, called *nucleosomes.* The term was given by *Oudet et al,* (1975). The nucleosome is made up of DNA and histone proteins. The proteins forms a *core particle* which is an octomer two molecules of each of the four histone proteins viz. H2a H2b, H3 and H4. The surface of core particle is surounded by 1.75 turns of DNA (200 base pairs). The DNA linking the core particle is called linker DNA. An another histone protein, H1, is bounded to the *linker DNA.* *(Kornberg* and *Thomas,* 1974). The core particle measures 40A° hight and 80 A° in diameter. The entire nucleosome measures 55A° in hight and 110A°in diameter.

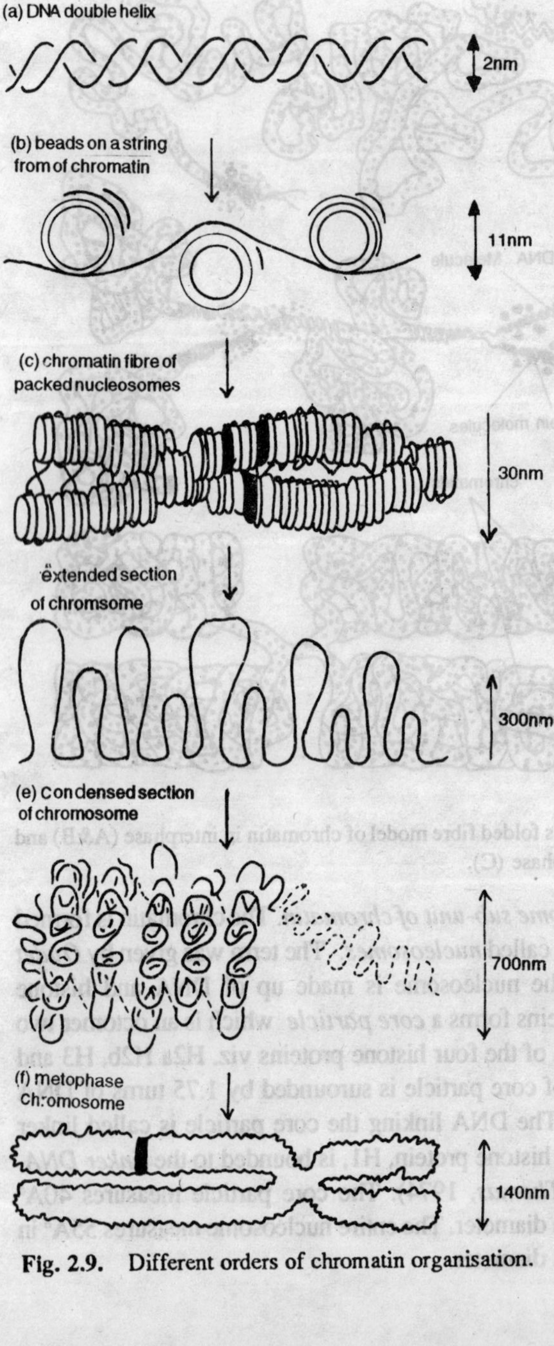

Fig. 2.9. Different orders of chromatin organisation.

Polytene Chromosomes

Balbiani in 1881 was the first to observe salivary gland chromosomes in the salivary glands of the *Chironomous larva.* This kind of giant chromosomes are strictly confined to certain type of somatic tissues in the insects belonging to the order Diptera. Usually they attain their largest size in the spherical nuclei of the larval salivary gland, but similar nuclei frequently exist in other tissues such as the lining cells of the gut and its derivatives, the Malpighian tubules as well as in muscles and fat cells, etc. Thus the term "polytene" chromosome of *Koller* is more preferable than the general term of salivary gland chromosomes.

Structure

The structure of the salivary gland chromosome is of great cytogenetic interest. Along the entire length of the chromosome there is a series of *dark* bands alternating into other clear zones called *interbands.* The dark bands stain intensly and are Feulgen positive. Further more, they absorb ultraviolet light at 600 A°. These bands may be considered as disks, since they occupy the whole diameter of the chromosome. They are of varying size. The longer bands have more complicated structure. They often form doublets, two bands located next to each other and of identical thickness and shape. The interbands are of fibrillar aspect, do not stain with basic dyes, are Feulgen negative and absorb very little ultraviolet light. Furthermore, they present a greater elasticity than the regions of the bands. The constancy in situation and distribution of the disks or bands in two homologous (paired) chromosomes is notable.

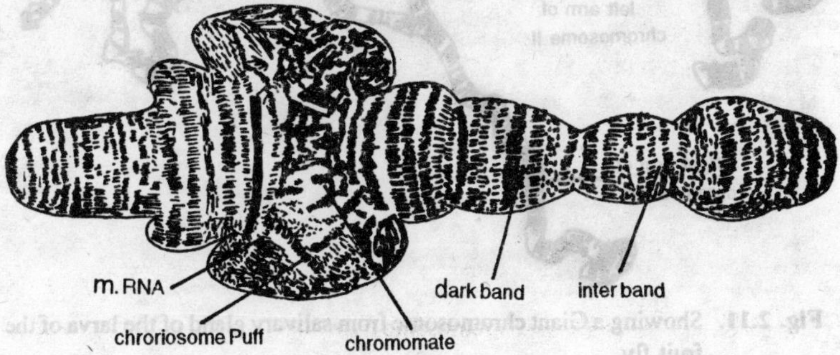

Fig. 2.10. Polytene chromosome of an insect

In the case of *Drosophila melanogaster*, the chromosomes of each polytene nucleus, when flattened, appear as fine long strands and one quite short, attached to a central mass known as the *chromocenter*, to which single lage *nucleolus* is also attached. The relationship between these strands are the light chromosomes of the ordinary mitotic set of this speices is not at first obvious. The explanation depends on two facts: (1) The two members of each pair of chromosomes are closely fused throughout their length; (2) The centromeres of all the chromosomes together with the heterochromatic segments adjacent to them are all joined together to form the chromocenter.

Thus, of the six strands the short one represent the two fused IVth chromosomes, and longer one represents the X chromosome, while the remaining four are the limbs of the 'V'-shaped second, and third chromosomes. In salivary gland nuclei from female larvae, the strand representing the 'X' is double, like the others, while in nuclei from male individual it is single. The V being quite small and almost completely included in chromocenter.

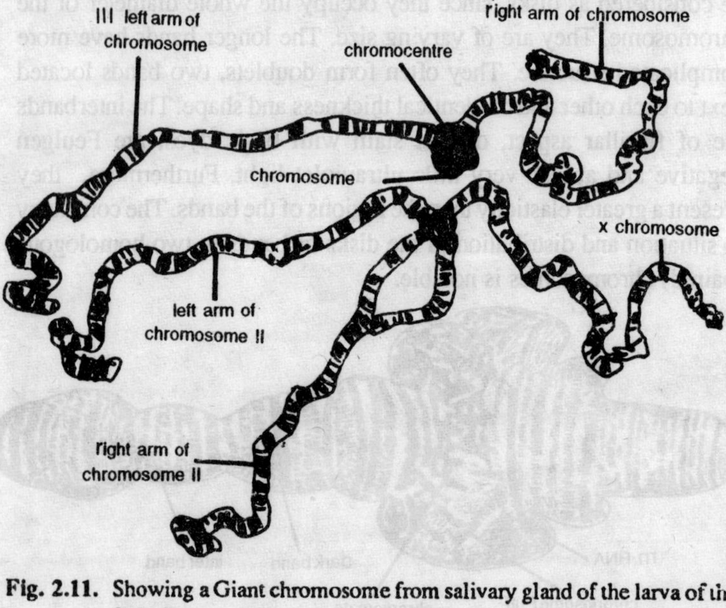

Fig. 2.11. Showing a Giant chromosome from salivary gland of the larva of the fruit fly.

Chromocenter occurs in all species of *Drosophila,* and its size depending on whether the proximal heterochromatic segments are extensive or not. In some other groups of Diptera, of the families 'Sciaridae' and 'Chromonidae' chromocenter is absent.

Puffs and Balbiani Ring and Gene activity. The most important morphological peculiarity of polytene gland chromosome is the presence of bands and interbands. *Brewer, Pavan, Beermann, McChelke* and others have found that at certain stages of larval development some specific bands of the polytene chromosome show an enlargement. These enlarged bands are considered as ultimate units of heredity—the *genes* at work. These active genes take the form of puffs scattered here and there along the salivary gland chromosomes. *Beermann* and *Clever* (1964) have found that the puffs produce RNA and the RNA made in one puff differs from the RNA of other puff. Observations of the puffs have shown the patterns of gene activity in several developing insects. It is also observed that certain hormones and other substances can start, stop and prevent some of these activities.

The fine structure of individual band can differ with respect to puffs that are in one location on a chromosome in one tissue and in another location on the same chromosome at other time or in another tissue. These localized modification in chromosome structure of various Diptera had been noted many years earlier but their possible significance was over-looked.

The coherence of chromosome filaments is loosened at the puffed regions. The loose ring always starts at a single band. In small puffs, a particular band simply loses its sharp contour and presents a diffuse, out of focus appearance in the microscope. At other loci or at other times a band may look as though it had "exposed" into a large ring or loop around the chromosomes. Such nut like structures are called *Balbiani-rings,* after *E.G. Balbiani,* who first described them in 1881; puffing is thought to be due to the unfolding or uncoiling of individual chromosome in a band. On observing that specific tissues and stages of development are characterized by definite puff pattern, *Beermann* (1952) postulated a particular sequence of puffs represents a corresponding pattern of gane activity. Of differential gene activation does in fact occur, one could predict that genes in a specific type of cell will regularly puff whereas the same gene in other tissue will not puff.

A gene of the same nature has been described in a group of four cells of salivary glands of *Chironomous*. *Chironomous pallidivittatus* produces a granular secretion. The closely related species *Chironomous tentatus* gives off a clear, non-granular secretion from the same cells. In hybrids of these two species this nature follows simple Mendelian laws of heredity. *Beermann* and *Clever* (1964) were able to localized the difference in a group of fewer that 10 bands in one of the chromosomes of *Chironomous* and that chromosome is designated as IV chromosome. The granule producing cells of *C. pallidivittatus* have a puff associated with this group of bands, a puff that is entirely absent at the corresponding loci of chromosome IV in *Chironomous tentatus*. In hybrids the puff appears only on the chromosome coming from the *C. pallidivittatus* parent; the hybrid produces a far smaller number of granules than the parent. Moreover, the size of the puff is positively co-related with the number of granules. This reveals quite clearly the association between the puff and a cellular (specific) product. This study demonstrates a specific relationship between puff-gene and the specific function of a cell.

Theories regarding the structure of polytene chromosome

There are three theories for explaining the structure of polytene chromosome. Out of them, the third explanation is actually the combination of first theories:

(1) Polytene chromosomes are the result of several cycle of intracellular chromosomal reproduction and consists of bundles of the folded ordinary chromosomes. This is the polytene theory sponsored by *Hertwig* (1935), *Cooper* (1938) *Painter* (1939) and *Beermann* (1952).

(2) Polytene chromosomes are paired chromosomes which have enormous length and width by addition or incorporation of extra material not present in ordinary chromosomes. This is the earlier alveolar concept of *Metz* (1935) and theory proposed by *Kodani* (1942) and *Darlington* (1949).

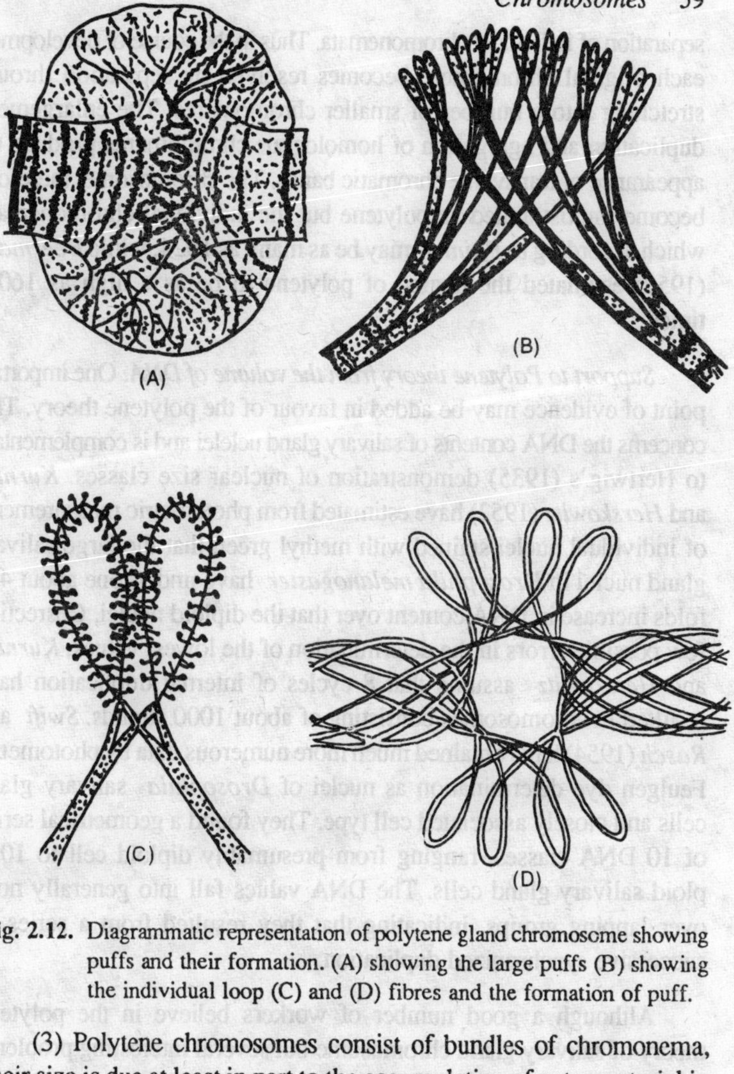

Fig. 2.12. Diagrammatic representation of polytene gland chromosome showing puffs and their formation. (A) showing the large puffs (B) showing the individual loop (C) and (D) fibres and the formation of puff.

(3) Polytene chromosomes consist of bundles of chromonema, their size is due at least in part to the accumulation of extra material in the center of chromosomes, and or to an actual growth in length of the chromonema (*Koltzof*, 1934; *Painter*, 1934; *Calvin et al.*, 1940; *Ris* and *Course*, 1954; *White*, 1945).

Polytene Theory. Painter (1941) believed that the increase in diameter is due to an enlargement and probably by a continuing duplication of the individual chromomeres but without a variable separation of individual chromonemata. Thus in the course of development each original chromomere becomes resolved by separation through stretching into a number of smaller chromomeres. The enlargement, duplication and agregation of homologous chromomeres produce the appearance of transverse chromatic bands. The chromosome, therefore, become multistranded as polytene but the individual chromonemata, which according to *Painter* may be as many as 1024, while *Beermann* (1952) estimated the degree of polytene to be as a high as 16000 times.

Support to Polytene theory from the volume of DNA. One important point of evidence may be added in favour of the polytene theory. This concerns the DNA contents of salivary gland uclelei and is complementary to Hertwig's (1935) demonstration of nuclear size classes. *Kurnick* and *Herskowitz* (1952) have estimated from photometric measurements of individual nuclei stained with methyl green that the large salivary gland nuclei in *Drosophila melanogaster* have undergone about 420 folds increase in DNA content over that the diploid nuclei. Cosrecting few possible errors in the determination of the lowest values, *Kurnick* and *Herskowitz* assume that 8 cycles of internal duplication have resulted in chromosomes consisting of about 1000 strands. *Swift* and *Rasch* (1954) have obtained much more numerous data by photometric Feulgen dye determination as nuclei of *Drosophila* salivary gland cells and closely associated cell type. They found a geometrical series of 10 DNA classes, ranging from presumably diploid cell to 1024 ploid salivary gland cells. The DNA values fall into generally non-over lapping groups, indicating that they resulted from a series of successive synchronized duplication.

Although a good number of workers believe in the polytene theory of salivary gland chromosome but several interesting problems remain unresolved. One of these is concerned with the question to what extent different chromosomes in one nucleus or even different regions within a chromosome, can become polytene to different degrees. The "sex ratio" phenomenon analysed by *Sturtevart* and *Dobzhansky* (1936) represented a well substantiated case of this type. Other class which suggest this possibility have been reported by *Schultz* (1941) and *White* (1948) who discovered that certain chromosomes appear to be out of steps with the rest of the set when judged by their sizes and stainability. It will be eventually determined by means of photometric DNA measurements whether these cases are due to unequal polytene or merely to abnormal states of contraction of certain chromosomes. This later possibility rises during their development. The mechanism of different regions of the same chromosome is still completely obscure. The cause of the break down of certain polytene chromosomes into their constituent fibrils are equally unknown.

Lampbrush Chromosomes

Throughout the vertebrate group of animals the somatic chromosomes present the usual structure, but within developing oocyte of those vertebrates which posses a yolky egg, and during the diplotene stage of meiosis, the same chromosomes undergo a remarkable change, especially characterized by an enormous increase in length and the appearance of radiating hairs or side loops which appear to organize themselves from the chromomeres. This gives a brush-like appearance during meiotic prophase. This type of chromosome was first described by *Flemming* (1882) and *Rukert* (1892) gave the famous name, 'Lampbrush.' *Ris* (1951) found similar chromosomes in sharks, birds, amphibians etc. Sometimes these chromosomes attain maxmimum size up to 800 to 1,000 m per chromosome.

The lampbrush chromosome possesses a central chromosomal axis from which project a series of lateral loops. The loops appear to project out from the dense region. Accoring to *Duryee,* each chromosome resembles a single plastic cylinder in which, at specific loci, are embedded chromatin granules. In a bivalent chromosome about 150-200 paired granules are present. These granules are of two sizes, viz., *smaller chromioles* and *larger chromioles,* later ones are ellipsoidal in appearance as though squeezed inside the matrix.

The lateral loops give rise to brush-like appearance. The loops are considered to be chromatin material synthesized for external utilization, and not an integral portion of the chromonemata extended in the form of a major coil as proposed by *Ris* (1945). *Duryee* (1941) hypothesis of lateral synthesis is supported by the fact that stretching of the chromosome by micro-manipulation or contractions by calcium

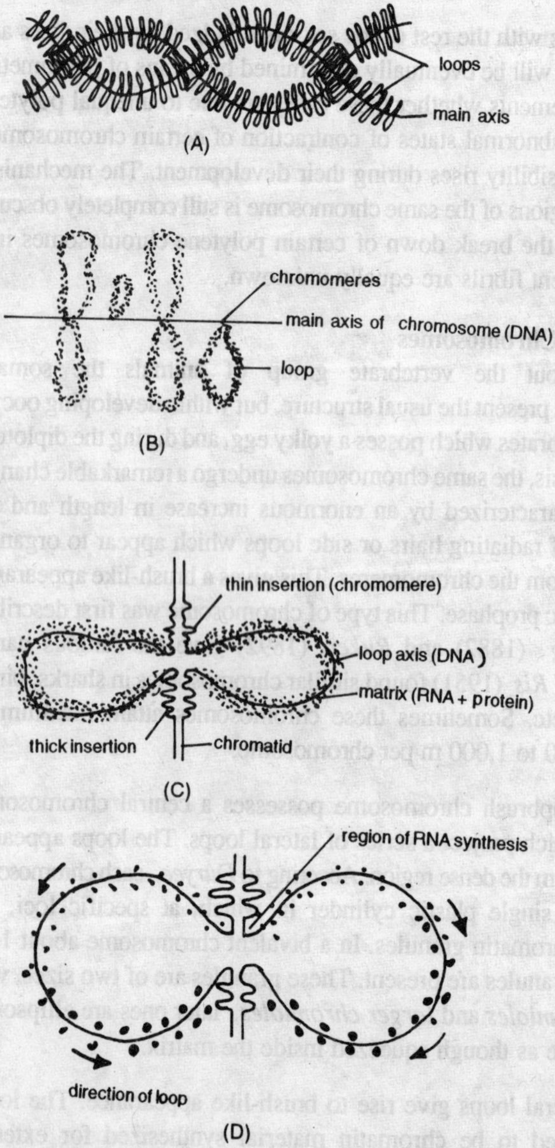

Fig. 2.13. Showing the lampbrush chromosome. (A) low magnification, (B) higher magnification (C) and (D) RNA synthesis in loop.

ions does not cause the loops to disappear or to be displaced and that dissolution of the loops by a variety of substances into granules does not effect the integrity of chromonemata. *Gall* (1956) showed quite conclusively through electron microscopy that loops are parts of the chromonemata and their apparent disappearance is due to the fact that prior to contraction they shed their coating of nucleic acid. The chromosome possesses a remarkable elasticity. The lateral loops extending from chromomeres are more fragile. *Gall* (1958) has interpreted that loops formation is a reversible physiological change which is probably non-genetic. However, he points out that loops exhibit variation in their morphology which also suggests, on the other hand, that perhaps each loop pair, represents a different genetic locus responsible for the formation of some particular cell product.

The lampbrush chromosome contains a continuous central main axis formed of DNA and protein. This varied from 30-50A° in diameter and is highly flexible. The loops axis is surrounded by the protein combined with RNA. Perhaps the loops are mainly concerned with synthesis of RNA, protein and yolk material.

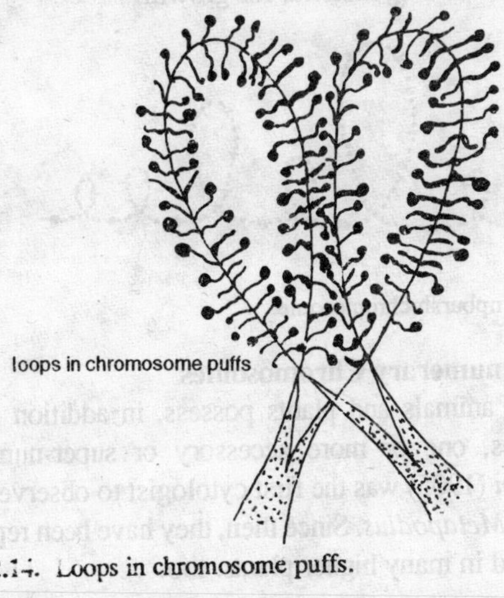

Fig. 2.14. Loops in chromosome puffs.

To explain the large size of lampbrush loops, *Callan* and *Lloyd* (1960) postulated that each consists not of gene, but of a number of duplicate copies, linearly arranged, of one gene. The chromomeres, basic units of the organization in lampbrush chromosomes exist in two forms. There is a 'master' copy of particular gene in a chromomere which resembles to its identical 'slave' copy of gene. In this way loop contains number of duplicate copies only. The genes are isolated by a double line and single transverse lines indicating the ends of the duplicate copies. *Gall* and *Callan* observed that the lateral loops always have one thin and one much thicker end at the point of insertion in the chromomere. It is also believed that loop spun out from the chromomere reunites to it at the thick end, which shows heavy accumulation or RNA. *Callan* further suggested that only in 'slave' copies take part in RNA synthesis.

This ensures possiblilities to synthesize large amount of ribose nucleic acid.

Nucleoli formation in lampbrush chromosome shows an unusual pattern. There may be several hundreds of nucleoli floating free in the nucleoplasm. The significance is not well-understood, but it is suspected that they must be synthesizing material for growth.

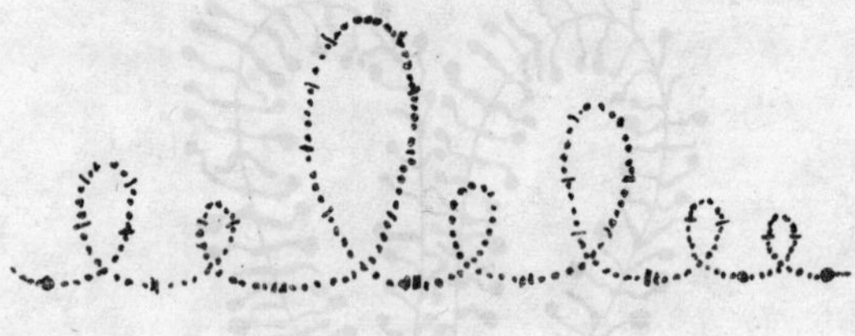

Fig. 2.15. Model of lampbursh chromosome.

Accessory or Supernumerary Chromosomes

Nuclei of some animals and plants possess, in addition to the normal chromosomes, one or more accessory or super-numerary chromosomes. *Wilson* (1905) was the first cytologist to observe them in hemipteran insect, *Metapodius*. Since then, they have been reported in several insects and in many higher plants. too.

In some cases, their nature and origin are certainly known. However, their ancestry is yet entirely unknown. The supermunerary chromosomes are usually of sizes smaller than those of there types. It is regarded that they perform some, as yet undermined, function, which is too slight to detect genetically.

3

Cell Division

There are two types of organisms-acellular and multicellular. The growth and development of an individual depends exclusively on the growth and multiplication of the cells. It was *Virchow* who first of all adequately stated the cell division. In animal cell the cell division was studied in the form of segmentation division or leavage by *Prevost* and *Dumas* in 1824. The mechanism of cell division was not precisely investigated until long afterward but *Remak* and *Kolliker* showed that the process involves a division of both the nucleus and the cytoplasm. The term *karyokinesis* was introduced by *Schleicher* (1878) to designate the changes of nucleus during division, and the term *cytokinesis* was introduced by *Whiterman* (1887) to designate the associated changes taking place in the cytoplasm.

Cell division is necessarily the avoidance of aging, and secondly for the segregation of an individual into semi-independent units which leads to efficiency. Thus, we see that cell division is a widespread phenomenon that is essential not only for the maintenance of life but also for the development of the organism itself.

Cell division can be conveniently described as:

(i) Direct division. Where the nucleus and cell body undergo a simple mass division into two parts. It is also called *amitosis*.

(ii) Indirect division. Here the nucleus undergoes complicated changes before it is divided into two daughter nuclei.

AMITOSIS

The amitosis or direct cell division is the means of asexual reproduction in acellular organisms like bacteria and protozoans and also a method of multiplication or growth in foetal membranes of some vertebrates. In amitosis type of cell division the splitting of nucleus is followed by cytoplasmic constriction.

During amitosis the nucleus elongates first and then assumes dumbell-shaped appearance. The depression or constriction increases in size and ultimately divides the nucleus into two nuclei. The division of nucleus is followed by the constriction of cytoplasm which divides

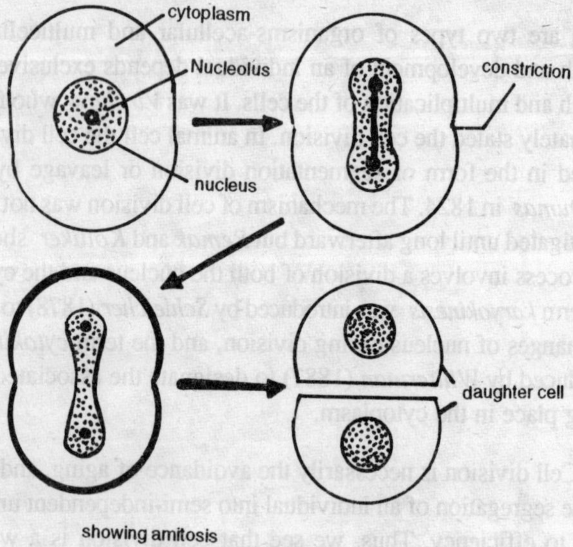

Fig. 3.1. Showing amitosis.

the cell into two equal or approximately similar halves. Therefore, without the occurrence of any nuclear event two daughter cells are formed.

MITOSIS

In mitosis, one cell divides into two, both of which are genetically identical to each other and to the parental cell. In other words, both, the chromosomes and the genes, are the same in all of the cells. This type of cell division is necessary if the organism and/or cell is to persist and survive. There are many facts to the need for cell division, and they vary depending on the particular biological function. For example, in liver tissue when some cells die or are damaged, others divide and provide new cell to replenish those that are lost. Other cells in the organism actually grow (increase in size), and it is possible that when they reach a point where there is too much cytoplasm far a given amount of nuclear material, they divide, and the whole process begins again. The phenomena of growth also involves an increase in numbers of cells. The increase in size of a tissue or organ is often due to numerical increase in cells rather than an increase in cell size. When subjected to the appropriate environmental and biochemical signals, these cells may be stimulated to differentiate upto a specific cell type. The sum total is that, as a rcsult of all division, a certain degree of plasticity and immortlity is provided to the organism. When the plasticity is lost, the organism undergoes the process of aging yet when the process of division is out of control, the organism literally "grows" to death!

The process of mitosis is a continuous one but on the basis of certain changes the same can be divided into several phases or stages.

Characteristic Features of the Mitotic Phases or Stages

In cell division, the first visible changes occur in the nucleus, when the chromonemata condense into *chromosomes* (*Gr;* coloured bodies). This stage is called the *Prophase* (*Gr;* early figure).

Then the nuclear boundary disappears and the chromosomes line up in or near one plane in the cell. This is *Metaphase* (*Gr;* middle figure).

Following this each chromosome separates into two parts, and these two parts migrate away from each other to the ends of the cells. This is *Anaphase (Gr;* up and down figure.)

Next the chromosomes at each end of the cell reconstitute a nucleus. This is *Telophase (Gr,* final figure).

Then the cytoplasm divides into two daughter cells, bringing about the *Interphase* again.

Interphase

The period of metabolic activity during which cell division is not in a process, has been called the 'interphase.' This is frequently referred to as the 'resting phase' but this term is not appropriate because the cell is metabolically most active at this stage that is why *Berril* and *Huskins* (1936) referred it as the *energy phases.* The interphase is the period between the telophase of one division and the prophase of the new cell division. During this phase the cell does everything except division. It is during interphase that the genes self-duplicate and carry on their function of supervising synthesis.

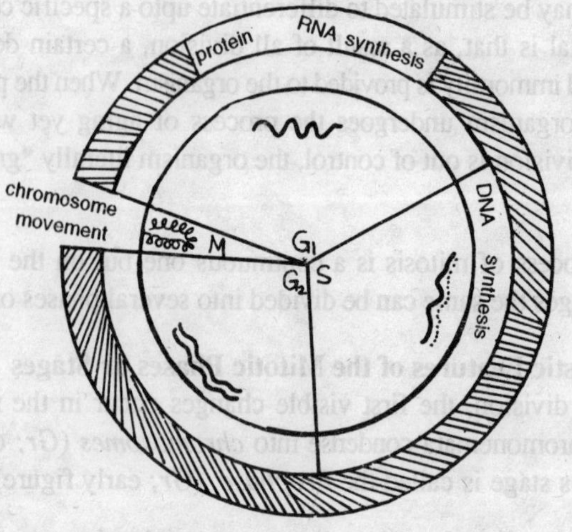

Fig. 3.2. The cycle of events in mitosis.

The chromatin granules in the nucleus are not readily distinguishable in the living cell, but may be brought out by treatment with chemicals which kill, fix and stain them. They appear at first glance to be scattered throughout the nucleus, but careful study has produced evidence that they are arranged in a definite pattern as long coiled strands and appear as a thread like net work in ordinary stained preparation.

"Nuclear sap" or, "Karyoplasm" fills the interstics between the chromosomes. One of more rounded bodies, the *nucleoli* are usually present. Between the nucleus and the surrounding cytoplasm, is the *nuclear membrane*. In the cytoplasm adjacent to the nucleus, there is a body, the *central body,* which consists of two granules or, after each granule has replicated, of two pairs of granules, the *centrioles.*

A typical cell cycle, including interphase lasts from 20-24 hrs. Interphase is the longest period in the cell cycle, and may last for several days in cells.

Interphase can be further divided into form sub-phases:

(1) G_1-phase
(2) S-phase
(3) G_2-phase
(4) M-Phase.

G_1-phase includes the synthesis and organization of the substrate and enzyme necessary for DNA synthesis. Therefore, G_1, is marked by the synthesis of RNA and protein.

G_1-phase is followed by the S-phase where the synthesis of DNA occurs.

During G_2-phase, all the metabolic activities are performed. M-phase is the period of chromosomal division.

The relative lengths of these phases differ in different organisms. A human cell in culture at 37°C, completes the mitotic cycle in about 20 hours and the M-phase lasts for only one hour. Temperature and cell environment plays an important role in determining the rate of cell division. Even the non-meristematic cells can sometimes be made to devide by changing the environmental conditions. Those cells which are not going to divide any more, have the mitotic cycle at the G_1 phase and start differentiating.

The cells shows following changes:

(1) The cell, as a whole, attains the maximum growth and possesses synthesized proteins for energy for various divisions and processes.
(2) The nuclear membrane is intact and the chromosomes are found in the form of more or less loosely coiled threads, somewhat closely appressed to the membrane. In this condition

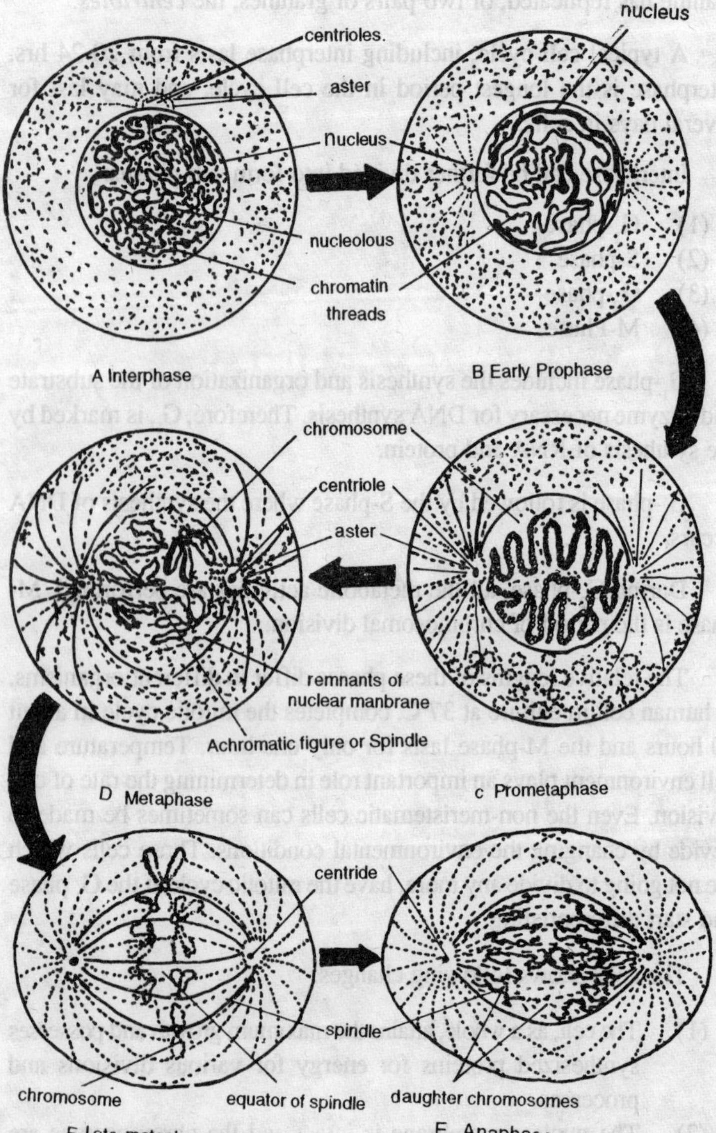

Fig. 3.3. Some early stages of mitosis.

of chromosomes, most of the cytologists regard them to be duplicated, while some workers are of the opinion that they are multipartite.

(3) The two centrioles, which are found at right angles to each other replicate into two each. *Mazia* (1961) has described that, if the replication is checked, then division will not take place.

(4) For the future spindle condensation of protoplasm into a coherent area of jelly-like consistency also takes place. The spindle also starts growing and pushes the centrioles apart.

(5) DNA synthesis occurs during autosynthetic interphase, when the chromosomes are dispersed.

(6) Chromocentre are also conspicuous during the interphase.

Prophase

The prophase is the longest in M-phase and may take about one to several hours. In the neuroblast cells of grasshopper it takes about 102 minutes. Various important changes in this stage are as follows :

(1) The cell tends to become spherical and increases its refractivity and turgidity by increasing its surface tension by removal of water.

(2) The nucleus imbibes water from cytoplasm, and the chromosomes get themselves organized differently. Each chromosome shows its peculiar structure with marked isolation of chromatids, each of which undergoes a regular cycle of coiling. At the beginning of division the chromosomes start to contract, get thickened and undergo their coiling. Partially this change appears to be associated with the development, at right angles to old gyres of the coils. *Sparrow* (1941) referred that total end to end contraction is approximately one fifth of the initial length. Physiologically, the coils are formed due to continuous condensation. The coils are of two types: smaller minor coils and larger somatic coils. The mode of twisting during the formation of coils and gyres is also categorized in two distinct varieties:

(i) *Plectonemic*. Twisted in such a way that it is not easy to isolate them, and

(ii) *Paranemic*. Coiled chromatids can be separated laterally.

(3) The chromatids are connected with centromere. The chromosomes are separately distributed in the nuclear cavity. The RNA and phospholipid contents gradually increase.

(4) The nuclear boundary becomes disrupted, the spindle apparatus begins to form, and the nucleolus and the centromere generally disappear. The spindle formation takes place in two ways, viz.,

(i) Single centriole divides into two daughter centrioles and on separation, astral rays appear as delicate filaments, called the spindle. The centrioles migrate along with asters until they become located at antipodal positions. This type of spindle is referred as to the *central spindle*.

(ii) The two centrioles are already polarized before the beginning of the division and the spindle formation takes place at metaphase. This type of spindle is known as the *metaphasic spindle*.

The mitosis in which achromatic figure and spindle are formed by centers is called as the *amphiastral mitosis* and whence centers are absent, the mitosis is called as '*anastral*'. The anastral mitosis takes place in plants.

Prometaphase

This stage follows the complete disappearance of the nuclear membrane, the chromosomes tend to aggregate in a central position in the cell, near the equator. The term prometaphase is described by *Coin* (1964). *White* (1963) defined it as the period during which the spindle is being formed and during which the chromosomes give the impression of struggling and jostling one another in an attempt to reach the equator of developing spindle. At least, in plants, this stage corresponds to the first appearance of an organised spindle. *Wilson* and *Hyppio* (1955) considered this positioning of chromosome to play an important role, both in the development and functional organization. The spindle fibers are tubular, elastic, fibrous and proteinous in nature.

Metaphase

The metaphase chromosomes are sharply defined and descrete bodies and are tightly coiled. In this stage the chromosome number can

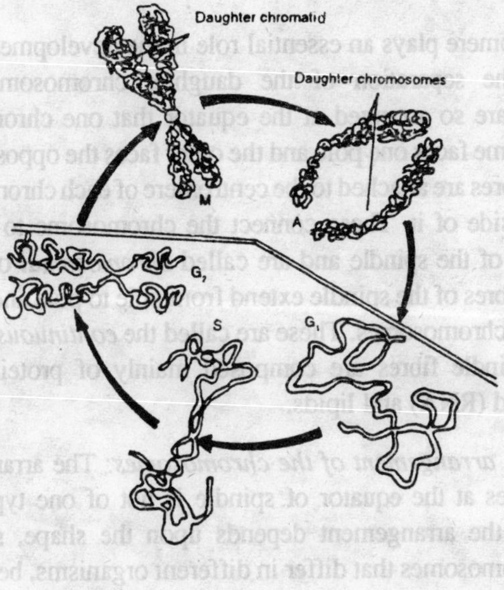

Fig. 3.4. Changes in chromosomes during cell cycle.

be easily counted and it is possible to recognise the different chromosomes from their sizes, shape and gross structure. In early metaphase characteristic spindle-shaped figure appears in the clear nuclear region. This consists of fibrous radiations which extend from the central broad-portion called the *equator* and converge at two points on the opposite ends of the nuclear region, called the *poles*. The chromosomes which were till now scattered in a random manner in the central nuclear portion of the cell begin to show peculiar movements and arrange themselves in a single plane midway between the two poles of the spindle, forming an equatorial plate. To be more precise it is the centromeres of the chromosomes that are lined up at the equatorial plate. Usually, the smaller chromosome occur near the center of the *equatorial plate* and the larger ones near the outer ends. It is, however, not necessary that the two homologus should be near each other on the equatorial plate. The position of each chromosome is indepedent of other on the equatorial plate. The metaphase chromosome is a visibly double structure and is clearly seen to be split longitudinally into two exactly identical chromatids.

The centromere plays an essential role in the development of the spindle and the separation of the daughter chromosomes. The chromosomes are so arranged at the equator that one chromatid of each chromosome faces one pole and the other faces the opposite pole. Two spindle fibres are attached to the centromere of each chromosome, one on either side of it. These connect the chromosome to the two opposite poles of the spindle and are called *chromosomal* or *tactile fibres*. Other fibres of the spindle extend from poie to pole and are not attached to the chromosomes. These are called the *continuous spindle fibres*. The spindle fibres are composed mainly of protein, some ribonucleic acid (RNA) and lipids.

Form and arrangement of the chromosomes: The arrangement of chromosomes at the equator of spindle is not of one type in all organisms, as the arrangement depends upon the shape, size and number of chromosomes that differ in different organisms, because in some organisms they (chromosomes) are thread like [Urodela and in insects (Orthoptera and Diptera)]. In Odonata, Coleoptera and Hamiptera they are shorter or rod shaped while in Orthopod they are rounded.

On the basis of their morphological study, one can divide them into three categories:

(i) *Straight rods or threads*. These chromosomes arise directly by shortening of the spireme thread.

(ii) *Loops, V's or hook forms*. These chromosomes are formed by a flexture at the middle point or near one end.

(iii) *Ovoidal or Spheroidal forms*. These forms arise by the extreme shortening of threads.

All the three forms mentioned above are double at metaphase due to their longitudinal splitting. The longitudinal splitting among rod shaped and thread shaped chromosomes, can be traced clearly but in spheroidal forms, it often appears as an apparently transverse constriction due to the extreme shortening of chromosomes.

The spindle attachment of the chromosomes. The mode of the attchment of the chromosomes to the spindle fibres based on the structure of chromosomes. But even though, the arrangement and the way of attachment is constant for every particular chromosome and that it is inherited from generation to generation. Chromosomes are attached to spindle fibres by centromeres. The chromosomes lacking centromere fail to attach with spindle fibres. The way of their attachment may be of two types:

(i) Terminal or telocentric
(ii) Non-terminal or atelocentric.

In terminal attachment the chromosomes may have their connection with spindle fibres at the free end. Non-terminal attachment may be at the middle point (*median*) or at an intermediate point *sub-median* or *sub-terminal.*

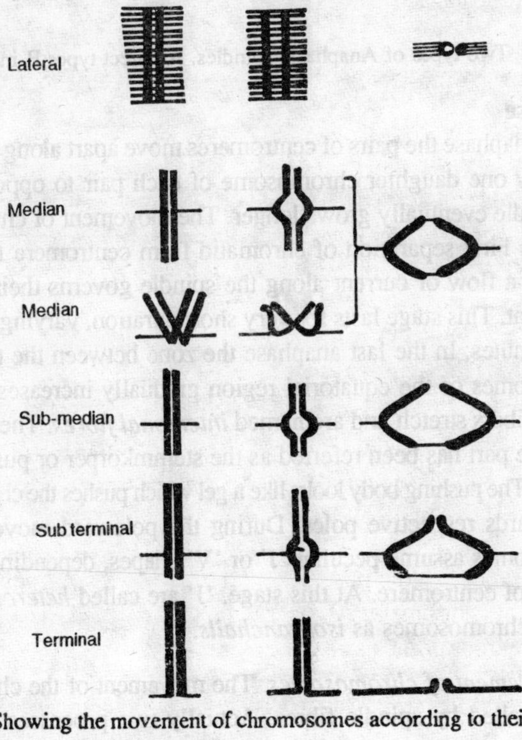

Fig. 3.5. Showing the movement of chromosomes according to their mode of attachment, during anaphase.

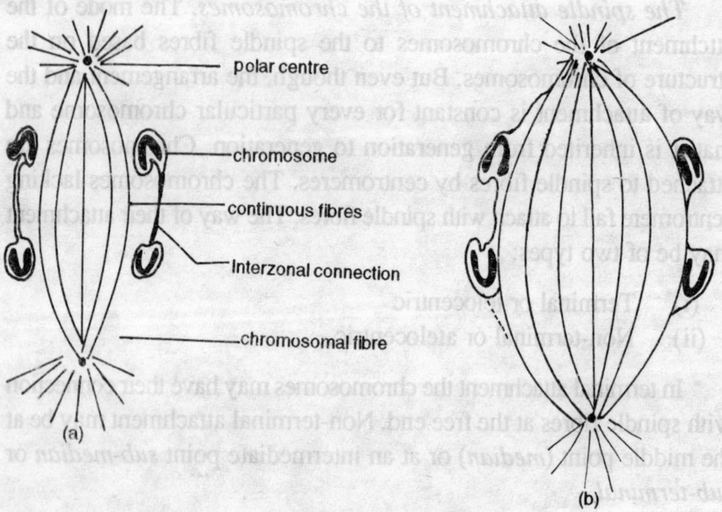

Fig. 3.6. Two types of Anaphasic spindles. A-Direct types B-indirect type.

Anaphase

In anaphase the pairs of centromeres move apart along the spindle and carry one daughter chromosome of each pair to opposite poles. The spindle eventually grows longer. The movement of chromatids is complex. First separation of chromatid from centromere takes place and then a flow of current along the spindle governs their poleward movement. This stage lasts for very short duration, varying between 6 to 12 minutes. In the last anaphase the zone between the two sets of chromosomes or the equatorial region gradually increases. In seems that the fibers stretch and are named *interzonal fibres*. The expansion of middle part has been referred as the stemmkorper or pushing body by *Bear*. The pushing body looks like a gel which pushes the chromosomal sets towards respective poles. During the poleward movements the chromosomes assume peculiar 'J' or 'V' shapes, depending upon the position of centromere. At this stage, 'J' are called *heterobranchials* and 'V' chromosomes as *isobranchails*.

Movement of chromosomes. The movement of the chromosome are controlled by spindle fibres. Actually two processes may be at work; a continuous enlargement and elongation of the spindle and the shortening of the chromosomal fibres. Since the fibres shorten without becoming thicker, the process probably involves the removal of water or other molecules from fibres. A band of 'interzonal fibres' is often seen for a time after the separation has been accomplished, connecting the chromosomes which have pulled apart, and often including a remnant of the spindle.

Forces involved in the movement of Chromosomes. Spindle fibres are responsible for the movement of chromosomes during anaphase i.e., from equator to the pole. Several models have been proposed by different workers to explain about the forces involved in the movement. Some of the them are given below:

(i) *Simple contraction model. Van Benden* (1883) suggested that chromosomes in the dividing cell are pulled towards the poles by the contraction of spindle fibres. *Swann* (1962) described that centromere secrete some substance which causes the contraction of spindle fibres. The main objection to the theory comes from the direct observations on chromosomes being carried even beyond the centriole to which the spindle fibres are attached. Further during cell division the whole cell elongates which is just opposite to the contraction model.

(ii) *Expansion Model. Watase* (1981) suggested that the spindle fibres exert pressure on the nucleus and chromosome become flattened on the metaphase plate. The chromosomal fibres become attached to the chromosomes, which are now pushed towards the opposite poles. For this reason, this is also known as *pushing model.*

(iii) *Contraction and expansion model. Belar* proposed that initial separation of daughter chromosomes is an autonomous process but further separation is the result of both contraction or expansion of various parts of spindle. The chromosomal fibres which extend from the centromeres of chromosomes to the poles of spindle contract and pull the attached chromosomes to the poles. The interzonal fibres present between the separating daughter chromosomes expand and push the daughter chromosomes towards opposite poles.

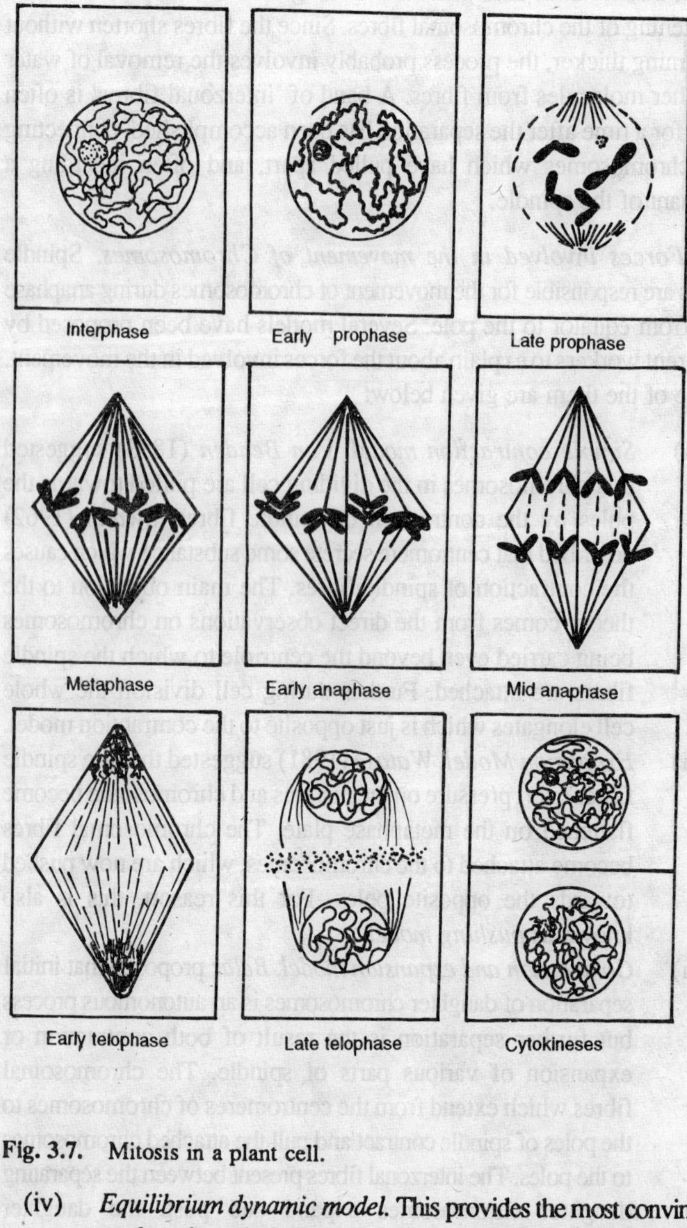

Fig. 3.7. Mitosis in a plant cell.

(iv) *Equilibrium dynamic model.* This provides the most convincing explanation about the possible mechanism of chromosome movement. *Inove* and *Sato* (1967) have described the occurrence of an equilibrium between a large pool of monomeres which constitute the proteins of microtubules. During polymerization some hydrophobic interactions occur between the non-polar groups of protein monomeres.

During chromosomal movement both contraction and elongation of spindle fibres occurs by the substraction or addition of new monomeres to the fibres. During anaphase chromosomal fibres contract by the delation of monomere from the polar end of the fibres and the continuous fibres increase in size by the assembly of new material at the polar ends. Thus the spindle elongates and distance between the poles increases.

(v) *Gliding model. Bajer. Ostergen* and other are of the opinion that the movement of chromosomes is an active process in which chromosomal fibres glide as sail boats between the continuous fibres. *Ambrose* has suggested that motile forces for these movements could be electrosmosis and electrohoresis. Break-down of ATP is supposed to provide the magnitude required for spindle movement.

(vi) *Ratchet model.. McInuosh, Helper, Van Wie* and others have described the presence of mechanical bridges between the microtubules similar to the ratchet connections in the muscles fibres. The chromosome fibres glide between the continuous fibres and draw the chromosomes apart. The energy is supplied by the break-down of ATP.

(vii) *Electrical model. Lillie* and *Coller* (1936) suggested that the change in the membrane potential is due to local changes in permeability occurs near the poles and around the nuclear membrane and produces electric field. The chromosomes are negatively charged in prophase and the charged chromosomes easily migrate in the electric field.

(viii) *Push Model.* According to this concept chromosomes are pushed apart by the colloidal component of cell cytoplasm. The colloid imbibes water, swells up and pushes the two chromatids of the chromosomes apart. The spindle fibres act as tracks that guide the movement of chromosomes to the poles, preventing their scattering throughout the cell.

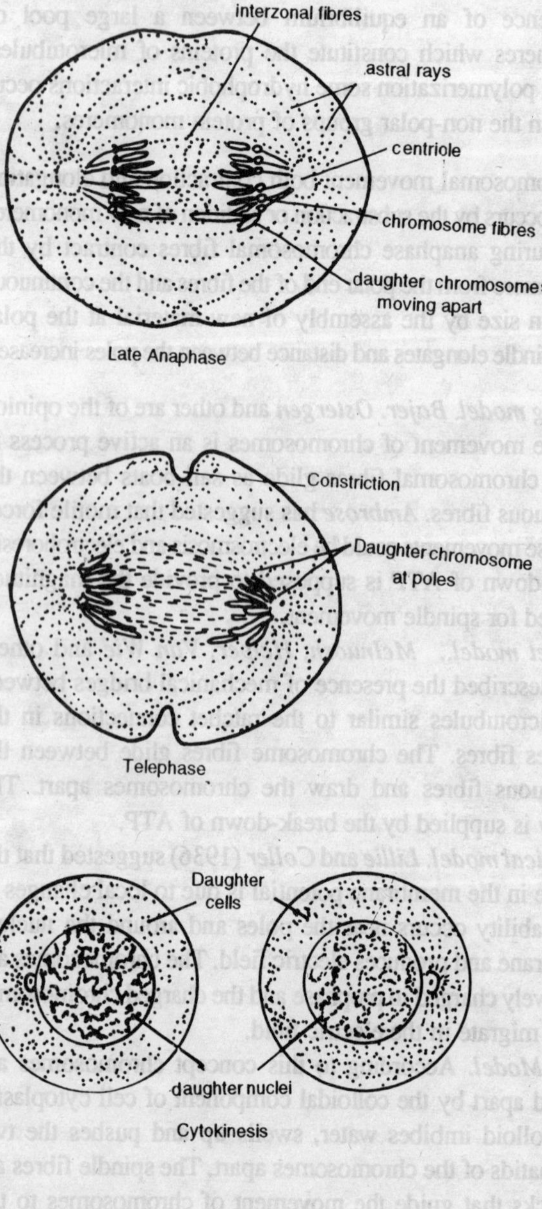

Fig. 3.8. Division of cell cytoplasm.

Telophase

The two groups of chromosomes aggregate at the poles and on arrival they progressively lose their chromaticity. Decondensation takes place. Reformation of the nuclear membrane by certain unknown process takes place. Perhaps new material is synthesized by RNA during telophase or it may be possible that the endoplasmic reticular system may be giving rise to a new membrane round the nucleus. All chromosomes become uncoiled. The spindle fibers are absorbed in the cytoplasm. The nucleolar organizer or SAT zone again forms the nucleolus. Finally, these change result into two daughter nuclei which are equivalent to the parent nucleus in all respects.

Cytokinesis in plants

Nuclear division or mitosis, as it is called is followed by the division of the cytoplasm. While the daughter nuclei are being organised at the poles, the mitotic spindle disappears except at the equator, where the continuous spindle fibres become more dense. This region is now called the *phragmoplast*. According to *Porter* and *Machade* (1960) the formation of cell plate is initiated by migration of tubular elements of the endoplasmic reticulum toward the interzonal region of the spindle where they spread out to form a close lattice along the mid line.

Theories of cytokines *Singh & Tomar* appears in phragmoplast and contain pectic substances which fuse to form in the middle of dividing cell a partition known as *cell plate*. More droplets are added to the cell plate to form the middle lamella which now begins to extend outwards until it reaches the outer walls of the original cell Primary cell wall material is now deposited on either side of the middle lamella. This divides the cytoplasm into two roughly equal parts. Cell division is now complete and two cells are formed. The division of the cytoplasm is called cytokinesis.

CYTOKINESIS

Cytokinesis in animal cells

Cytokinesis starts by the appearance of a shallow furrow in the cytoplasm at the equator of the spindle. Slowly and slowly the furrow deepens and constricts the cytoplasm and the cell into two daughters. A number o theories have been proposed for the formation of furrow.

These are as follows:

1. *Contractile Ring Theory.* According to *Swann* and *Mitchison* (1958) the cytoplasm around the equator of the spindle contains some contractile proteins. These proteins form a sort of ring at the equator. As the dividing cell elongates, the contractile ring contracts which results into the furrow formation.

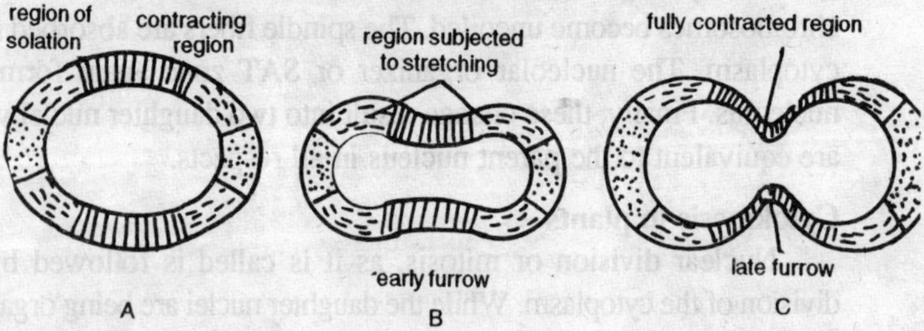

Fig. 3.9. Diagrams showing contractile ring theory of Marsland. A, Early telophase. B. Early furrow. C. Late Furrow.

2. *Expanding surface theory. Mitchison* (1922) proposed that a nuclear material is liberated by chromosomes which is responsible for the cellular expansion at the poles. As the polar areas expand, the equator contracts which results in the appearence of furrow. The furrow divides the cell into two daughter cells. There are certain examples, showig that the furrow formation takes place in the absence of nuclei or chromosomes (*Nachtway*, 1965). This indicates that exciting material might have originated from other than nucleus.

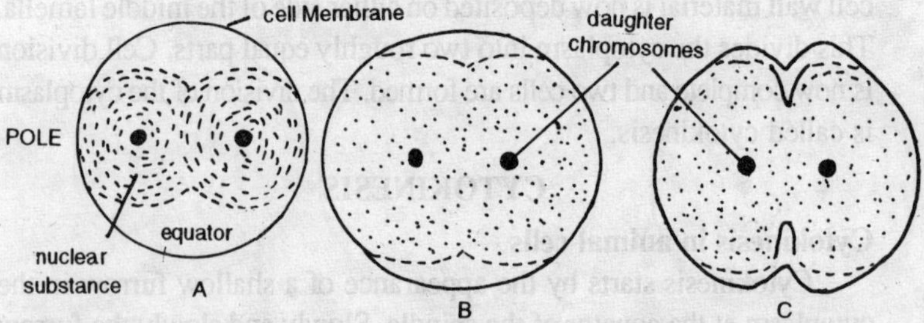

Fig. 3.10. Diagram to illustrate the expanding surface theory.

3. *Spindle elongation theory*. This theory was proposed by *Dan* and *Dan* (1947) and *Dan* (1958). According to them the spindle and asters are responsible for the cytokinesis. During experiment, kaolin particles were attached with the egg membrane. It was observed that elongation of the cell at anaphase is accompanied by shrinkage of equatorial plane results two kaolin particles on either side of equator come close to each other. The driving force is believed to be the elongation of microtubules of the spindle which push the centres apart.
4. *Astral relaxation Theory*. *Wolpert* (1960, 1963) proposed that the ourface of a call is under uniform tension. During cell division, as the astral rays reach the poles they lower in the surface tension at the poles. The surface tenson at the equator remains same. Because of low surface tension, the polar regions expand and cause the appearence of furrow at the equator.

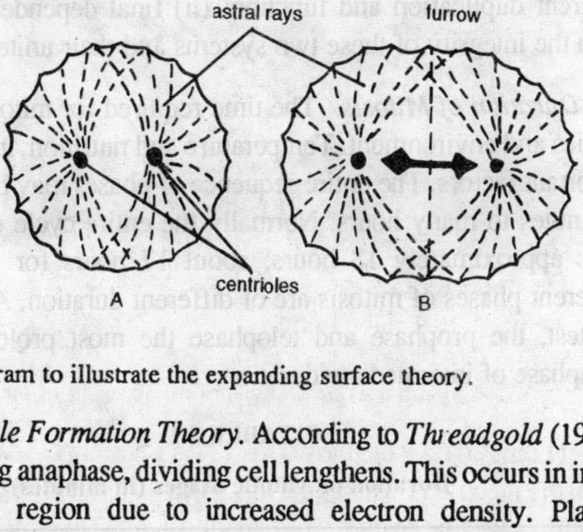

Fig. 3.11. Diagram to illustrate the expanding surface theory.

5. *Vesicle Formation Theory*. According to *Threadgold* (1968), during anaphase, dividing cell lengthens. This occurs in inter-zonal region due to increased electron density. Plasma membrane also exhibits high electron density in the furrow simultaneously, the continuous spindle fibres, present in the inter zonal region, continue to increase its density and finally form an adielectronic fibrillary plate in the equator. As furrow progresses, a large, empty membrane-bound vesicle appears on each side of this plate. Later on small vesicles are added

in the equatorial plane. At final stage, fusion occurs of all the large and small vesicles to form a deep furrow, leaving the dauthter cells to be connected by an intercellular bride. Further, a row of tiny vesicles arises which causes final separation of two daughter cells.

Role of Centriole During Mitosis. The centriole acts as an epicentre for the development of the central spindle. Whether the centriole produces the spindle fibres or serves to direct their formation or on the other hand is entirely passive in the proess, cannot be uniquestionable deduced from all the informations at hand. According to *Cleveland* (1957) the spindle fibres and astral and chromosomal fibres are recognizable first in the close vicinity of the centrioles. The completely developed achromatic apparatus then functions to separate daughter chromatids towards respective poles. With relation to the mitosis, *Cleveland* concluded two important points: (i) relative temporary independence of the achromatic and chromatic apparatus as regards apparent duplication and function; (ii) final dependence of the cell upon the integrity of these two systems and their united function.

Duration of Mitosis. The time required for mitosis differs with species and environment. Temperature and nutrition, in particular, are important factors. The entire sequence of phases may be completed in 6 minutes to many hours. Normally the entire cycle of cell division takes approximately 18 hours; about 17 hours for the interphase. Different phases of mitosis are of different duration. Anaphase is the shortest, the prophase and telophase the most prolonged, and the metaphase of intermediate duration.

TABLE 3.1

Duration of Mitotic Stages (in minutes)

Example	*Temp °C*	*Prop-hase*	*Meta-phase*	*Ana-phase*	*Telo-phase*
Mouse spleen	38	21	13	5	4
Chicken (mesen-chime cells in tissue culture)	39	30-60	2-10	3-7	2-10

Frog (Fibroblasts in culture)	20-24	32	20-29	6-11	
Grosshopper (Neuroblasts)	38	102	13	9	57
Sea Urchin Embryo	12	19	17	2	18
Onion (root tip)	20	71	6.5	2.4	3.8
Pea (root tip)	20	78	14.4	4.2	13.2

Significance of Mitosis

1. *Equal distribution of chromosomes.* The essential feature of mitosis is that chromosomes are distributed equally among the two daughter cells. With every cell division there is a division of chromosomes. The constant number of chromosomes in all the cells of the body in because of mitosis.

2. *Surface-volume ratio.* Mitosis restores the surface-volume ratio of the cell. A small cell has a greater amount of surface available in relation to volume than a large cell. As the cell increases in size the available surface area in relation to increased volume becomes less. By undergoing division the cell becomes smaller in size and the surface-volume ratio is restored.

3. *Nucleoplasmic index.* The growth of a multicellular organism is due to mitosis. A cell cannot grow in size to a large extent without disturbing the ratio between the nucleus and the cytoplasm. After a particular size has been reached the cell divides to restore the nucleoplasmic index. Thus growth takes place by an increase in the number of cells, rather than by increase in the size of the cells.

4. *Repair.* The repair of the body takes place because of addition of cells by mitosis. Dead cells of the upper layer of the epidermis, cells of the lining of the gut, and red blood corpuscles are constantly being replaced. It is estimated that in the human body about 500,000,000,000 cells are lost daily.

Abnormalities in Mitosis

Under unfavourable conditions the mitotic divisions become

defective and this various abnormalities are formed when exposed to physical and chemical agents like temperature, radiations, nacrotics and enzyme inhibitors etc. Some abnormal mitotic divisions are given under the following heads:

1. *C-Mitosis (abnormal spindle formation). Brachet* (1975) has described that colchicine inhibits mitosis by disorganizing spindle formation. The results are the formation of polyploid cells, and reduplication of chromosomes. Besides, during mitosis, sometime cell nucleus is either deformed into a single shperoid compact mass (pycnosis) or becomes broken down (karyorrhexis).

2. *Cytasteral mitosis. Wilson* (1901) noticed the formation of numerous asters in the cytoplasm of unfertilized eggs called cytasters by placing them in hypertonic sea water. In the eggs of marine invertebrates, production of many small cytoplasmic asters is of frequent occurrence. The intermixing of nucleoplasm and hyaloplasm seems to be necessary condition for the aster formation. *Costello* (1940) showed that asters can be produced only if germinal vesicle has broken down.

3. *Multipolar and catenar mitosis.* Mitotic division showing several spindles and centrosomes are common in many protozoans and fish eggs. Multipolarity is usually caused by uneven division of centrosomes as well as irregular distribution of chromatids of the different spindles. It results in the formation of cells as aneuploidy (uneven chromosome number). Catenar mitosis have been described by *Dalcq* and *Simon* (1932) in amphibian eggs. During this, dividing cell forms a large number of asters without spindles for chromosomes to link them. It suggests that asters are capable of dividing in the absence of spindle or nucleus in an autonomous way.

4. *Achrosomal motosis.* Sometimes, mitosis is induced by various agents without use of nucleus or chromosomes. *Briggs et al.* (1951), in fertilized eggs with heavily x-rayed sperms, removed the egg's maturation spindle by pricking and sucking. In this way, he obtained very nice non-nucleate blastulae. Similaly, *Stauffer* (1945) obtained normal non-nucle ated Axolotl blastula.

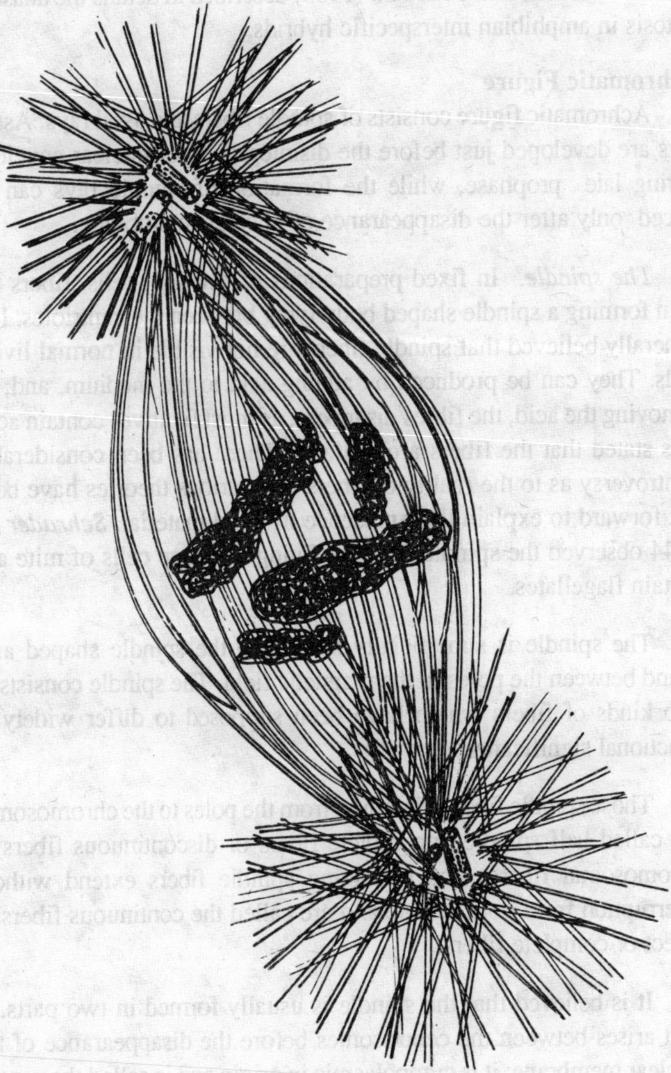

Fig. 3.12. Showing achromatic figure.

5. *Anastra mitosis.* In plant cells and many oocytes, generally asters around the centrosomes are absent, thus producing anastral

mitosis. *Bataillon and Tchou Su* (1933) described in details the anastral mitosis in amphibian interspecific hybrids.

Achromatic Figure

Achromatic figure consists of spindle fibers and asral rays. Astral rays are developed just before the disintegration of nuclear envelope during late prophase, while the formation of spindle rays can be traced only after the disappearance of nuclear rmembrane.

The spindle. In fixed preparations of dividing cells, fibers are seen forming a spindle shaped body lying between the centrioles. It is generally believed that spindle fibers are not visible in normal living cells. They can be produced by adding acid to the medium, and, on removing the acid, the fibres disappear; as most fixatives contain acid, it is stated that the fibers are artifacts. There has been considerable controversy as to the reality of fibers and various theories have been put forward to explain their presence in fixed material. *Schrader* in 1944 observed the spindle fibers in living cleavage cells of mite and certain flagellates.

The spindle is superficially similar to the spindle shaped area found between the poles in the magnetic field. The spindle consists of two kinds of fibers which have been supposed to differ widely in functional significance.

Those spindle fibers extending from the poles to the chromosomes are called half spindle, incomplete fibers or discontinuous fibers or chromosomal fibers. The remaining spindle fibers extend without interruption from pole to pole and are called the continuous fibers or direct or complete fibers.

It is believed that the spindle is usually formed in two parts. A part arises between the centrosomes before the disappearance of the nuclear membrane; it is cytopolasmic in origin and is called the central spindle. As the nuclear membrane disappears the central spindle moves into nuclear region. Then the nuclear material undergoes gelation around the central spindle to from the 1/2 spindle fibers or chromosomal fibers. In the metaphase the chromosomes are attached to the 1/2 spindle fibers at the equator. *Schrader* suggested that the spindle may consist of protein molecules arranged in parallel lines.

It appears that the spindle is organized from a pre-existing pool of protein. It is postulated that protein monomers aggregate to from microtubules of the spindle fibres. The spindle fibres are thus oriented polymers. It has been proposed that the continuous fibres are elongated by the insertion of subunits within the fibre. This leads to the elongation of the spindle. Conversely, the shortening of chromosomal fibres takes place by loss of subunits. This depolymerization of fibres takes place in the region of the spindle poles. Thus the fibres do not contract like a rubber band, but shorten by loss of material at one end. This is in agreement with observations that no change in the thickness of fibres takes place during shortening and elongation.

There is evidence that the microtubules of the chromosomal fibres originate at the kinetochore. Micotubules are also found in cells which hve no centrioles, e.g., those of higher plants. Here the microtubules end in the pole region and are not attached anywhere.

The Asters

Astral rays are developed by the centrosome present in the viscinity of nuclear region. Astral rays together with the spindle from the *amphiaster.* In early stages the astral rays are straight, simple and few in number. But in latter stages they rapidly elongate in all directions, and increase in number. In certain cases the astral rays of both asters cross one another in the equatorial region outside the spindle. Sooner or later, the crossing of the rays disappear and rays adjust themselves in such a way that their free ends move towards the opposite asters and meet in the equatorial plate thus giving the impression of continuous fibres. These continuous fibres which are formed by the fusion of astral rays are lying in the region outside the spindle. It has been suggested that the long protein molecules of the protoplasm become arranged in rows parallel to the lines of protoplasmic flow and thus constitute the astral rays.

REVIEW QUESTIONS

1. There are 40 chromosomes in somatic cells of the house mouse. (a) How many chromosomes does a mouse receive from its father? (b) How many autosomes are present in a mouse gamete? (c) How many sex chromosomes are in a mouse ovum? (d) How many autosomes are in somatic cells of a female?
2. Name. each stage of mitosis described. (a) Chromosomes line up in the equatorial plane, (b) Nuclear membrane reforms and cytokinesis occurs. (c) Chromosomes become visible, spindle apparatus forms. (d) Sister chromatids move to opposite poles of the cell.
3. Identify the mitotic stage represented in each of the following diagrams of isolated cells from an individual with a genome consisting of one metacentric pair and one acrocentric pair of chromosomes.

4. Identify the meiotic stage represented in each of the following diagrams of isolated cells from the germ line of an individual with one pair of acrocentric and one pair of metacentric chromosomes.

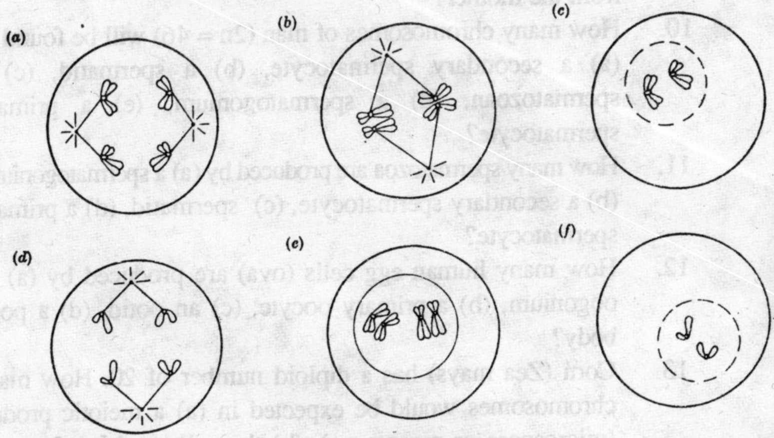

5. How many different types of gametic chromosomal combinations can be formed in the garden pea (2n = 14)? *Hint:* See problem 1.2.

6. (a) What type of division (equational or reductional) is exemplified by the anaphase chromosomal movements shown below?

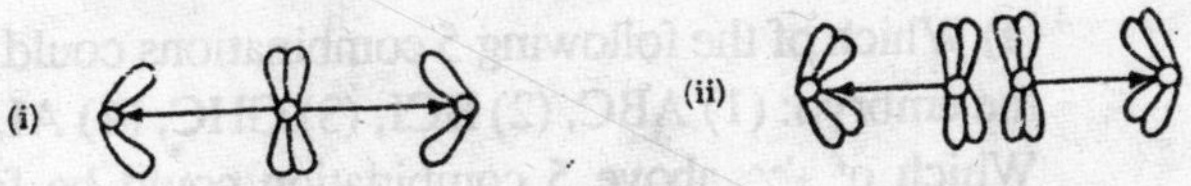

(b) Does the movement shown at (i) occur in mitosis or meiosis?

(c) Does the movement shown at (ii) occur in mitosis or meiosis?

7. What animal cells correspond to the three megaspores which degenerate following meiosis in plants?
8. What plant cell corresponds functionally to the primary spermatocyte?
9. What is the probability of a sperm cell of a man (n = 23) containing only replicas of the centromeres which were received from the mother?
10. How many chromosomes of man (2n = 46) will be found in (a) a secondary spermatocyte, (b) a spermatid, (c) a spermatozoan, (d) a spermatogonium, (e) a primary spermatocyte?
11. How many spermatozoa are produced by (a) a spermatogonium, (b) a secondary spermatocyte, (c) spermatid, (d) a primary spermatocyte?
12. How many human egg cells (ova) are produced by (a) an oogonium, (b) a primary oocyte, (c) an ootid, (d) a polar body?
13. Corn (Zea mays) has a diploid number of 20. How many chromosomes would be expected in (a) a meiotic product (microspore or megaspore), (b) the cell resulting from the first nuclear division (karyokinesis) of a megaspore, (c) a polar nucleus, (d) a sperm nucleus, (e) a microspore mother cell, (f) a leaf cell, (g) a mature embryo sac (after degeneration of non-functional nuclei), (h) an egg nucleus, (i) an endosperm cell, (j) a cell of the embryo, (k) a cell of the pericarp, (l) an aleurone cell ?
14. A pollen grain of corn with nuclei labeled A, B and C fertilized an embryo sac with nuclei labeled D, E, F, G, H, I, J and K as shown below.

    (a) Which of the following 5 combinations could be found in the embryo: (1) ABC, (2) BCI, (3) GHC, (4) AI, (5) CI? (b) Which of the above 5 combination could be found in the aleurone layer of the seed? (c) Which of the above 5 combinations could be found in the germinating pollen tube? (d) Which of the nuclei, if any, if the pollen grain would contain genetically identical sets of chromosomes? (e) Which of the nuclei in the embryo sac would be chromosomally and genetically equivalent? (*f*) Which of the nuclei in these two gametophytes will have no descendants in the mature seed?

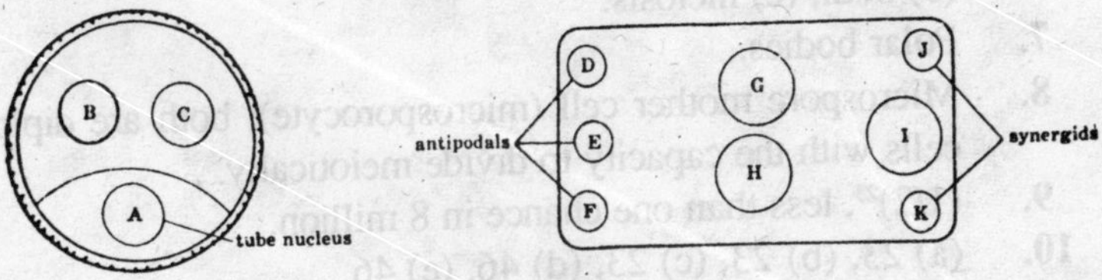

15. A certain plant has eight chromosomes in its root cells: a long metacentric pair, a short metacentric pair, a long telocentric pair, and a short telocentric pair. If this plant fertilizes itself (self-pollination), what proportion of the offspring would be expected to have (a) four pairs of telocentric chromosomes, (b) one telocentric pair and three metacentric pairs of chromosomes, (c) two metacentric and two telocentric pairs of chromosomes?
16. Referring to the preceding problem, what proportion of the meiotic products from such a plant would be expected to contain (a) four metacentric pairs of chromosomes, (b) two metacentric and two telocentric pairs of chromosomes, (c) one metacentric and one telocentric pair of chromosomes, (d) two metacentric and two telocentric chromosomes?
17. How many pollen grains are produced by (a) 20 microspore mother cells, (b) a cluster of 4 microspores?
18. How many sperm nuclei are produced by (a) dozen microspore mother cells, (b) a generative nucleus, (c) 100 tube nuclei?

ANSWERS

1. (a) 20, (b) 19, (c) 1, (d) 38.
2. (a) Metaphase, (b) telophase, (c) prophase, (d) anaphase.
3. (a) Metaphase, (b) prophase, (c) telophase, (d) anaphase.
4. (a) 1st anaphase, (b) 1st metaphase, (c) 2nd prophase or end of 1st telophase, (d) 2nd anaphase, (e) 1st prophase, (f) 2nd telophase (meiotic product).

5. 128.
6. (a) (i) is an equational division; (ii) is a reductional division; (b) both; (c) meiosis.
7. Polar bodies.
8. Microspore mother cell (microsporocyte); both are diploid cells with the capacity to divide meiotically. .
9. $(1/2)^{23}$, less than one chance in 8 million.
10. (a) 23, (b) 23, (c) 23, (d) 46, (e) 46.
11. (a) 4, (b) 2, (c) 1, (d) 4.
12. (a) 1, (b) 1, (c) 1, (d) 0.
13. (a) 10, (b) 20, (c) 10, (d) 10, (e) 20, (f) 20, (g) 30, (h) 10, (i) 30, (j) 20, (k) 20, (l) 30.
14. (a) 5; (b) 3; (c) 1; (d) A, B, C; (e) D, E, F, G, H, I, J, K; (f) A, D, E, F, J, K.
15. (a) 0, (b) 0, (c) all.
16. (a) 0, (b) 0, (c) 0, (d) all.
17. (a) 80, (b) 4.
18. (a) 96, (b) 2, (c) 0.

4

Meiosis

Mitosis is a device of increasing the number of cells. During asexual reproduction it is the means of multiplying the individual. Therefore, in a sexually reproducing organisms, set of characters are transferred unchanged from parents to offspring, except of course accidents. There is uniparental inheritance due to mitosis. But in sexually reproducing organisms, gametes are produced as a result of meiosis. Two gametes unite to give rise to a zygote which divides mitotically to form the individual. More often than gametes from different parents unite to form a zygote, thus ensuring biparental inheritance. This result into a blending of chrarcters from two parents into their offsprings and ensures variation which is important for evolution. Gametes contain half the number of chromosomes, as compared to somatic cells. Union of two gametes restores the somatic number of chromosomes in the zygote. Gametic union results into the mixing of cytoplasms of two cells which is followed by the fusion of two nuclei. During meiosis or at the time of gamete formation, therefore, halving of the number of chromosomes or the amount of DNA is achieved by two successive divisions of cytoplasm and nuclei (meiosis I and meiosis II) accompanied by only one replication of chromosomes. For example, each sometic cells of man has 46 chromosomes. Somatic cells are derived from the zygote due to mitotic divisions. Zygote is the result of a fusion of two gametes, each of which contain only 23 chromosomes. Both the male and female gametes contain the same amount of DNA and equal number of morphologically similar chromosomes. The zygote and somatic cells derived therefrom, therefore, contain 23 pairs of homologous chromosomes, one of each pair derived from the male parent through the male gamete and the other from the female through the female gamete. The members of each pair are homologous to each other and non-homologous to the other chromosomes. In some organisms, the chromosome complements of male and female cells are slightly different and this forms the basis of sex differentiation. Those cells which possess two sets of chromosomes, are called *diploid* and they have *2n* or *dipoid* or *unreduced* or *zygotic* number of chromosomes. Gamets, on the other have n, or *haploid* or *reduced* or *gametic* number of chromosomes. In sexual reproducing organisms the diploid and haploid phases of the life cycle alternate and this phenomenon in called *alternation* of *generation*. In plants, there is an alternation of sporophytic (2n) and gametophytic (n) generation. Reduction division heralds the beginning *of* gametophytic generation and fusion of gametes that of the sporaphytic phase. Thus, meiosis can be studied at the time of gamete formation which occurs in specialised organs i.e., the embryo sac and the pollen sac. A single cell undergoes meiosis in the embryo but in the pollen sac a large number of spore mother cells divide meiotically, almost synchronously, to produce hapoid pollen grains. It is, therefore, easier to study meiosis is anthers, similarly testes in animals.

Meiosis consists of two divisions which take place one after the other, during which the number of chromosomes is halved. The two divisions are known as the first meiotic division or *heterotypic division* and the second meiotic division or *homotypic division*. The first meiotic division has the same stages as mitosis. These stages are known as prophase I, prometaphase I, metaphase I, anaphase I and telophase I. Except for prophase, the other stages are similar to those of mitosis. Following telophase I is a period of interphase or rest. The second meiotic division consists of prophase II, metaphase II, anaphase II and telophase II.

I. HETEROTYPIC OR IST MEIOTIC DIVISION

This is the first half of meiosis and is true reductional in nature. Here the diploid cell divides into two haploid daghter cells. As already stated that the phases of this division one called prophase I, metaphase I, anaphase I and telophase I.

Prophase I

It is the longest stage and its duration makes from a few hours to a few days. The nucleus starts increasing in volume and this increase is much than the mitotic increase to some extent. This is due to an increase in hydration.

Nobel and *Ruttle* (1936) have called this stage as the *premeiotic spiral prophase.* Prophase I is Furtner divided into six substages.

(i)	Pre-leptotene	(iv)	Pachytene
(ii)	Leptotene	(v)	Diplotene
(iii)	Zygotene	(vi)	Diakinesis

(i) *Preleptotene.* As noted previously, replication of DNA has already taken place during interphase. The chromosomes at the beginning of meiosis are, therefore, note single but double strands. The beginning of meiotic prophase is marked by an increase in nuclear volume. During preleptotene the chromosomes are very thin and cannot be easily seen, except sometimes for the sex chromosomes.

(ii) *Leptotene ("slender thread") or Leptonema.* During this stage the chromosomes become more distinct, and their double nature is seen in many organisms. Leptotene extends from the time single unpaired chromosomes are first seen to the time when they begin to undergo homologous pairing. Under the light microscope the chromosomes appear as slender threads bearing a series of granule-like structures called *chromomeres*. Whole mount electron microscope preparations show that leptotene chromosomes have an axial filament to which chromatin fibres are attached as a series of lateral loops. In certain regions these loops appear in bunches, and from the chromomeres of light microscopy. The axial filaments are apparently composed of two independent protein fibrils, one for each chromatid. They later become transformed into the lateral elements of the synaptonemal complex. In the leptotene cells of the quail there is evidence of premeiotic pairing. Two homologous chromosomes become apposed to each other, chromomere for chromomere, even though no synaptonemal complex is formed. In leptotene cells there is a tendency for the ends of homologous chromosomes to be associated with each other at the nuclear membrane, where they are attached.

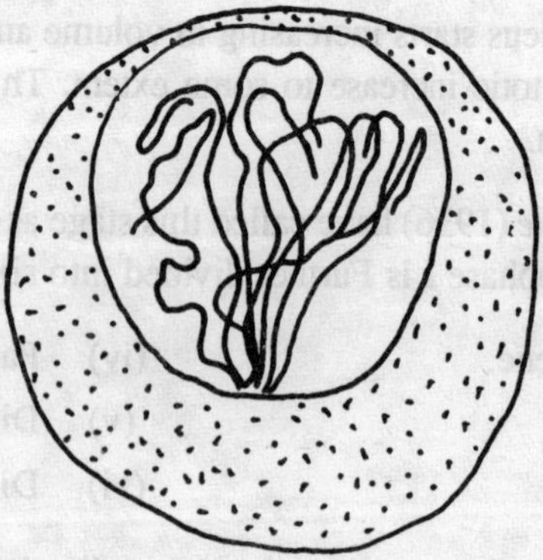

Fig. 4.1. Leptotene chromosomes arranged in a 'bouquet.'

Leptotene chromosomes may be irregularly arranged, or may be polarized toward the centrioles forming a 'bouquet.' Electron microscopic studies have shown that bouquet formation results when a group of chromosomes is attached close together on the nuclear membrane.

During leptotene the cytoplasm has many polyribosomes, but endoplasmic vesicles are few. In maize *(Zea mays)* the nucleus increases in size during leptotene, reaching its maximum size during pachytene. This increase in size is correlated with synthesis of RNA and protein.

(iii) *Zygotene ("mating thread") or Zygonema.* The chromosomes becomes much shorter and thick and number of coils is also increased. Homologous chromosomes begin to pair by force of attraction. In synapsis or pairing, two homologous chromosomes come close to each other precisely chromomere to chromomere. The pairing starts from one point and proceeds onward in a zipper like fashion. Out of the two synaptic chromosomes, one is of paternal origin while the other is maternal. These are called *bivalents.* Since each chromosome is made of two chromatids, each pair of chromosomes consists of four chromatids thus representing *'tetrad'* of chromatids at this stage.

The pairing of zygonemal stage is completed in three different ways, as follows.

(i) *Proterminal synapsis.* The two homologous chromosomes start pairing at the terminals, which gradually then progress towards the centromere.

(ii) *Procentric synapsis*. The pairing starts near the centromere and progresses onwards on either side.

(iii) *Random synapsis*. The pairing may be random at many points which then finally fuse together.

Two views have been advocated to explain the mode of initiation of synapsis. *Darlington* (1937) proposed precocity theory which explains that the chromosomes undergo synapsis due to singleness, but this theory does not explain the extra synthesis of DNA and duplication at leptotene stage. Another view is proposed by *Sax* (1932) which is referred as the retardation theory. It explains that pairing takes place due to the retardation or cessation of metabolic activities of the cell. During zygonema, the nucleus remains intact and the centriole moves apart initiating the aster formation.

(iv) *Pachytene "thick thread" or pachynema*. It generally begins when the pairing of chromosomes ends. It represents one of the longer substages of prophase I and is characterised by the following important events:

When the pairing has completed, the apposition of homologues becomes so tight that it is difficult to identify two separate chromosomes. Within the major coils, and at right angles to them smaller minor coils appear. At the height of pachytene, the pairing chromosomes twist about one another as relational coils and each soon longitudinally splits if no splitting had occured. This coiling is further complicated by a coiling of two chromatids of each chromosome around each other. This coiling put the chromosomes under considerable strain. This means that at this stage, each pacytence element consists of four chromatids. The chromatids of each homologue are called *sister chromatids*. Formerly this element of four chromatids was called a *tetrad*-a name which has now been replaced by *bivalent*. When the chromosomes have longitudinally splitted, transverse breaks ocour at the same level in two of the adjacent homologous chromatids. Soon both segments of the chromatids interchange and fuse together with those of homologues. In this process which is called *crossing over*, portions of the two chromatids with their genes are interchanged and two remain intact. The new chromatids thus interchanged will be mixed.

After this interchange, the parts begin to separate, repalling each other. However, this separation is not at all complete since the homologous chromosomes remain joined at their point of interchange. This point, where the two are joined, is called the *chiasma* (pl. chiasmata). This chiasma is an immediate consequence of crossing over and are found in all plants and animal with few exceptions (e.g. males and females silkworm). The number of chiasmata is variable but for one bivalent atleast one chiasma is found. Bivalents with 2, 3 and 4 chiasmata are rare. The highest number of chiasmata has been observed in the long chromosome of the broad bean, *Vicia faba,* where there are probably 12 chiasmata in a bivalent. The average number of chiasmata is known as *chiasma frequency*. This number seems to depend upon the length of the chromosome. Short acrocentric chromosomes have a greater number of chiasmata than metacentric chromosomes. In small chromosomes, chiasmata are localized at the distal ends. Nucleus remains as such, being somewhat larger than the proceeding substage.

Cytomixis. It is a botanical term which refers to the passage of chromatin from one microsporocyte (pollen mother cell) to another during early prophase. This phenomenon is so widespread in the normal species of hybrids that one wonders if it is a general phenomenon.

(v) *Diplotene or Diplonema*. The separations of homologous chromosomes initiates diplotene. The synaptic forces of attraction between them lapse due to breakage at one or more point, so that the homologous chromosomes uncoil and start separating. But the separation is incomplete since the homologues are in, contact at one or more where the crossing over has already taken place. These points of contact are points known as chiasmata (sing-chiasma; meaning cross) which present crosshaped appearance. The chiasma is the morphological equivalent of genetic crossing over. Their number and position varies in the same pair of chromosomes and in different cells of the same individual.

By the end of diplotene, the chiasmata begin to move along the length of chromosomes from the centromere towards the end. This displacement of chiasmata is termed as terminalization. When the terminalization of chiasmata is complete the homologous chromosomes are kept in contact by terminal chiasmata. The degree of terminalization is generally expressed as a coefficient of terminalization (T).

$$T = \frac{\text{Number of terminal chiasmata}}{\text{Total number of chiasmata}}$$

The average number of chiasmata in bivalents is known as frequency of chiasmate (fq).

$$\text{Frequency (fq)} = \frac{\text{Total number of chiasmata}}{\text{Total number of bivalents}}$$

According to *Darlingtion* two types of repelling forces operate on the chromosomes at diplotene. One of the forces is electroegatively charged and operates on the surface of the chromosome throughout its length and the other with electropositive charge is localized on the centromere. The former controls the repulsion of the chromosomes and the later causes distal movement of the chiasmata.

Diakinesis. At this stage the chromosomes are best seen as very short and thick bivalents. The contraction is nearly maximum and the chromosomal pairs are well spread throughout the cell as thought by mutual repulsion. The chromatids of each chromosome are very close to each other and their identity is not distinguishable. Terminalization may be completed in the substage. The bivalents migrate to the periphery of the nucleus. The nucleus generally disappears during this stage but may persist, in reduced size until anaphase. The nuclear membrane disrupts and eventually disappears thus all the chromosomes are set free in the cytoplasm. The spindle fibres are organized with an establishment of the pole.

Prometphase I

Spiralization reaches its virtul maximum in this stage; the nuclear membrane disappears, and the chromosomes become arranged at the equator of the cell to begin the metaphase. Nuclear membrane disappears and spindle and astral rays are formed during this stage.

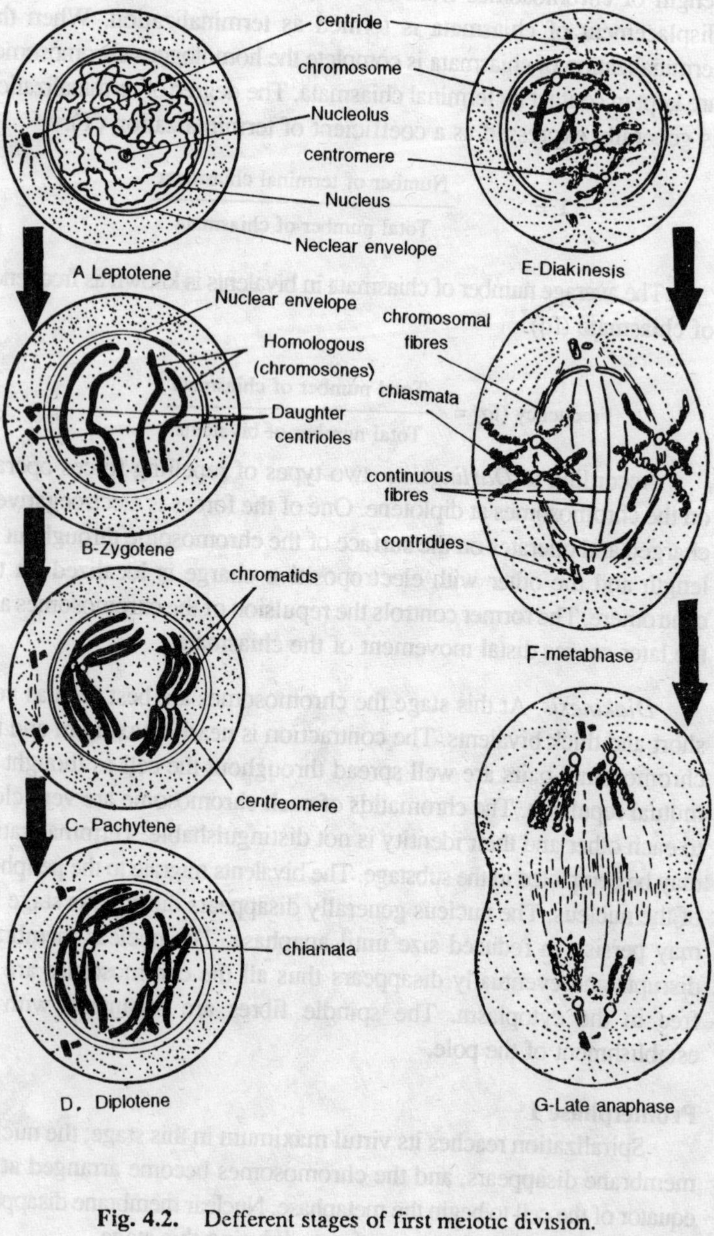

Fig. 4.2. Defferent stages of first meiotic division.

Metaphase I

The bivalent chromosomes arranged themselves in the equatorial plane so that the two members of each homologous pair are found with their centromeres directed toward oppposite poles. Whereas in mitosis there is a single centromere which unites the chromosomes, in meiosis there are two centromeres. Each homologous chromosome has a centromere to which the daughter chromatids are united. The repulsion of the centromere is accentuated, and the chromosome is found on the verge of dividing.

Anaphase I

In this the chromosomes are pulled apart by the movement of their centromeres which remain undivided, with the result that each tetrad breaks up into two component pairs of chromatids or dyads. The dyads separate and move along the tactile spindle fibres to the opposite poles. This is the true reduction division.

Telophase I

The group of chromosomes at each pole forms a nucleus. Each daughter nucleus has half the number of chromosomes as compared to the parent nucleus. Unlike the mitotic telophase each chromosome consists of two distinctly separated chromatids united only at their centromeres. This is followed by the formation of a cell plate between the separated sets of chromosomes, dividing the mother cell into two daughter cells. Sometimes, cytokinesis fails to occur after the first meiotic division. In such cases the second meiotic division takes place in the common cytoplasm of the mother cell.

Interkinesis

After telophase, two haploid daughter nuclei or cells sometimes undergo typical resting stage as in mitosis. This intervening stage between first telophase and the beginning of second prophase is termed as interphase or interkinesis. This stage is either very short or may be entirely absent, if present, the coiling of the chromosomes persists. No increase in DNA takes place between the I and II meiotic divisions. The occurrence of an interphase between the two division suggests that additional RNA and protein synthesis is required to carry meiosis through to completion.

SYNAPTONEMAL COMPLEX

The behaviour of chromasomes during meiosis is more easily followed with the light microscope than with the electron microscope. However, distinctive structures known as synaptonemal complexes are visible with the electron microscope in the nucleus of cells during I prophase of meiotic division. *Moses* (1950) first of all discovered this complex.

Synaptonemal complex is a tripartite structure found between the two paired homologous chromosomes of each bivalent nuclei undergoing meiosis. Complete synaptonemal complexes are seen at zygotene stage in the region of pairing however, at pachytene they are even more conspicuous under electron microscope. The synaptonemal complex appears consisting of three parallel dense lines. These dense lines are separated by less dense areas. The two lateral elements are composed of fibres that are slightly wider than 10 nm. These are called *Synaptomeres*. The central element has a ladder-like configuration in the centre of the synaptonemal complex. The transverse elements have electron dense filaments that interconnect the central element with the lateral elemens. The lateral elements varies in width from 20nm to 30 nm, and the two lateral elements are situated 100 nm to 120 nm apart.

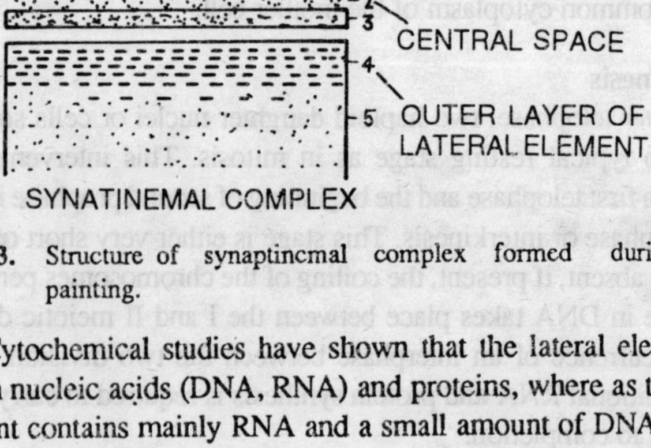

Fig. 4.3. Structure of synaptinemal complex formed during meiotic painting.

Cytochemical studies have shown that the lateral elements are rich in nucleic acids (DNA, RNA) and proteins, where as the central element contains mainly RNA and a small amount of DNA.

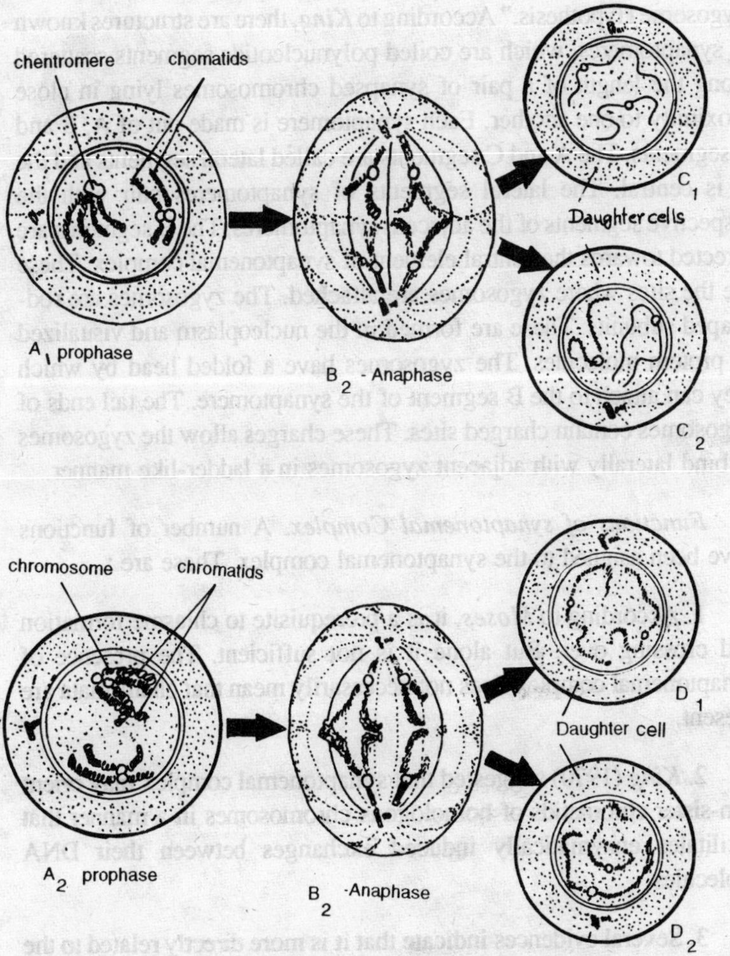

Fig. 4.4. Different stages of second meiotic division.

According to *Comings* and *Okada* (1971) a series of lateral loops of chromatin arise from the lateral elements. These loops make up the substance of chromosome. Besides another series of smaller loops also arise from the lateral elements. These smaller loops fuse to form the central element of synaptonemal complex.

In 1970, *Robert King* proposed a hypothesis for the formation of synaptonemal complex. This hypothesis is known as, "The synaptomere-

Zygosome Hypothesis." According to *King*, there are structures known as synaptomeres, which are coiled polynucleotide segments scattered along the length of a pair of synapsed chromosomes lying in close proximity to one another. Each synaptomere is made out of A, B and C segments. The A and C segments are called lateral segments and the B is central. The lateral segments of synaptomeres pair with the respective segments of the adjacent synaptomeres. The B segments are directed towards the central element of synaptonemal complex. These are the sites where zygosomes are attached. The zygosomes are rod-shaped subunits. These are formed in the nucleoplasm and visualized as protein molecules. The zygosomes have a folded head by which they can attach to the B segment of the synaptomere. The tail ends of zygosomes contain charged sites. These charges allow the zygosomes o bind laterally with adjacent zygosomes in a ladder-like manner.

Functions of synaptonemal Complex. A number of functions ave been asigned to the synaptonemal complex. These are :

1. According to *Moses*, it is a prerequisite to chiasme formation nd crossing over. But alone it is not sufficient. The presence of synaptonemal complex does not necessarily mean that chainsmata are present.

2. *King* (1970) suggested that synaptonemal complex may orient non-sister chromatids of homologous chromosomes in a manner that facilitates enzymatically induced exchanges between their DNA molecules.

3. Several evidences indicate that it is more directly related to the process of recombination. It has been observed that a small amount of DNA is synthesized at meiotic prophase. If this synthesis is inhibited, the function of synaptonemal complex is arrested.

II. HOMOTYPIC OR 2nd MEIOTIC DIVISION

Before the start of this division, most of the activities of heterotypic division cease for sometime. This resting stage is termed as the interkinesis. 2nd Meiotic division involves following staps:

Prophase II

This stage superficially resembles that of the mitosis prophase with following exceptions:

At the early prophase II, the two chromatids of each dyad looks like X, since they are conjoined by a common centromere. The four arms are widely separated. There is no rational coilling. The X-shaped dyads are quite longer than at telophase. The chromonemata are still not completely coiled. The genetic constitution of the two chromatids of each dyad depends upon the kind and number of cross overs which took place in the phase I. If there is no crossing over, the dyads would consist of two identical sister chromatids of either paternal or maternal origin. At the end of the prophase II the nucleolus disappears. The nuclear membrane disappears and the acromatic figure is develoed.

Metaphase II

As usual, all the chromosomes arrange themselves on the equaor for a short duration. The centromeres touch the equator but their arms radiate out in different directions. Later on the centromere in each dyad divides into two sister centromeres.

Anaphase II

Sister centromeres separate to the poles, pulling with them the chromatids to which they are attached. The chromatids, however, become much thick and stout.

Telophase II

The chromosomes at each pole uncoil and thin out to form the nuclear net. Each group gets surrounded by a nuclear membrane. Nucleolus reappears. Thus two nuclei are recognised in each cell. This is soon followed by cytokinesis and two cells are formed from each haploid daughter cell. Thus as a result of meiosis four cells are produced, each with a haploid set of chromosomes, i.e., each contains just one members of each homologous pair.

Cytokinesis

Sometimes meiosis I is followed by cytokinesis and sometimes it is deferred until the end of meiosis II. The details of the process are same as during mitosis. Cell plate formation results into a tetrad of cells with reduced chromosome numbers. The arrangement of cells in a tetrad is different in different organisms, but is characteristic of the species.

Like mitosis, meiosis too is a dynamic process and its different stages merge into subsequent ones. There is no sharp demarcation between different stages. DNA synthesis takes place during interphase, prior to prophase I. Duplication of chromosomes takes place during interphase itself and the chromosomes that appear during leptotene are double and not single. Number of chromosomes is reduced during meiosis I. If the reduction process is examined more closely, it becomes clear that there is qualitative reduction during meiosis I of those parts of chromosomes which are not involved in exchange. The homologous chromosomes or the maternal and paternal chromosomes segregate or undergo qualitative reduction during meiosis I. Their quantitative reduction takes place during meiosis II, when their centromeres divide and the sister chromatids separate. But the reverse is true of those parts of chromatids which are exchange due to the formation of chiasmata. For these parts quantitative reduction takes place during meiosis I and qualitative reduction during meiosis II. Thus, complete of both qualitative and quantitative reduction takes place as a result of meiosis I and II and none of them alone can bring about complete reduction. It is, therefore, not proper to say that meiosis I is reductional and meiosis II equational. Duration of meiosis varies a great deal in different organisms. So does the duration of different stages in an organism. Prophase I lasts for the longest duration.

Types of Meiosis

There are three types of meiosis, *zygotic or initial meiosis, gametic or terminal meiosis* and *sporic or indeterminate meiosis.*

(1) Zygotic or initial meiosis (haplotone pattern) occurs in lower plants. Fertilization is immediately followed by meiosis, giving rise to cells with the haploid chromosome number. The zygote is the only diploid stage in the life cycle.

(2) Gametic or terminal meiosis (diplotonic pattern) occurs in animals. The cells resulting from the two meiotic divisions are transformed into sperms or eggs without further cell division. The process is as follows: Primordial germ cells divide repeatedly to give rise to many generations of daughter cells, spermatogonia (in males) or oogonia (in females). The spermatogonia and oogonia are transformed into primary spermatocytes and primary oocytes, respectively. Each primary spermatocyte undergoes the first maturation (meiotic) division and forms two secondary spermatocytes. Each secondary spermatocyte undergoes the second maturation (meiotic) division to give rise to two spermatids, which are then transformed into spermatozoa. In the female during the first meiotic division the primary oocyte transformed into a large secondary oocyte and a smaller cell, the first polar body. The secondary oocyte undergoes the second meiotic division and divides into an ootid and a second polar body. The first polar body may or my not divide into two. The ootid is the ovum proper.

(3) Sporic or indeterminate meiosis (diplohaplotonic pattern) is characteristic of higher plants and some thallophytes, but is not known in animals. In the diploid sporophyte, meiosis takes place at some indeterminate point between the zygote and gamete formation. As a result of meiosis haploid megaspores and microspores are formed. The megaspores and microspores undergo repeated mitotic divisions and form the male and female gametophyte, respectively. Certain cells of the gametophyte differentiate and from the gametes. In plants the cells that undergo meiosis are the microspore mother cell and the megaspore mother cell.

Significance of Meiosis

The process of meiosis keeps the chromosome number constant from one generation to other. This also involves reshuffling the hereditary characteristic which arises from reassortment of the parental chromosomes though crossing over. *Carothers* (1913) and others mentioned that in the metaphase I of meiosis it is usually a metter of chance whether the centromere of the paternal or the bivalent lies towards the particular pole. This results in random origin of new set on speration. There may be many new combinations. In fact, since each chromosome can go to either pole, the total possible number of arrangements will be 2n where n is the haploid number of chromosomes.

Darlington (1931) has suggested that the meiosis can be regarded as a precocious mitotic, and that the force which retains daughter chromatids together in mitotic prophase also retains the homologous chromosomes pairs in the earlier stages *(Ambrose)*.

Role of Centriole During Meiosis

The centrioles perform precisely controlled synchronous function during meiosis. Where meiosis occurs in two steps, that is, two nuclear divisions take place in sequence, the centriole reproduces and functions to from two achromatic figures. In the first division of meiosis, the centromeres do not duplicate but rather homologous chromosomes pair together and move to opposite poles, while in the second meiotic division the centromeres of chromatids are separated. A scheme for relating various key events is presented briefly as under:

(i) Non-duplication of centromeres in meiosis-I,
(ii) Non-duplication of chromosomes in meiosis-II, and
(iii) Duplication and function of centrioles in both divisions.

By a complete repression of the duplication, simultaneous function of the chromatic appearatus brings about the reduction. The cetriolar function is super-imposed upon the duplication process of the chromatic apparatus in a sequence which leads to the loss of a generation of chromatic apparatus. When meiosis is gametic and fertilization follows after, the immediate formation of zygote as a cell might contain the maternal as well as paternal centrioles, having endowed as it were, with the extra centriole generation. According to *Daniel Mazia* the kinetochore is an essential parts of the chromosomes so far as the mitosis is concerned. In gametogenesis there is a gain in total chromatic material, since these types of cell behaviour lead to duplication and subsequent reamassing of chromosomes. The extra set of centriole is rejected normally in fertilization or autogamy. Thus a cell results in a chromatic apparatus. This is accomplished by meiosis: one division meiosis gives the centriole a chance to 'catch up' in a single duplication.

TABLE 4.1.

Showing Differences between Mitosis and Meiosis

	Mitosis		*Meiosis*
1.	Prolonged resting stage.	1.	This stage is very short.
2.	The prophase is short without any substages.	2.	The prophase is subdivided into six stages, viz., Proleptonema Leptonema, Zygonema, Pachynema, Diplomena and diakinesis.
3.	The chromosomes split longitidinally into sister chromatids during the preceding resting stage. From early prophase they appear as double tetrads.	3.	The chromosomes do not split longitudinally.
4.	The split homologues are not attracted towards each other.	4.	At zygonema or succeding stages, in prophase of first division, the homologous chromosomes are attracted towards each other and they from bivalents.
5.	Synapsis and tetrad stages do not occur.	5.	The chromosomes split after synapsis to from sister chromatids which are arranged in tetrads.
6.	No chromosomal crossing over takes place.	6.	The crossing over is a prominent feature which results in exchange of genic material.
7.	The chromatids are coiled in plectonemic manner.	7.	Coiling is paranemic type, and much evident.
8.	Doubling of chromosomes takes place early in prophase.	8.	Doubling of chromosomes occurs in pachytene stage of prophase.
9.	Nucleolus and nuclear membrane disappear later on and the formation of nuclear spindle is completed with the commencement of metaphase.	9.	The nuclear membrane and nucleolus disappear and the formation of acromatic figure is completed with diakinesis simultaneously.

Mitosis	Meiosis
10. The nucleus enlarges to some extent.	10. The nucleus is greatly enlarged.
11. At metaphase, the centrosome of each chromosome divides and the sister chromatids move towards the opposite poles. The somatic number of chromosomes is maintained without any reduction.	11. At metaphase I, the centrosome does not divide but the homologous chromosomes move towards opposite poles. The chromosome number is reduced to half by reduction division.
12. The chromosomes are seen in dyad consisting of two chromatids.	12. The chromosomes are in the form of tetrads having four chromatids
13. Each chromosome of paternal and maternal origin is longitudinally divided and the separation to opposite side results in identical cells at the end of mitosis.	13. At metaphase I, the oricentation of chromosomes at the equatorial plate is in such pattern that either paternal or maternal chromosome goes to either pole. The orientation of one pair is independent of the other.
14. During orientation of dyads the centromeres face towards the equator and the arms directed towards the poles.	14. During the orientation of tetrads, the centromeres are placed equidistant and face the pole while arms are directed towards the equator.
15. The centromere of each bivalent undergoes division at metaphase.	15. No division of centromere occurs, but homologous chromosomes simply separate to form two dyads.
16. The initial movement of chromosomes at late metaphase is due to repulsion between the centromeric parts.	16. The main repulsive force lies between the centromeres of homologous chromosomes.
17. At anaphase the chromosomes are less thick and short.	17. The chromosomes are very thick and short.
18. At anaphase further separation of chromosomes is due to the formation of stem body.	18. Most of the changes are similar to mitosis.

19.	At telophase two daugher nuclei, each with diploid number of chromosomes, are reconstituted.	19.	At this phase two daughter nuclei are haploid.
20.	There is no evidence of second division in mitosis.	20.	There are two divisions for each meiotic division The second division follows after a short gap, interkinesis. The second division is homotypic by nature, and results into four cells, in the form of tetrads.
21.	In the mitosis neither the chromosomal number nor the quality is changed and the daugher cells carry 2n chromosomes.	21.	At the end of meiosis each of tertrads is haploid and quality is changed.
22.	This division takes place in growing somatic cells.	22.	This division takes place only in gonadial cells and result in the formation of gametes.

5

The Genetic Material

Now-a-days it is well-known fact that genes are found in a linear fashion on the chromosomes. There may be as many as 1000 to 10,000 genes on a chromosome depend's upon the length of chromosome bearing genes. Various attempts were made by early molecular geneticists to identify the physical and chemical nature of genes. But, the genes were found so minute structure that their physical identity remained almost impossible. However, the extensive chemical analysis of chromosomes of different organisms have revealed that chromosomes contain proteins and nucleic acids (DNA and RNA) and it was thought that genes might have either proteins or nucleic acids as their component molecules. In 1953, it was universally accepted that DNA is the genetic material. Later on RNA was also found to be the genetic material in some viruses.

Requirements for the Genetic Material

The genetic material is of central importance to cell function and therefore must fulfill a number of basic requirements:

1. It must contain the information for cell structure, function, and reproduction in a stable form. This information is encoded in the sequence of basic building blocks of the genetic material.

2. It must be possible to replicate the genetic material accurately such that the same genetic information is present in descendant cells and in successive generations.

3. The information coded in the genetic material must be able to be decoded to produce the molecules essential for the structure and function of cells.

4. The genetic material must be capable of (infrequent) variation. Specifically, mutation and recombination of the genetic material are the foundations for the evolutionary process.

The nucelic acids, *deoxyribonucleic acid* (DNA) and *ribonucleic acid* (RNA), meet all these requirements.

Evidence that DNA is the Genetic Material

Many lines of evidence strongly indicate the DNA is the genetic material in many organisms. Three examples are given here:

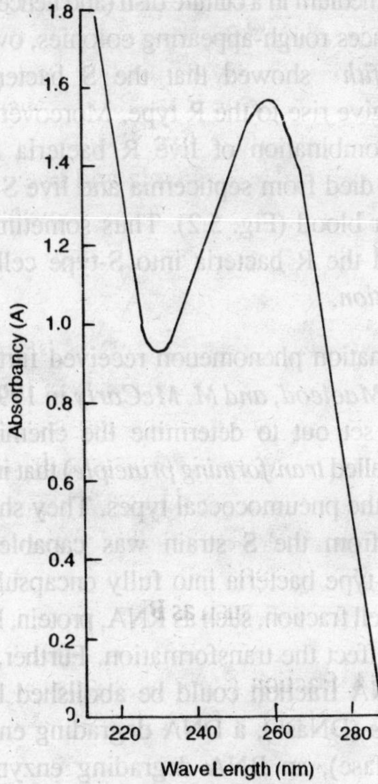

Fig. 5.1. Ultraviolet light absorbancy spectrum of DNA showing the maximal absorbancy.

1. The nucleic acids show maximal absorbance of ultraviolet light at a wavelength of 260 nm (Fig. 5.1), and this correlates exactly with the wavelength a which maximal mutagensis of cells can be achieved by ultraviolet irradiation. This observation provided further

evidence that nucleic acids and not proteins (which show maximal absorbance of light at 280 nm) are the genetic material.

2. In 1928 *F. Griffith* discovered that the S strain of the bacterium *Diplococcus pneumoniae*(pneumococcus), when injected into mice, causes death by septicemia (blood poisoning). Another strain, the R strain, had no effect on the same mice. The distinction between the two strains lies in the fact the S strain bacteria have a polysaccharide capsule around them, resulting in a smooth colony appearance when they grow on solid medium in a culture dish (and hence the S designation). The R strain produces rough-appearing colonies, owing to the lack of the capsule. *Griffith* showed that the S bacterial could mutate spontaneously to give rise to the R type. Moreover, when mice were injected with a combination of live R bacteria and heat-killed S bacteria, the mice died from septicemia and live S bacteria could be isolated from their blood (Fig. 5.2). Thus sometimes from the dead bacteria converted the R bacteria into S-type cells; this process is called *transformation.*

The transformation phenomenon received further scrutiny from *O.T. Avery, C.M. Macleod, and M. McCarty* in 1994. In some classic experiments they set out to determine the chemical nature of the substance (the so-called *transforming principle*) that induced the specific transformation of the pneumococcal types. They showed that a DNA fraction isolated from the S strain was capable of transforming unencapsulated R-type bacteria into fully encapsulated S-type cells. None of the other cell fraction, such as RNA, protein, lipid, crbohydrates, etc., was able to effect the transformation. Further, the transforming activity of the DNA fraction could be abolished by treatment with *deoxyribonuclease* (DNase), a DNA degrading enzyme, but nor by *ribonculease* (RNase), an RNA degrading enzyme. These results strongly indicated *that DNA* was the genetic material. However, Avery's work was criticized because the nucleic acids isolated were not completely pure and contained proteins.

The ultimate proof that DNA is genetic material came from other experiments such as the performed by *A.D. Hershey* and *M. Chase*, which is described next.

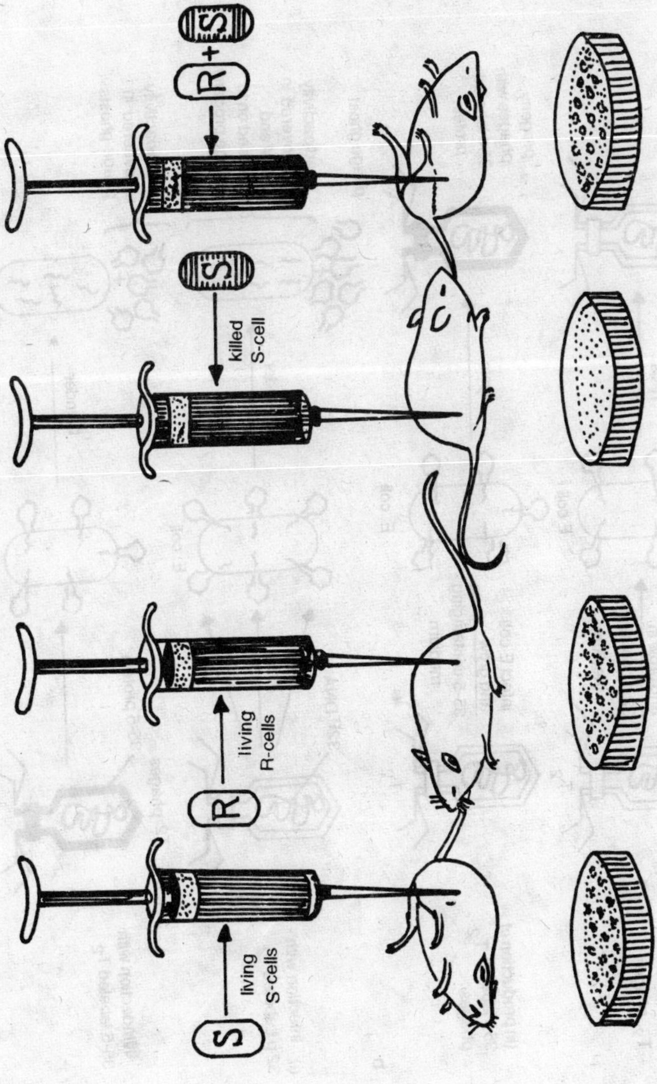

Fig. 5.2. Transformation experiment of Griffith.

3. *Hershey* and *Chase* in 1952 studied the replication of bacteriophages (bacterial viruses) in their bacterial hosts. Bacteriophages consist of two components: DNA and protein. *Hershey* and *Chase*

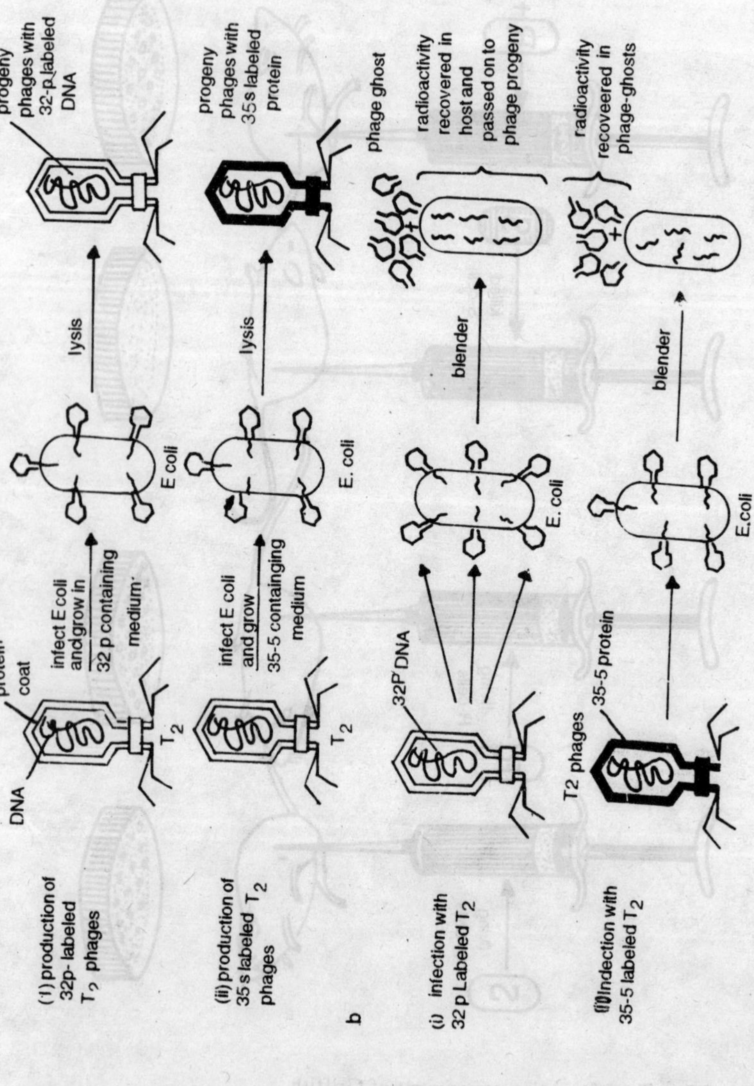

Fig. 5.3. The Hershey and Chase experiment. (a) Production of T2 either with (i) ^{32}P-labeled DNA or with (ii) ^{35}S-labeled protein (b) Experiment that showed DNA and not protein was the genetic material of T2. In (i) *E. Coli is infected with ^{32}P-labeled T2, and in (ii)* E. coli *is infected with ^{35}S-labeled T2.*

infected *Escherichia coli* cells growing in media containing either a radioactive isotope of phosphorus (^{32}P) or a radioactive isotope of sulfur (^{35}S) with bacteriophage T^2 and collected the progeny phages that were produced (Fig. 5.3).

TABLE 5.1

Approximate DNA Content per cell in Some Organisms Including Viruses

	Species	*DNA amount Pg/ per cell*	*Number of nucleotide pairs (in millions)*
1.	Bacteriophage-T4	0.00024	0.7
2.	Bacteria	0.002-0.06	2
3.	Fungi	0.02- 0.17	20
4.	Sponges	0.1	100
5.	Molluscs	1.2	1,100
6.	Crustaceans	3	2,800
7.	Fishes	2	2,000
8.	Amphibia	7	6,500
9.	Reptiles	5	4,500
10.	Birds	2	2,000
11.	Mammals	6	5,500

Since DNa contains phosphorus but no sulfur and protein contains sulfur but no phosphorus, the DNA in the phages produced in ^{32}P-containing medium was radioactive and the protein in the phages produced in 35S-containing medium was radioactive. Since each T^2 phage consists of approximately half DNA and half protein, one of these two components must be the genetic material. To determine which component it was. *Hershey* and *Chase* next infected unlabeled *E. coli* with the two types of radioactively labeled T^2 and obtained the results depicted in Fig. 5.3. When the infecting phage was ^{32}P-labeled (Fig. 5.3), most of the radioactivity was found within the bacteria soon after infection, and after lysis of the cells some of the label was found in the progeny phages. Very little label was found in the protein parts of the phages (called the phages ghosts) that were released from the cell surface by mixing the cells in a kitchen blender. When the infecting phage was ^{35}S-labeled (Fig. 5.3), essentially none of the radioactivity appeared either within the cell or in progeny phages, whereas most of the radioactivity was found in the phage ghosts released after the blender treatment.

We know that the genetic material in most organisms in DNA. In some viruses, however, the genetic material is RNA, as will be discussed later.

Biochemical Evidences for DNA as Genetic Material

Strong supporting evidences that DNA is hereditary material come from biochemical study:

(i) The amount of DNA in any given species of cell or organism is remarkably constant and cannot be altered by environmental circumstances or by changes in the nutrition etc.

(ii) The amount of DNA per cell appears to be in proportion to the complexity of the cell and the amount of genetic information it contains. Table 5.1 presents the amount of DNA in some organisms. This shows that the greater is the amount of DNA, contains about 0.02 pg. (1 pg=picrogram=1×10^{12}gm) per cell whereas tissues of higher animals contain about 6 pg. DNA per cell. Since bacteria and viruses contain relatively little genetic information and so relatively few genes and consequently small amount of DNA.

EVIDENCES FOR RNA AS THE GENETIC MATERIAL OF SOME VIRUSES

The demonstration that RNA is the genetic material in RNA containing viruses came in 1956, when *A. Gierer* and *G. Scharmm* showed that tobacco plants could be inoculated with purified RNA from the tobacco mosaic virus (TMV) and TMV-like lesions could later be identified on the tobacco leaves. A different approach was taken by *H. Fraenkel Conrat* and *B. Singer* in experiments published in 1957. They first separated the RNA from the protein of TMV viruses, much as *Hershey* and *Chase* had done with the DNA and the protein of T^2 bacteriophage. They then developed techniques for forming "reconstituted" viruses containing the protein from one mutant strain of TMV and the RNA from another, or vice-versa. Such hybrid viruses were allowed to infect tobacco leaves, and the progeny were examined. In all cases the progeny were the parental RNA type and not the parental protein type.

In a much improved technique *N. Pace* and *S. Spiegelman* in 1966 purified RNA from two different mutant strains of the RNA phage QB which had quite distinct base compositions. The isolated RNAs were then incubated separately in the presence of an *E. coli* cell extract containing and enzyme capable of RNA replication. The new RNA synthesized was in each case identical in base composition to the particular phage RNA presented to the *in vitro* system, thus indicating that the phage of RNA can serve as a template for its self-replication.

6

Nucleic Acid

The fundamental investigations which led to the discovery of the nucleic acids were made by *Friedrich Miescher* (1844-95) who may be regarded as the founder of our knowledge of the chemistry of the cell nucleus. In early work carried out in 1868, in the laboratory of Hoppe Seyler in Tubingen, he isolated the nuclei from pus cells obtained from discarded surgical bandages and showed the presence in them of an unusual phosphorus containing compound that he called '*nuclein*' and which we now know to have been nucleoprotein. It was subsequently shown that nucleic acids were normal constituents of all cells and tissues and Miescher's investigations of the nucleic acid were continued by *Altmann,* who in 1889 described a method for the preparation of protein free nucleic acids from animal tissues and from yeast. The term *nucleic acid* was introduced by *Altmann. Albrecht Kossel, P.A. Levine* and *Walter Jones* described the chemical nature of nucleic acid. According to them the nucleic acid is composed of phosphoric acid, a sugar and nitrogenous bases. *Franklin W. Stahl* presented first evidence that nucleic acid forms the genetic material. *Chargaff* (1951) described the occurrence of nitrogenous bases in equal proportion. *Dotty* (1961) has emphasized much on the physical properties of DNA. In 1962 *Willkins, Wlatson* and *Crick* proposed a model to explain the double helicular structure of DNA. They also shared the Nobel Prize for the same.

Deoxyribonucleic Acid or DNA

DNA is found in the cells of all living organisms except plant viruses, where RNA forms the genetic material and DNA is absent. In bacteriophages and viruses there is a single molecule of DNA, which remains coiled and is enclosed in the protein coat. In bacteria, mitochondria and plastids of eukaryotic cells DNA is circular and lies naked in the cytoplasm. In the nuclei of eukaryotic cells DNA occurs in the form of long spirally coiled and unbranched threads. The number of DNA molecules is equivalent to the number of chromosomes per cell. In them DNA is found in combination with proteins forming nucleoproteins—the *chromatin* material and is enclosed in the nucleus.

DNA is a polymer, the monomeres of which are the deoxyribonucleoside monophosphates. Most DNA have high molecular weight, varying from 10^6 to 10^9 or more. Chemical analysis of DNA molecules from different kinds of organisms suggest that in one kind of DNA the number of adenine (A) nucleotides was always the same as the number of thymine (T) nucleotides. Similarly the number of guanine (G) nucleotides was always the same as the number of cytosine (C) nucleotides. This evidence suggests that nucleotides occur in pairs in the DNA molecule. Further, quantity of A present in one organisms's DNA (and it varies from species to species) is equal precisely to the amount of T. Similarly the amount of G present is always equal to the amount of C.

Physical Structure of DNA

According to the model proposed by *Watson* and *Crick*, the DNA contains two polynucleotide chains running antiparallel and twisted around each other in the form of a regular double helix. The diameter of the helix is about 20A° and each chain makes a complete turn every 34A°. Each chain is polynucleotide in which the sugar of each nucletide is linked by a phosphate group to the sugar of the adjacent nucleotide. There are 10 nucleotides on each chain in every turn of the helix. The two helices are coiled in such a way that they cannot be separated without unwinding. The two chains of a helix are of opposite polarity. If one chain runs 3'-5' direction, then the other will run as 5'-5' direction.

The molecular weight for purified DNA is found in the range 1000,000 to 120 x 10^6. The amount of DNA in a mammalian cell corresponds to about 25,000 molecules of molecular weight 120 x 10^6, a total of about 10,000 million nucleotides.

Monomeric Components of DNA

Before any account is given of the structure of DNA proper it is necessary to describe the structures of its component parts. The DNA molecule is a polymer consisting of several thousand pairs of nucleotide monomeres. Each nucleotide consists of the pentose sugar, *deoxyribose,* a *phosphate* group and a *nitrogenous base,* which may be either a *purine* or a *pyrimidine.*

1. *Deoxyribose suguar. Levine* (1909) identified ribose sugar as a main constituent of nucleic acid structure. *Mori* (1929) isolated sugar from the guanine nucleoside and showed that it is deoxyribose sugar. It is pentose type which in DNA molecule shows the absence of one oxygen molecule from carbon-2 position of ribose sugar. Both deoxyribose and ribose (pentose sugars of nucleic acids) have a pentagonal ring with five carbons, among which two (i.e. 3' and 5') are attached to phosphoric acid and three (1') to the base. This sugar is called *deoxyribose* and it simply acts as a support column to which bases are attached. When the pentose sugars occur in nucleic acids nucleotides the carbon atoms an numbered as 1', 2', 3' etc., to avoid confusion with the numbering of the ring atoms of the bases.

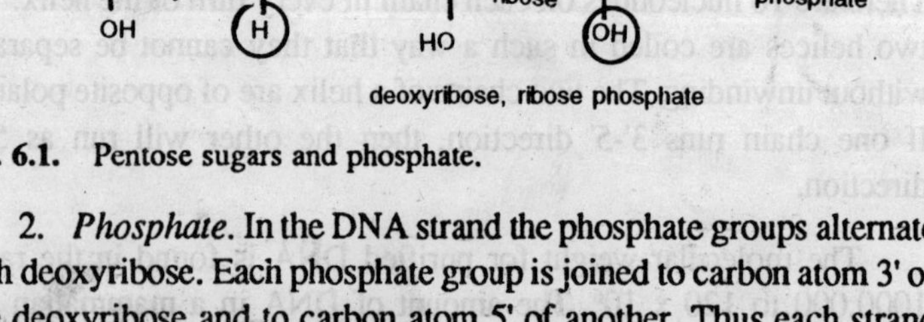

Fig. 6.1. Pentose sugars and phosphate.

2. *Phosphate.* In the DNA strand the phosphate groups alternate with deoxyribose. Each phosphate group is joined to carbon atom 3' of one deoxyribose and to carbon atom 5' of another. Thus each strand has a 3' end and a 5' end. The two strands are oriented in opposite directions. The 3' end of one strand corresponds to the 5' end of the other. Consequently the oxygen atoms of deoxyribose point in opposite directions in the two strands. *Phosphodiester bonds* are formed between

the sugars of two different nucleotides and a phosphate group.

3. *Nitrogenous Bases*. The nitrogenous bases of nucleic acid are of two types: *Purine* and *Pyrimidine*. Purine bases comprise mainly adenline (A) and guanine (G) which are common to both DNA and RNA while pyrimidine bases comprise cytosine (C) and thymine (T) in DNA and in the cases of RNA, thymine (T) is replaced by uracil (U).

These bases are arranged linerly as the back bones of chains which (back bones) are composed of alternating sugar and phosphate. The bases have specific way of combining, if C is found in one chain G will be opposite to it and in other if A is on one T must be on the other. Or in other words adenine (A) is always paired with thymine (T) and guanine (G) with cytosine (C).

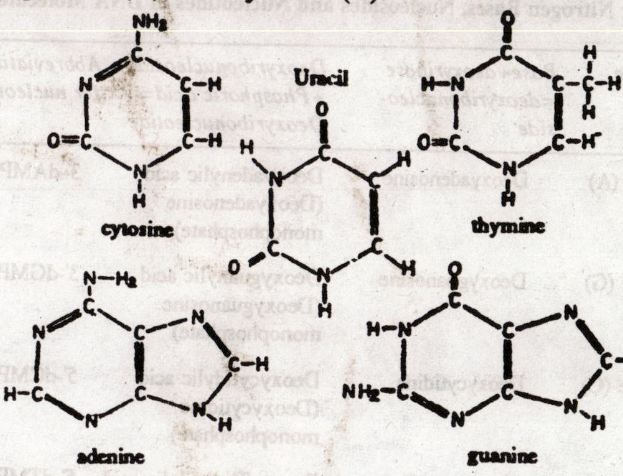

Fig. 6.2. Common nitrogenous bases of nucleic acids.

Among bases A-T and G-C, two hydrogen bonds are formed between A and T and three hydrogen bonds are formed between C and G. This hydrogen bond formation precludes A-C or G-T pairs. Only these arrangements A-T and G-C are possible, for two purines would occupy too much space to allow a regular helix and correspondingly, two pyrimidines would occupy too little. The stickness of these pairing rules result in a complementary relation between the sequences of bases on the two interwined chains. For example, if we have a sequence

ATGTC on one chain, the opposite chain must have a sequence TACAG. The hydrogen atom with its positive charge is shared between an oxygen atom and a nitrogen atom, both with slight negative charges. Although hydrogen bonds are weak, the fact that there are so many gives stability to the DNA molecule. The weak hydrogen bonding enables the two strands of the DNA to separate during replication.

The bases are joined to the pentoses by N-C *glycosidic bonds.* For the purine, the glycosidic bond is between the C_1 position of the pentose sugar and the N_9 position of the bases. For the pyrimidines the linkage joints the C_1 and N_3 positions.

TABLE 6.1

Four Nitrogen Bases, Nucleosides and Nucleotides of DNA Molecule

Nitrogen base	*Base+deoxyribose =deoxyribonucleoside*	*Deoxyribonucleoside +Phosphoric acid= Deoxyribonucleotide*	*Abbreviation for nucleotide*
1. Adenine (A)	Deoxyadenosine	Deoxyadenylic acid (Deoxyadenosine monophosphate)	3'-dAMP
2. Guanine (G)	Deoxyguanosine	Deoxyguanylic acid (Deoxyguanosine monophosphate)	3'-dGMP
3. Cytosine (C)	Deoxycytidine	Deoxycytidylic acid (Deoxycytidine monophosphate)	5'-dCMP
4. Thymine (T)	Deoxyhymidine	Deoxy Thymidylic acid (Deoxythymidine monophosphate)	5'-dTMP

Nucleosides

When a purine or a pyrimidine base is linked to ribose (in case of RNA) or deoxyribose (in case of DNA) the resulting compound is known as a *nucleoside*. Thus adenine condenses with ribose to form the nucleoside adenosine, guanine forms guanosine, cytosine forms cytidine and uracil form uridine. These ribonucleosides can be formed on partial hydrolysis of RNA. The ribonucleoside from hypoxahtnine is named inosine. The nucleosides derived from 2-deoxyribose are known as deoxyribonucleosides-deoxyadenosine, deoxyguanosine, deoxycytidine, deoxythymidine and so on.

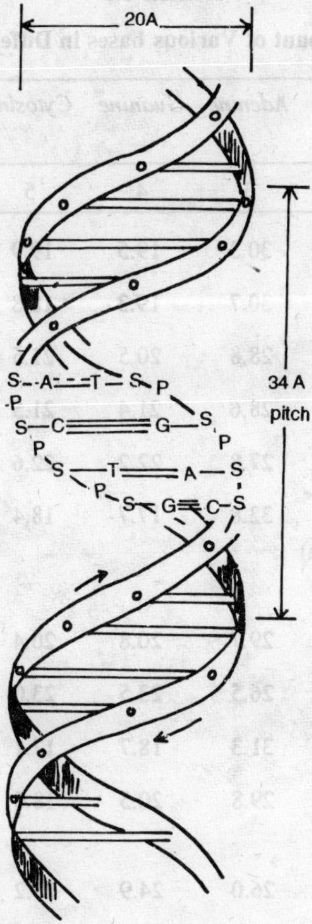

Fig. 6.3. The Watson-Crick base pair in DNA, Adenine-Thymine, and Guanine-Cytosine.

Nucleotides

When a nucleoside is linked to phosphoric acid the resulting compound is known as a *nucleotide*. They are all phosphoric acid esters of the nucleosides. Those derived from ribonucleosides are usually referred to as *ribonucleotides* and those from deoxyribonucleosides as *deoxyribonucleotides*. These terms are sometimes abbreviated to ribose, ribotide, deoxyriboside and deoxyribotide.

TABLE 6.2

Showing Amount of Various bases in Different Tissues

S.No.	*Source*	*Adenine*	*Guanine*	*Cytosine*	*Thymine*	*A + T* / *G + C*
1	2	3	4	5	6	7
1.	Human Liver	30.3	19.5	19.9	30.3	1.53
2.	Human sperm	30.7	19.3	18.8	31.2	1.62
3.	Hen red cells	28.8	20.5	21.5	29.2	1.38
4.	Rat bone marrow	28.6	21.4	21.5	28.4	1.33
5.	Herring sperm	27.8	22.2	22.6	27.5	1.23
6.	*Paracentrotus lividus* (sea urchin) sperm	32.8	17.7	18.4	32.1	1.85
7.	Salmon	29.7	20.8	20.4	29.1	1.43
8.	Wheat germ	26.5	23.5	23.0	27.0	1.19
9.	Yeast	31.3	18.7	17.1	32.9	1.79
10.	*Diplococcus pneumonial*	29.8	20.5	18.0	31.6	1.59
11.	K-12 *Escherichia Coli*	26.0	24.9	25.2	23.9	1.00
12.	*Trycobactarium tubercubsis*	15.1	34.9	35.4	14.6	0.42
13.	Bacteriophage T_2	32.5	18.2	16.7	32.6	1.86

In addition to the nucleosides listed above several others are found in very small amounts in certain classes of nucleic acids.

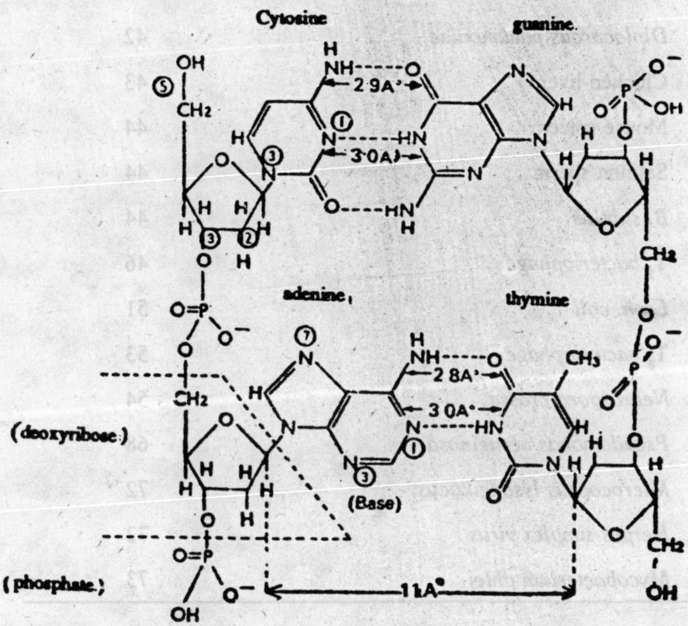

Fig. 6.4. Segment of a DNA molecule.

TABLE 6.3

The Relative (G+C) Content of DNA's from Various Sources

Source of DNA	*Per cent (G+C)*
Dictyostelium (slime mould)	22
M. pyogenes	34
Bacillus cereus	37
B. megaterium	38
Haemophilus influenzae	39
Saccharomyces cerevisiae	39
Calf thymus	40
Rat liver	40
Bull sperm	41

Diplococcus pneumoniae	42
Chicken liver	43
Mouse spleen	44
Salmon sperm	44
B. subtilis	44
T_1 *bacteriophage*	46
Esch. coli	51
T_3 *bacteriophage*	53
Neurospora crassa	54
Pseudomonas aeruginosa	68
Micrococcus lysodeikticus	72
Herpes simplex virus	72
Mycobacterium phlei	73

The Primary Structure of DNA

As stated already that nucleic acids are polymers made up of hundreds, thousands or millions of nucleotides coupled together by *phosphodiester linkages.* In the case of DNA where C-4' in the sugar is occupied in ring formation and C-2' carries no hydroxyl group, only the hydroxyl groups at positions 3' and 5' are available for internucleotide linkages. The absence of a hydroxyl group at C-2' makes cyclic phosphote formation impossible and DNA is therefore not hydrolysed by alkali in the same way as in RNA. The primary structure of the polynucleotide chain in DNA is shown in the following Fig. 27.5.

The representation of polynucleotide chains by complete formulae is clumsy and it has become customary to use the schematic system illustrated in the above figure, where the chain is abbreviated as on the right. The vertical line denotes the carbon chain of the sugar with the base attached at C-1'. The diagonal line from the middle of the vertcal line indicates the phosphate link of C-3' while that at the end of the vertical line remote from the base denotes the phosphate link at C-5'. This system may be used for either RNA or DNA.

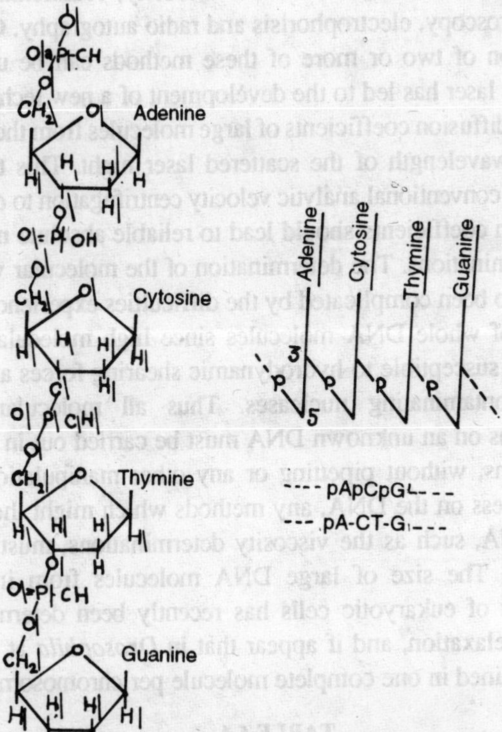

Fig. 6.5. Part of the polynucleotide chain in DNA.

Molecular Weight of DNA

The molecular weight of DNA molecules are very difficult to determine accurately by the methods of classic chemistry since they range from 10^6 to more than 10^{10}. Conventional analytical equilibrium ultracentrifugation is unsatisfactory since the available instruments are not stable at speeds low enough to balance centrifugal forces against diffusion forces. The best absolute method of DNA molecular weight determination are light scattening on low angle instruments equillibrium analytical ultracentrifugation in Caesium chloride gradient and viscoelastic relaxation. Most other methods have an empirical bases and rely on the few light scattering experiments which have been performed for their calibration. These more empirical methods include the measurement of intrinsic viscosity, sedimentation rate, electron microscopy, electrophorisis and radio autography. Generally a combination of two or more of these methods can be used. The advent of the laser has led to the development of a new technique for determining diffusion coefficients of large molecules from the Doppler shift in the wavelength of the scattered laser night. This technique together with conventional analytic velocity centrifugation to determine sedimentation coefficients should lead to reliable absolute molecular weight determinations. The determination of the molecular weight of DNA has also been complicated by the difficulties experienced in the preparation of whole DNA molecules since high molecular weight DNA is very susceptible to hydrodynamic shearing forces and to the action of contaminating nucleases. Thus all molecular weight determinations on an unknown DNA must be carried out in nuclease free conditions, without pipetting or any other manipulation which puts shear stress on the DNA, any methods which might themselves shear the DNA, such as the viscosity determinations, must be used with caution. The size of large DNA molecules from individual chromosomes of eukaryotic cells has recently been determined by viscoelastic relaxation, and if appear that in *Drosophila* at least the DNA is contained in one complete molecule per chromosomes.

TABLE 6.4

DNA Molecular Weights

Source	*Mol. wt.*	*Length*	*Number of nucleotide pairs*	*Conformation*
E. coli				
chromosome	2×10^9	1 mm	3×10^6	Cyclic, duplex
H. influenzae				
chromosome	8×10^{-3}	300μm	12×10^5	—
Mycoplasma PPLO				
strain H-39	4×10^8	150 μm	6×10^5	—
Bacteriophage				
T2 or T4	$1.3. \times 10^3$	50μm	2×10^5	Linear, duplex

Bacteriopage λ	33 x 10^6	13μm	0.5 x 10^5	Linear, duplex
Bacteriophage ϕX174	1.6 x 10^6	0.6*μ	—	Cyclic, single-stranded
Polyoma virus	3 x 10^6	1.1μm	4.6 x 10^3	Cyclic, duplex
Drosophila melanogaster chromosome	43 x 10^9	20 mm	65 x 10^6	—

In more recent years if has become customary to describe the size of DNA molecules in terms of kilobase pairs (kbp), rathee than daltons or molecular weight. A molecu!ar weight of one million corresponds approximately to 1.5 kilobase pairs.

Secondary Structure of DNA

Using early information obtained by X-ray diffraction and chemical observations, *Watson* and *Crick* in 1953 put forward the view that the DNA molecule is *double stranded* and in the form of a right handed helix with the two polynucleotide chains would round the same axis and held together by hydrogen bonds between the bases. By using scale models they were able to show that the bases could fit in if they were arranged in pairs of one pyrimidine and the purine. The equivalence of adenine and thymine, and guanine and cytosine in most naturally occuring DNA molecules. First observed by *Chargaff* suggested that the most likely hydrogen bonding configuration of base pairs of these two types. These base pairing arrangements have been confirmed as the only ones possible in the B-form of DNA. Although other base pairing arrangements have since been suggested they have only been shown to occur in RNA where they are involved in secondary and tertiary structure stabilization.

The most important consequence of the base pairing configuration formed in DNA is that the order in which the bases occur in one chain automatically determines the order in which they occur in the other, *complementary chain*. Apart from this essential condition, there are no restrictions on the sequence of the bases along the chains.

The shorthand method of describing two DNA strands has been

to have the strand of 5'—>3' polarity on the top line of the sequence with the complementary strand of opposite polarity lying below. For example :

5' - AGGTC - 3'

3' - TCCAG - 5'

The pairs of bases are flat and may be stacked on above the other like a pile of plates so that the molecule is readily represented as a spiral staircase with the base pairs forming the treads. The two polynucleotide chains are of opposite polarity in the sense that the internucleiotide linkage in one strand is 3'—>5' while in the other it is 5'—>3'. The two helices are right handed (i.e. (a) turn in a clockwise direction as viewed from the near end or (b) turn in the direction of the fingers of the right hand when the thumb indicates the line of sight) and cannot be separated without unwinding. The pitch of the duplex is 3.4 nm and since there are 10 base pairs on each turn of the helix, there is a distance of 0.34 nm between each base pair. The diameter of the helix is 2 nm.

The base sugar linkage is normally *anti* with the C8 of the purine or the C6 of the pyrimidine ring over the sugar. The sugar puckev is variable but a base pair usually has the sugar attached to the pyrimidine in the C3' *endo* configuration. There is lot of water associated with DNA. Upto 72 molecules of water per 12 base pairs are found largely in the *minor groove* where they form a spine.

The stacked bases form a hydrophobic core but the amino and keto groups point upto the grooves and allow interaction with the solvent. The sugar phosphate backbone points out to give hydrophilic ridges with one negative charge per phosphate.

While the basic model put forward by of *Watson* and *Crick* remains close to the accepted structure of the DNA molecule in solution, the more refined X-ray diffraction studies of *Wilkins* and his colleagues demonstrated that DNA fibres can have three possible structures with the B structure corresponding closely to the original Watson-Crick model. The A and C structures are also right handed double helices but differ in the pitch and in the number of bases per turn. In both A and C conformations the bases are not flat but are titled. Their biological significance is not clear but the A conformation is believed to be very close to the structure adopted by double stranded RNA and by DNA - RNA hybrids. Because of the presence of extra 2'hydroxyl group, RNA appear to be unstable to adopt the B conformation. Thus when it is engaged as a template for making RNA, the DNA molecule must adopt the A conformation. Various organic solvents and proteins have been suggested to force the DNA from the B to the A form, and the transition from the B conformation to the C conformation appears to occur in concentrated salt solution and ethylene glycol.

Various forms of DNA

A
B
C
D
Z.

The Tertiary Structure of DNA

The DNA from many sources, particularly cellular chromosomal DNA, cannot be isolated in the form of intact molecules because of its large size and consequent sensitivity to shear and enzynatic degradation. However, the DNA from many viruses from mitochandria and chloroplast and from a few other sources (e.g. bacteral plasmid DNA) can be isolated in an ungraded form. The examination of such ungraded DNA has shown that the double helical secondary structure of DNA can in some instances be further contained giving rise to tertiary structure such as super coiled and open cyclic molecules. Linear DNA occurs naturally and is also a product of the degradation or cyclic DNA. Single stranded DNA, both in linear and cyclic forms, is found in some viruses, and can be formed artifically by denaturing double stranded DNA.

TYPES OF DNA'S

Eukaryotic DNA

It is the filamentous type of DNA found in all eukaryotic cells of animals and plants. In these cells, DNA synthesis takes place in a restricted time period during the cell cycle. In mitosis the cells divide and form two daughter cells. This is then followed by a gap period designated G1 during which no DNA synthesis occurs. The length of the various gap periods and indeed the length of the entire cell cyclic depends very much upon the cell type. The synthesis of DNA occurs in S phase and then follows a second shorter gap period known as G_2. The cells then go into mitosis again.

Prokaryotic DNA

Also called the *circular* a *superhelicular DNA*. The DNA molecules of prokaryotes and most viruses are circular. A circular molecule may be a covelently closed circle, which consists of two unbroken complementary, single strands, or it may be a *nicked circle*, which has one or more interruption (nicks) in one or both strands. With few exceptians, covalently closed circles are twisted. Such a circle is said to be a *superhelix* or a *supercoil*.

Extranuclear DNA

Also known as non-chromosomal or cytoplasmic DNA. In cytoplasm this DNA is associated with some of cell organelles as given belows.

(a) Mitochondrial DNA : Mitochondrial DNA is usually found in cyclic double stranded, supercoiled molecules, the exceptions being the linear mitochondrial DNAl molecules from *Tetrahymena* and *Parameciui*. Marmialian mitochondrial DNA molecules are not packaged into nucleosomes. They are about 15 kbp long and can therefore code for 15 to 20 proteins. However, some yeast ones are considerably larger and *Saccharcmyces* mitochondrial DNA is 75 kbp long. Plant mitochondrial DNA is much longer still. Yeast mitichondrial DNA molecules have been widely studied since they are readily amenable to genetic analysis. Mitochondrial DNA can only codes for a small proportion of mitochondrial proteins.

Mitochondrial DNA from several mammai, *Xenopus* and *Drosophila,* has been sequenced and the sequence analysed. There are genes for two ribosomal RNAs, 22 tRNAs and for 13 protein most or all of which are involved in electron transport. The only region of mammalian mitochondrial DNA which is non-coding is the *D-loop region* involved in the initiation of DNA replication. This is also the region at which transcription of both strands is initiated. Transcription continues uninterrupted around the cyclic molecule and the transcripts are the processed to give individual messenger, ribosomal and tRNAs. The tRNA forms the puncturation between the various protein coding regions and provides the signals for the processing enzymes.

Plant mitochondrial DNA is apparently much more complex than animal mitochondrial DNA. It consists of per mutation of basic structure related to each other by recombination.

(b) *Chloroplast DNA* : Chloroplast DNA is in general much larger than mitochondrial DNA, being in the molecular weight range 100 million (150 kbp). In contrast to mitochondria which have from one to ten molecules of DNA per organelle, chloroplasts tend to have a very large number of copies of the DNA molecules in each organelle, in some cases greater than one hundred. Like mitochondrial DNA, the DNA in chloroplasts carries the coding information for essential membrane components, tRNA and rRNA. All known chloroplast DNA molecules are cyclic and supercoiled.

(c) *Kinetoplast DNA*: The kinetoplast is part of highly specialized mitochordrion found in certain groups of flagellated protozoa such as trypanosome. Its DNA (kDNA) consists of an interlinked series of many thousands of cyclic DNA molecules which vary in size from 0.6 kbp to 2.4 kbp depending from the type of trypanosomes from which they are obtained. These components, which are known as *minicircles*, are further interlinked with a much smaller number of larger circular DNA molecules of above 30 kbp in size known as *maxicircles* in the majority of system analyzed. While maxicircles appear to perform the connectional functions of mitochondrial DNA in trypanosomes, minicircles are microheterogeneous in sequence and size and there is no evidence to suggest that they are ever transcribed so that it is possible that they may fulfil some structural rather than coding role.

Unique DNA

In human cells it is generally classified as that DNA which has a

$Cot_{1/2}$ value of about 1000 moles of nucleotide seconds litre^{-1}. It comprises about half of the total haploid DNA content and is thought to consist of the sequences coding for most enzyme functions for which there is only one or a small number of genes per haploid genome. The genes coding for the various chains of globin or the enzyme glucose 6-phosphatas fall into this category.

Repetitive DNA

In this type of DNA certain bases or nucleotides are repeated many times. Among the repetitive DNA is a fraction which reanneals with a Cot value of between 100 and 1000 moles of nucleotides seconds litre^{-1}. The sequences represented in this group are generally thought to be those coding for proteins which form major structural components of the cell such as the histones. The genes for rRNA and ʻRNA also fall into this category.

Satellite DNA

At the other extreme DNA with a $Cot_{1/2}$ value as low as 10^{-3} moles of nucleotide seconds litre^{-1} consists largely of *satellite DNA*. This represents highly repeated sequences of which there may be a million or more copies per haploid genome, which are usually quite short and are arranged in tandem arrays. The origin of the name satellite relates to the method of its isolation on caesium chloride buoyant density gradients of sheared DNA where it will sometimes form a satellite band separate from main DNA band, due to its differing content of adenine and thymine residues. The simplest known DNA is poly [d(A-T)], which occurs in certain crabs. Other satellites can have any number upto several hundred base pairs which are repeated in tandem fashion along the genome.

The distribution of satellite DNA among chromosomes varies. Some chromosomes have virtually no satellite sequences while others are largely composed of satellite sequences. In general, satellite DNA appears to be concentrated near the centromere of the chromosomes in the heterochromatin fraction. DNA sequence analysis has shown that the basic repeat unit of satellite DNA is itself made up of sub repeats. For example, the major momse satellie has a repeating structure of 234 base pairs made up of four related 58 and 60 bp segments each in turn made up of 28 and 30 bp sequences. The satellite between and within the related species are themselves related in an evolutionary sense by cyclic rounds of multiplication and divergence of an initial short sequence.

Fold-back DNA

Also known as *palindromic DNA*. It is a special class of DNA sequences comprising 3-6% eukaryotic DNA. The size range is from 300 to 1200 base pairs and the molecules have a $Cot_{1/2}$ value of less than 10^{-5} moles of nucleotide second $litre^{-1}$. This palindromic DNA is represented in all frequency classes and is widely distributed throughout the metaphase chromosomes. Some palindromic DNA arises when two copies of the *Alu* sequences are present close to one another in opposite orientations and it was as a result of the ability of such DNA to renative instantaneously.

Z-DNA or Left Handed DNA

On the basis of the arrangement of pentose sugar forming chains of DNA two forms of DNA have been reported. Usually the sugar chain has the pentose sugar with ascending sequence of carbon atoms. This is the right handed DNA or B-DNA. Recently in 1979, another form of DNA have pentose sugar with ascending arrangement of carbon atom has been reported and named Z-DNA or left handed DNA.

Tautomeric form of DNA

An important chemial feature of DNA is the position of the hydrogen atoms in the purine and pyrimidine bases. Before 1953, many chemists thought that some of these hydrogen atoms randomly moved from one ring nitrogen or oxygen atoms to another and so could not be assigned a fixed location. Now we realize that although such movements, called *tautomeric shifts* do occur, these hydrogens have preferred atomic locations. The nitrogen atoms attached to the purine and pyrimidine rings are usually in the *amino* (NH_2) form and only rarely assume the *imino* (NH) configuration. Likewise, the oxygen atoms attached to the C6 atoms of guanine and thymine normally have the *keto* (C-O) form and only rarly take up the *enol* (COH) configuration.

It is essential to the biological functioning of DNA that the hydrogen atoms have relatively stable locations. If the hydrogen atoms had no fixed positions, adenine could often pair with cytosine and guanine with thymine, and sequence of the bases on the two intertwined chains would not necessarily be complementary. If that were the case, the DNA could not function as a genetic molecule, for it is the complementary relationship between the opposing chain sequences that gives DNA its capacity for self-replication. Replication of the

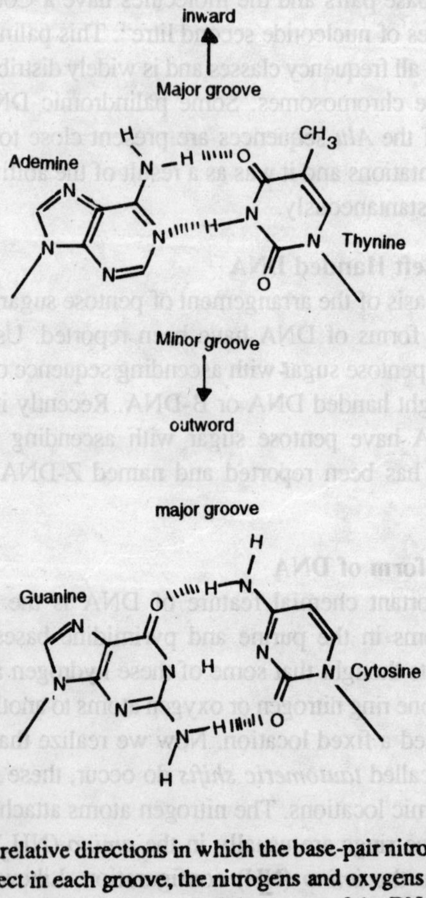

Fig. 6.6. **The relative directions in which the base-pair nitrogens and oxygens project in each groove, the nitrogens and oxygens of the floor of the major groove project toward the center of the DNA molecule, while those of the minor groove project toward the outer edge of the DNA molecule.**

double helix involves strand separation followed by formation of complementary DNA chains using the now free single strands as templates to attract the appropriate partner bases dictated by AT and GC base-pairing rules. The imino forms adenine and cytosine and the enol forms of guanine and thymine must indeed occur only very rarely compared to their amino and keto alternatives. Otherwise, the many errors (mutations) incurred during DNA replication would be incompatible with orderly cell growth and division.

Denaturation and renaturation of DNA

Single-stranded DNA

Single-stranded DNA molecules rarely occur naturally. The best known example is the DNA molecule from the small spherical bacteriophage ϕX174 first isolated by *Sinsheimer* in 1959 but it is also found in the filamentous bacteriophages such as fd. Among the animal viruses single-stranded DNA is found in the parvo viruses. Some of these naturally occurring single-stranded DNA's are circular and other linear. The former can be distinguished by their insensitivity to exonucleases. In general they are comparatively small molecules with less than 10,000 nucleotides. The molar proportions of the bases do not show the usual equivalence of A and T, and G and C, required for double helix formation and the bases react readily with formaldehyde since the amino groups are not protected by hydrogen bondings as they are in double stranded DNA. Single-stranded DNA molecules are very much more flexible than double-stranded molecules of the same size since they lack the rigid double helical structure. In general they behave in solution in a manner similar to RNA molecules.

The helix-coil transition

When double stranded DNA molecules are subjected to extremes of temperature or pH, the hydrogen bonds in the double helix are ruptured and the DNA collapses into two single-stranded molecules. If heat is used as the denaturant, the temperature at which this collapse occurs is known as the T_m or transition temperature. The absorption at 260 nm of any polynucleotide is due to that of its component bases. However, this absorption tends to be suppressed in the double-stranded DNA molecule where the bases are stacked above one another and

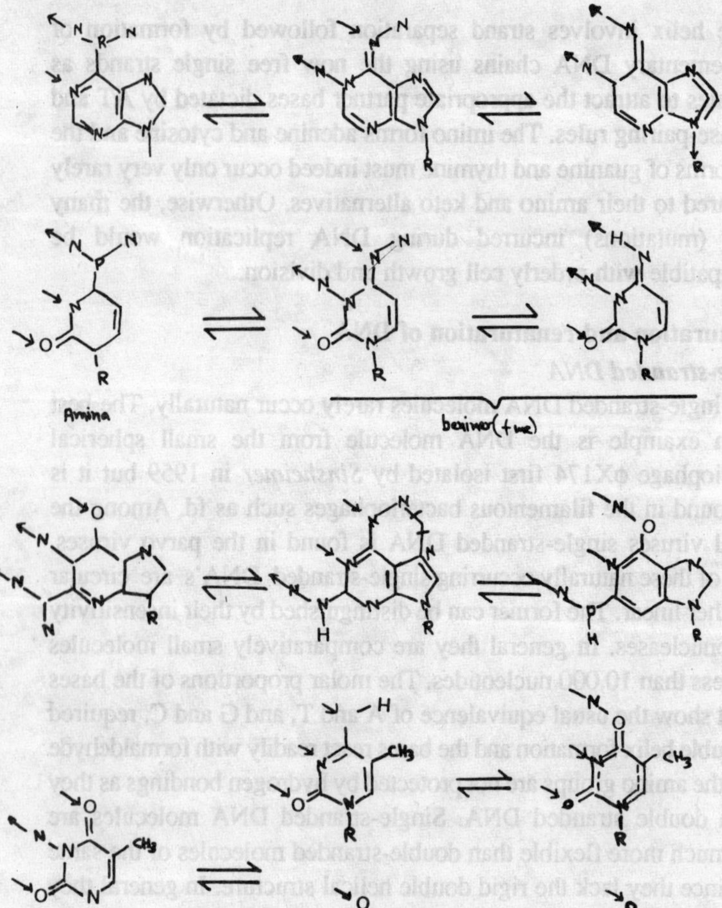

Fig. 6.7. Amino=imino and keto = enol tautomerism. (a) The A and C nuclotide bases are usually in the amino form but rarely form the imino configuration. (b) The bases G and T are usually in the keto form but are rarely found in the enol configuration. Atoms shown in colour may participate in hydrogen-bond formation.

inhibited from swinging out freely in solution in hydrogen bonds, and consequently it is very much lower than that of equimolar amounts of the component bases or nucleotides free in solution. This inhibition of absorption is to some extent relieved when the DNA molecule goes through the helix-coil transition and the bases are no longer so rigidly stacked although they still interact to some extent. As a consequence of this absorption of DNA solutions rises by about 20-30 per cent as the DNA undergoes the melting process, and the transition process is usually measured by ultraviolet spectroscopy. The increase in absorbance on melting is known as the *hyperchromic effect*.

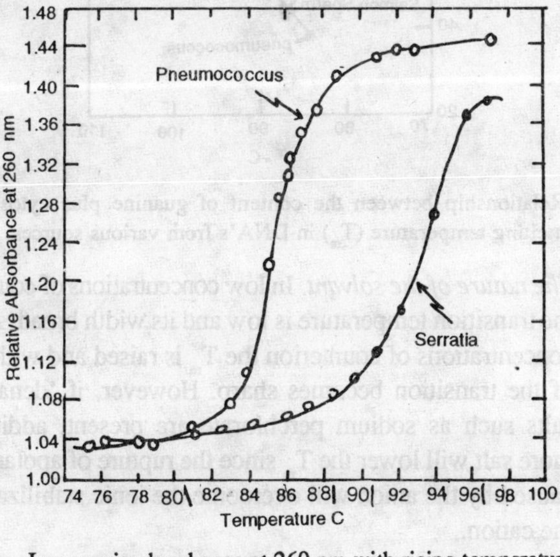

Fig. 6.8. Increase in absorbance at 260 nm with rising temperature for DNA from two different sources.

Such a transition can be characterized by several factors :

(1) *The nature of the DNA*. Homogenous DNA, such a viral DNA, melts over a short temperature range but heterogeneous DNA melts over a longer range.

(2) *The (G+C) content of the DNA*. The melting temperature of any DNA can be related to its (G+C) content since this base pair confer extra stability on the molecule. In DNA molecules such as those from bacteriophage λ, in which some regions

are richer in (G+C) than others, the transitions of both regions can be observed.

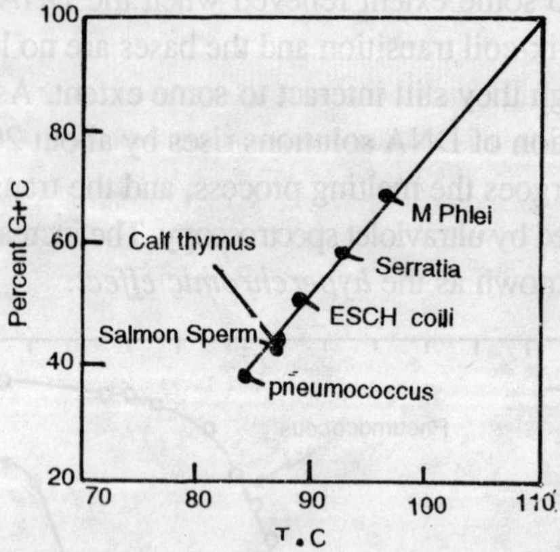

Fig. 6.9. Relationship between the content of guanine plus cytosine and melting temperature (T_m) in DNA's from various sources.

(3) *The nature of the solvqnt.* In low concentrations of counterion the transition temperature is low and its width broad. At high concentrations of counterion the T_m is raised and with width of the transition becomes sharp. However, if 'denaturing' salts such as sodium perchlorate are present, addition of more salt will lower the T_m since the rupture of apolar bonds caused by the anion will overcome the ionic stabilization of the cation.

The helix-coil transition is also associated with a change in density of the DNA molecule, the single-stranded DNA being more dense than the equivalent double stranded form with the same (G+C) content.

The renaturation of DNA

When two DNA strands are returned from the extreme conditions which caused them to melt to their original state they may reassociate to form a doule helix again. However, whether or not they do so depends on a variety of factors. In principle the process should follow simple second-order kinetics but, because of the high probability of two sequences which were not previously matched forming an imperfect union which is then difficult to dissociate, the degree of renaturation can vary from less than 1 per cent according to the nature of the sample and the conditions of renaturation.

(1) *The nature of the sample.* Simple sequence DNA molecules such as $d(G)_n d(C)_n$ have no difficulty finding the appropriate sequence with which the reanneal and do so without difficulty. However, if a eukaryotic DNA sample from a large genome is reannealed, clearly some of the sequences will have to encounter many other non-complementary sequences in solution before they find the correct partner. Thus if most of the DNA from eukaryotic cells is quickly cooled to a low temperature, so that the diffusion of the DNA in solution is inhibeted, few of the single strands will reassociate.

(2) *The temperature of the reassociation process.* At very low temperatures (4°C) not only a diffusion limited but, for a DNA molecule which has become mismatched with a strand having only a few complementary bases, the opportunities to break away and continue its search for the correct complementary strands are reduced. Consequently DNA which is heated beyond the T_m and then quickly cooled to a low temperature will be denatured whereas solutions maintained at high temperatures below the T_m may renature.

(3) *The size of the DNA fragments.* Large string-like fragments of single-stranded DNA encounter diffusion problems and frequently cannot reanneal correctly since this effectively reduces their opportunities for finding complementary srands. For this reason DNA is often sheared for reannealing experiments.

(4) *The ionic strength of the solution.* Two highly charged DNA molecules are likely to repel one another, and the presence of salt is necessary to mask this repulsion in renaturation.

(5) *The concentration of the DNA.* Clearly, at higher concentrations the probability of two complementary strands encountering each other is raised.

(6) *The time allowed for reannealing experiments.* If renaturation is allowed under ideal conditions, two DNA samples of identical

concentration should take different times to reanneal according to the genome size. For this reason the term *Cot value* has been defined for the study of reannealing of DNA and also for the study of the formations of DNA-RNA hybrids. *Co* represents the DNA concentration and *t* represents time in seconds. This is again monitored by ultraviolet spectroscopy or, more frequently, by hydroxyapatite chromatography since this can be used to separate double- and single-stranded DNA molecules and can be used at higher concentrations and hence for shorter times.

Single-Stranded DNA or Viral Chromosomes

At first it was thought that all DNA molecules are double stranded except during replication, when a small region around the replicating fork is temporarily in a non-hydrogen-bonded, single-stranded form. It therefore came as quite a surprise when experiments revealed that the DNA of several groups of small bacterial viruses exists as single-stranded molecules in which the amount of A is not equal to the amount of T and the amount of G does not equal the amount of C. Among these single-stranded phages are the spherically shaped ϕX174 and S13 viruses and the rod-shaped f 1 and M13 viruses. The chromosomes of the parvoviruses, minute viruses that infect many vertebrate organisms, are also single stranded DNA molecules.

When these single-stranded viral chromosomes enter their host cells, they serve as templates for the formation of complementary strands. The resulting double helices in turn serve as templates for new single strands that then become incorporated into new virus paricles. Thus, the fundamental mechanism for orderning nucleotides during the synthesis of single-stranded DNA is basically the same as that used for double-hellical DNA. Nuclotide selection always occurs by attaction of the complementary base. Single-stranded DNA replication usually differs from double-helical replication in that it uses only one of the two complementary strans (the—strand) as a template for the progeny (+) strand.

Although single-stranded DNA can serve as the chromosome of a small virus, it would not be very effective in storing genetic information within the chromosomes of cells. In bacteria and in many eukaryotic

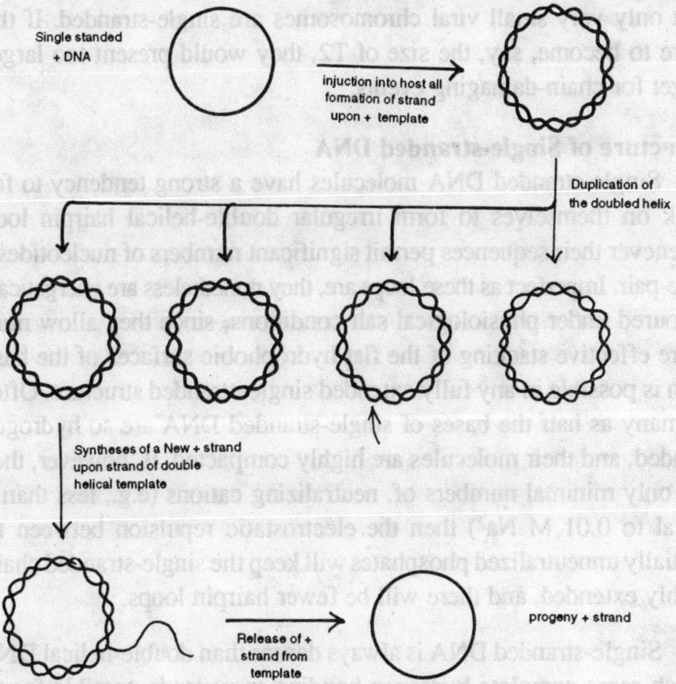

Fig. 6.10. The replication of a single-stranded DNA molecule. Each double helix generally servers as the template for the formation of a large number of new + strands.

cells (e.g., the haploid phase of yeast cells); there usually is only one copy of a given gene. If interphase chromosomes contained single-stranded DNA, with the double helix existing only briefly during mitosis, then the daughter cells would contain two completely different sets of genetic informations. This follows from the fact that the two complementary chains do not have identical base sequences and hence would code for entirely different amino acid sequences. The double-strand ensures that each daughter cell will contain the same genetic information. The double-stranded form also permits effective DNA repair mechanisms to exist. For example, when one strand is damaged by exposure to X-rays, the remaining strand can provide the genetic information to rebuild the damaged section. It is probably no accident that only very small viral chromosomes are single-stranded. If they were to become, say, the size of T2, they would present too large a target for chain-damaging events.

Structure of Single-stranded DNA

Single-stranded DNA molecules have a strong tendency to fold back on themselves to form irregular double-helical hairpin loops whenever their sequences permit significant numbers of nucleotides to base-pair. Imperfect as these loops are, they nonetheless are energetically favoured under physiological salt conditions, since they allow much more effective stacking of the flat hydrophobic surfaces of the bases than is possible in any fully extended single-stranded structure. Often, as many as half the bases of single-stranded DNA are so hydrogen-bonded, and their molecules are highly compacted. If, however, there are only minimal numbers of, neutralizing cations (e.g., less than or equal to 0.01 M Na') then the electrostatic repulsion between the partially unneutralized phosphates will keep the single-stranded chains highly extended, and there will be fewer hairpin loops.

Single-stranded DNA is always denser than double-helical DNA. Much more complete hydrogen bonding to water is possible for the atoms of the highly regular double helix than for the atoms of the inherently irregularly shaped single-stranded DNA. In the absence of base-pairing hydrogen bonds, Van der Waal's interactions take over and bring the atoms of DNA closer together than would be the case if they formed regular hydrogen bonds with each other. An analogy can be made with the behaviour of water, which becomes less dense as it forms the regular, hydrogen-bonded lattice of crystalline ice.

Rigorous Crystallography

X-ray diffraction patterns from parallel-oriented fibers of DNA provided the first clues that the polynucleotide chains of DNA have helical conformations. They also told us that DNA was a multichained molecule composed of two or three polynucleotide chains. However, the double-helical nature of DNA did not directly emerge from X-ray diffraction analysis alone. Equally important was the use of model building in which the question was asked. Given the chemical features of a single DNA chain, into what three-dimensional helical conformations could it fold? The simplest models were those in which the sugar-phosphate backbones were on the outside with the bases stacking in the center. The double helix emerged when it was realised that thymine and guanine had keto, not enol, confiurations and that by base-pairing adenine to thymine and guanine to cytosine, a stereochemically pleasing, regular helical molecule resulted (i.e., the AT base pairs had the same "shape" as the GC base pairs).

They were originally discovered through their antimicrobial and anticancer attributes. Each time these intercalating agents become inserted into a DNA molecule, they necessarily *increase* the spacing of successive base pairs along the helical axes to roughly 7 Å, almost the distance between phosphate atoms in a fully extended polynucleotide chain. In this situation, very little rotation is possible around the helical axis, and so intercalating agents not only extend double helices, but extensively unwind them. For example, when a molecule of either ethidium bromide (an inhibitor of DNA synthesis) or actinomycin D (a powerful inhibitor of RNA synthesis) intercalates, the rotation angle between the two adjacent base pairs is reduced from 36° to 10°. The fact that intercalation occurs of readily indicates that it must be energetically favoured, with the van der Waals bonds holding the inserted molecules to the base pairs being stronger than those found between conventionally stacked base pairs. Intercalation is additional evidence for the *metastability* of the double-helical structure—its ability to temporarily assume many inherently unstable configurations that normally quickly revert back to the standard B conformation. One such metastable variant must be short stretches of extended chains with gaps separating adjacent base pairs, into which an intercalating agent may bind.

The Chromosomes of Viruses, *E. coli,* and Yeast are Single DNA Molecules

The first estimates of the average molecular weights of DNA centered at about a million, the size needed to encompass an average-size bacterial gene coding for some 300 to 400 amino acid. Therefore, it seemed natural to equate single DNA molecules with single genes. However, these early reports were inaccurate owing to DNA breakage during its isolation and study. Now it is clear that virtually all undegraded

DNA molecules contain the information of at least several genes. The most certain molecular weight values come from DNA-containing viruses. Regardless of whether the DNA content is relatively small or large, each virus particle contains a single DN molecule. For example, all the DNA of the small monkey (simian) virus SV40 is present within a single molecule of molecular weight ~ 3×10^6 daltons (5×10^3 hase pairs, or 5 kbp), while the DNA molecule of the large bacterial virus T2 has a molecular weight of 1.2×10^8 daltons (2×10^3 kbp). In these cases, the entire viral chromosome, rather than separate genes, corresponds to a single DNA molecules. Likewise, the chromosome of an *E. coli* cells is a single DNA molecule whose molecular weight is about 2.5×10^9 daltons (about 4×10^3 kbp) and whose extended length is, roughly 1 mm.

DNA molecules of the yeast *Saccharomyces cerevistiae* range in size between the T2 and *E. coli* DNAs, with their average size being that expected if each of the yeast's 17 chromosomes contain one DNA molecule. Thus, the centromeres of yeast chromosomes (if not those of all eucaryotic chromosomes) do not represent discontinuities between separate DNA molecules, but instead are specialized regions of DNA evolved to interact with the spindle bodies upon which chromosomal segreation occurs during cell division. As of yet, there is no direct proof that the very much larger chromosomes of higher plants and animals (often 50 times larger than the *E. coli* chromosomes) also contain only one DNA molecule. However, we now expect that the one chromosome - one DNA molecule rule will hold for the chromosomes of all organisms. In any case, some DNA molecules are larger than any other biological molecules by several powers of ten.

Circular Versus Linear DNA Molecules

Our evidence, which came largely from autoradiography and electron microscopy, initially suggested that all DNA molecules are linear and have two free ends. But when it became possible to take a better look at undegraded DNA, many DNA molecules were found to be circular. For instance, the small monkey DNA virus SV40 has a 5000-base-pair circular double-helical chromosome; the similarly short chromosomes of single-stranded phages are likewise circular, as are almost all autonomously replicating plasmid DNAs. Most, if not all, bacterial chromosomes are also circular. We suspect that the DNA found in the rare circular chromosomes of higher cells are likewise circular molecules. Circular shapes were initially puzzling, since they seemed to present obstacles to the untwisting of double helices during DNA replication. Only when specific enzymes that first snip and then join DNA chains were discovered did the untwisting dilemma disappear.

Equall important is the realization that linear DNA molecules do not have the uncomplicated structures first envisioned for them. As we see in the next chapter, replicating the ends of DNA molecules is not a straightforward process. Thus, all linear DNA molecules have evolved special tricks to replicate their ends, as well as to prevent their free ends from either being nibbled back by cellular enzymes or being enzymatically ligated together to form even larger chromosomes. For example, the ends of the linear adenovirus chromosomes have specialized proteins covalently attached to the 5' ends of strands; and poxvirus chromosomes have closed hairpin loops at their ends, with their basic structures being that of largely self-complementary circular single strands. Still other linear viral chromosomes (e.g., T2 and T7) have the same sequence at both ends, allowing the ends of the chromosome to recombine together during DNA replication to form very long, end-to-end aggregates (concatamers).

Moreover, some DNA molecules that are linear when isolated from a virus particle (e.g., phage λ) are found as circles inside the host cell. This tells us that the linear and circular forms of such DNA molecules are interconvertible. Starting with a closed circle, a specific enzyme introduces two breaks, one in the + strand and one in the — strand. As these breaks are very close to each other, the intervening hydrogen bonds occasionally are broken by thermal agitation, and then the circle unfolds. Most importantly, the resulting linear form contains single-stranded ends that have complementary nucleotide sequences (*"stickly ends")*. Thus, at a later time the linear form can base-pair and resume a circular configuration. If the mission phosphodiester bonds are then reformed, the covalent circle is regenerated.

There is solid evidence that phase λ DNA interconverts between its circular and linear forms using precisely this mechanism. Later, we shall see that the duplication of such "sticky" DNAs involves circular replicative intermediates. Thus, the linear form is most likely an adaptation for injection of the viral chromosome through the narrow phase tail.

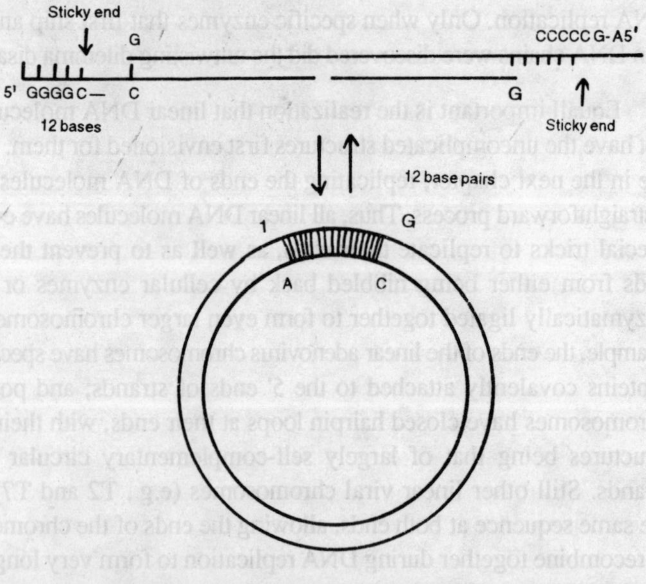

Fig. 6.11. Interconversion of the linear and circular forms of λ DNA.

Supercoiling of Circular DNA

As DNA is a very flexible structure, its exact molecular parameters are a function of both the surrounding ionic environment and the nature of the DNA-binding proteins with which it is complexed. Because their ends are free, linear DN molecules can freely rotate to accommodate changes in the number of times the two chains of the double helix twist about each other. But once the two ends become covalently linked to form circular DNA molecules, the absolute number of times the chains twist about each other (the *linkage number*) cannot change. For the most part, changes in the average number of base pairs per turn of the double helix will necessarily be accommodated by the formation of the appropriate number of supercoils in the opposite direction. *Supercoiling* refers to the further twisting of double-helical DNA molecules. Untwisting of the double helix usually leads to supercoiling in the negative (left-handed) direction while overtwisting leads to positive (right-handed) supercoiling.

The supercoiled state is inherently less stable than uncoiled DNA, and the cutting, or *nicking*, of a single strand instantly converts a supercoiled molecule into its simple (*relaxed*) circular state. Because circular DNA molecules become increasingly compacted as they become more supercoiled, relaxed DNA is easily distinguished from supercoiled DNA by its slower sedimentation in a centrifugal field and by the longer time it takes to move through an agarose gel. At the present time gel electrophoresis affords the most direct way to measure supercoiling, since it can separate molecules that differ by only single turns.

Protein Containing Supercoils

Just because isolated DNA molecules would energetically prefer to be in the relaxed state does not mean that the "natural state" of DNA is relaxed. In fact, virtually all DNA within both procaryotic and eucaryotic cells exists in the negative supercoiled state. As such it seldom, if ever, occurs as free DNA, but is complexed with specific DNA-binding proteins to form compacted molecules called *chromatin*. How this compaction occurs is best understood for eucaryotic chromatin, whose most prominent DNA-binding components are the histones. *Histones* are relatively small, positively charged, arginine- and lysinerich proteins that aggregate together to form discrete ellipsoid-shaped packets (histone cores) around which the DNA supercoils. The resulting *nucleosomes* give to chromatin its beaded appearance.

Since the same collection of histones binds to nearly all sections of DNA, histones are thought to play essentially structural roles, as opposed to enzymatic or regulatory roles. Only about half the mass of chromatin proteins is histone. The remaining proteins are not nearly as well characterized with their exact functions yet to be determined.

Viral DNA chromosomes are also complexed with proteins, both when multiplying within cells and when packaged into virus particles. The SV 40 viral chromosome is at all time compacted into 24 histone-containing nucleosomes identical it structure to those existing in the

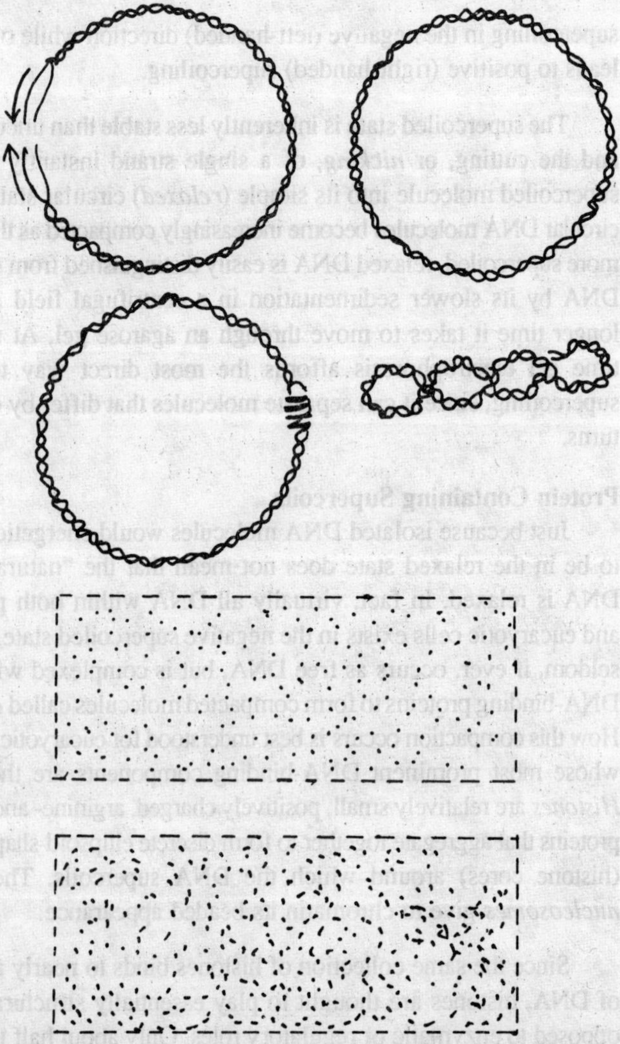

Fig. 6.12. Generation of supercoiled DNA by cutting and ligating. In a circular DNA molecule, the number of times the DNA strands twist about each other is fixed. If both strands are cut (1) and rotated in the opposite direction from that of the twist of the helix (2), and if the cut ends are then rejoined (3) then the DNA molecule negatively supercoiled. (4) The upper electrons micrograph is a relaxed (non-supercoiled) DNA molecule of bacteriophage PM2. The lower electron micrograph shows the phages in its supertwisted form.

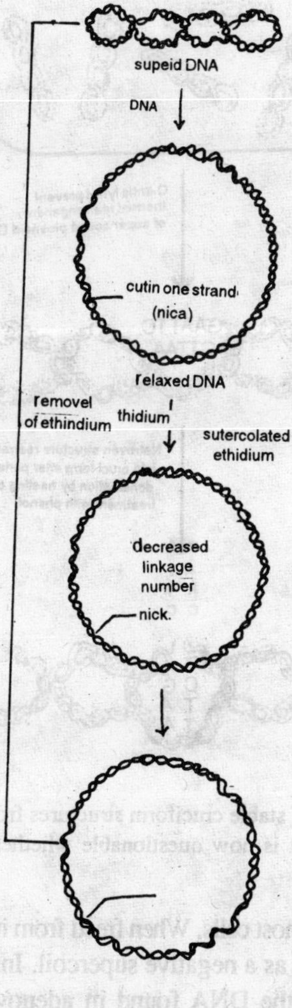

Fig. 6.13. Use of nicking and intercalation to interconvert supercoiled and relaxed DNA. When one strand is nicked, a supercoiled circular DNA molecule relaxes. Ethidium intercalated into the DNA molecule increases the spacing of base pairs, unwind the double helix by 26°, and thus decrease the linkage number. DNA ligase can rejoin the free ends. When ethidium is removed, the DNA molecule supercoils negatively.

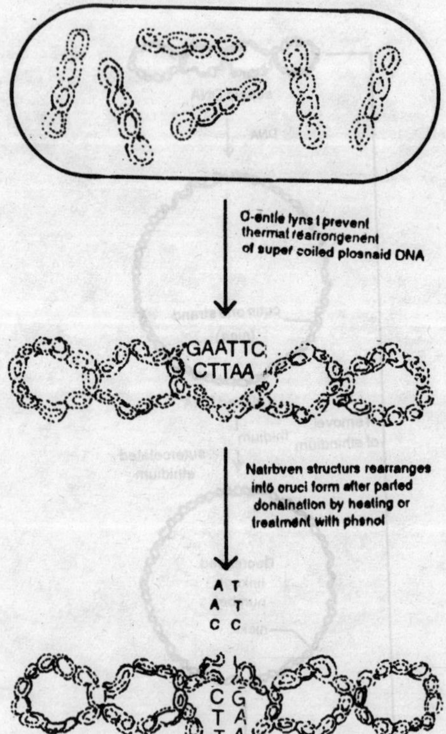

Fig. 6.14. Formation of stable cruciform structures from inverted palindromic sequences. It is now questionable whether these do indeed form within cells.

chromosomes of their host cells. When freed from its protein components, SV 40 DNA is found as a negative supercoil. In contrast, no histones are complexed with the DNA found in adenovirus particles. These vertebrate viruses encode a unique DNA binding protein to help package their DNA within their protein capsules. When, however, an adenovirus DNA functions within a cell, its own DNA-binding protein is released and replaced with the same histone components that bind to all other cellular DNA.

DNA Supercoils Twice Around Each Nucleosome

In forming the nucleosomes of eucaryotic cells, 200-base-pair-long segments of DNA negatively (left-handedly) supercoil twice around the histone cores. The energy lost by supercoiling may be partially compensated by the energy gained from the ionic and hydrogen bonds formed between the histones and the DNA. The number of supercoils in solution need not be the same as found in chromatin. In chromatin, the double helix is slightly more twisted than in solution, with 10 base pairs per helical turn in chromatin versus 10.5 base pairs per helical turn in naked DNA in solution. As a result, there are fewer superhelices found in solution than found in chromatin.

Histone-like DNA-binding Proteins in Prokaryotes

For many decades, it was believed that bacterial DNA, unlike eucaryotic DNA, occurs naked and does not have a compacted chromatin-like arrangement. But recently, using more genetle preparative procedures, condensed *E. coli* chromosomal sections have been seen that are organized into bead-like packets from which small, basic proteins analogous to the histones can be extracted. That bacterial DNA is complexed with proteins that have histone-

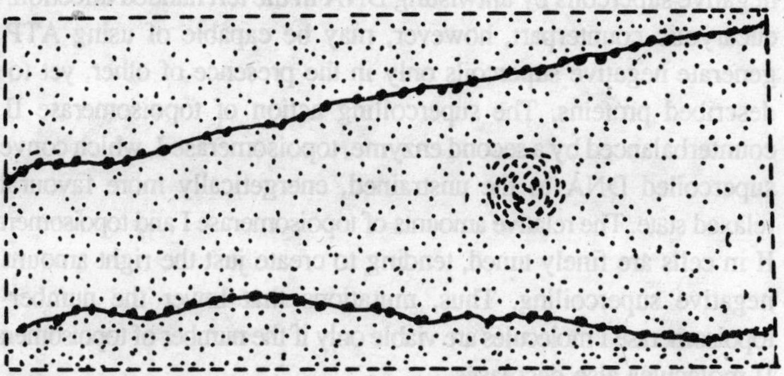

Fig. 6.15. Nucleosomes give chromatin strands a beadlike appearance. Magnification 100,000.

like properties undoubtedly reflects its need to be regularly compacted in order to function. Crystallization of one of these proteins, DNA-binding protein II, reveals that its 9500-dalton chains associate in pairs as dimers containing extended arginine containing arms able to interact with the phosphates of one turn of the DNA backbone. So it is likely that dimers sited adjacently along procaryotic DNA are oriente into a helical arrangement with the DNA bound on the outside. Interestingly, the average *degree of supercoiling* (i.e., the number of super-helical twists per ten base pairs) is about 0.05 for all naturally occurring DNA supercoils, regardless of their source. This may in part be because the underlying chromatin structure is so similar within both procaryotic and eucaryotic cells.

Topo-isomerases Change the Linkage Numbers of Supercoiled DNAs

That supercoi!ing not only occurs but is biologically very important is affirmed by the existence of the enzyme topoisomerase II, which specifically generates the negative supercoils that characterize so much of chromosomal DNA. *Topoisomerases* are a group of enzymes that convert (isomerize) one topological version of DNA into another. They do so by changing the linkage number, which is the number of times two DNA chains twist around each other. Procaryotic topoisomerase II (sometimes called *gyrase*) uses the energy of ATP to generate negative supercoils by untwisting DNA in the left handed direction. Its eucaryotic counterpart, however, may be capable of using ATP to generate negative supercoils only in the presence of other, yet to be described proteins. The supercoiling action of topoisomerase II is counterbalanced by a second enzyme, topoisomerase I, which converts supercoiled DNA to the unstrained, energetically more favourable relaxed state. The relative amounts of topoisomerase I and topoisomerase II in cells are finely tuned, tending to create just the right amount of negative supercoiling. Thus, mutations that lower the number of topoisomerase I molecules are viable only if the number of topoisomerase II molecules also decreases.

Both topoisomerase I and II work by catalyzing the breakage and rejoining of DNA phosphodiester bonds. Unlike virtually all other enzymes, their action does not lead to new patterns of covalent bonds. Rather, their role is to create temporary gaps in polynucleotide chains. When so acting topoisomerases do not create free ends but instead become themselves covalently attached to one of the two broken ends (depending on the specific enzymes, either the 3' and 5' end). Through such catalysis, a tyrosine group on the topoisomerase becomes linked to the terminal phosphate group of the cut polynucleotide chain. The free energy present in the original phosphodiester bond is thus preserved so that it can be used to rejoin the broken chain.

When a DNA chain is broken by a topoisomerase, its broken ends do not fall apart but are held together by the topoisomerase. At no time do the broken ends have the capacity to rotate freely. If that happened, topoisomerase would be limited to only completely relaxing double helices. Instead, individual topoisomereases function to make discreate changes of either one positive turn (topoisomerase I) or two negative turns (topoisomerase II) in the linkage number. Such discrete steps result from DNA chains passing through either transient single-stranded breaks (topoisomerase I) or transient double-stranded breaks (topoisomerase II).

Topoisomerases are constructed in such a way that they can undergo reversible conformational changes that create cavities through which DNA chains can pass. Thc ATP requirement for bacterial topoisomerase II (gyrase)-induced supercoiling most likely reflects the use of ATP in mediating the necessary conformational changes that lead to chain passage. Topoisomerase I, however, relaxes supercoiled DNA without requiring a cyclical input of energy. Complete understanding of how topoisomerase I and II work can occur only when their precise structures are determined through X-ray crystallographic procedures. Even now, however, the prediction can be made that double helices wrap themselves around topoisomerase II in positive supercoils prior to the start of the cutting-rejoining cycle. Without this restraint on the orientation of the DNA, topoisomerase II would break as well as make negative supercoils, and the net result of its action would be the eventual relaxation of the double helical substrate to the uncoiled state.

Looped and Supercoiled Structure

The fact that linear DNA molecules have free ends would seem to preclude their possession of supercoiled segments. Yet, examination of gently isolated linear eucaryotic chromosomes shows that their chromatin is organized into a large number of successive looped domains organized along a scaffold containing two major DNA-binding proteins. Somehow, attachment to the scaffold prevents the rotation of one domain from being transmitted to adjacent sections, since the DNA within each of these loops independently supercoils and may be under different torsional strain. Now there is much evidence both from *Drosophila's* giant salivary chromosomes and from the looped lamph rush chromosomes of the salamander that the individual loops represent functional units of chromatin, with given loops usually being transcribed into one or more very long RNA molecules. How the protein scaffold might maintain independently supercoiled loops was a complete mystery until recently. Now it is less so, following the discovery that a major component of the scaffold is none other than topoisomerase II.

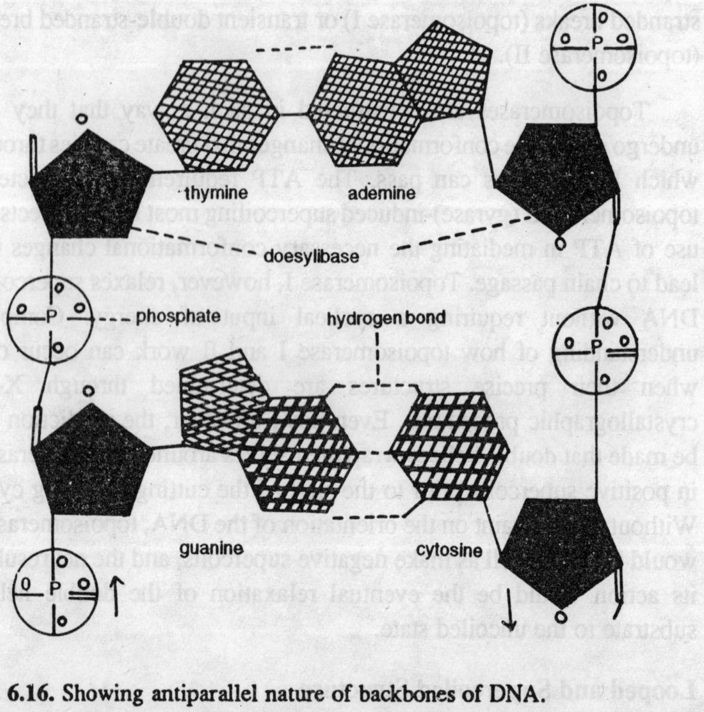

Fig. 6.16. Showing antiparallel nature of backbones of DNA.

Not only are the long, linear chromosomes of higher organisms divided into looped, functional domains, but the circular *E. coli* DNA chromosome also is somehow divided into some 50 looped domains (each about 20 mm long) whose supercoiling is independently controlled. No scaffold has yet been isolated from any bacterial chromosome.

Indeed, the best way now to search for it may be to look for topoisomerase II activity in very gently ruptured cells.

Biological Replication of DNA

Study of replication of DNA is one of the interesting aspect of molecular control of heredity. Various study of the duplication of genetic material was presented from time to time in different organisms. But most of them are discarded because no concrete ideas have been proposed till *Watson* and *Crick* proposed a detailed scheme for the molecular structure of DNA. Later on *Korenberg, Stahl* and *Taylor* have confirmed the Watson-Crick hypothesis. They also made a suggestion for its mode of replication which follows in a natural and attractive fashion from the proposed structure. During the replication, the DNA duplicates exactly into two chains. In the beginning both the strands of double stranded DNA, apart from each other by the disolution of hydrogen bonds and thus uncoiling of the chains occur. And then new chains, "the complementary chains" are built on them so that the newly formed daughter chains are the mirror images of each other.

Mechanism of DNA Replication. The entire process of DNA replication involves following steps in *E. coli* :

1. *Recognition of the initiation point.* Replication of DNA begins at a specific point, called initiation point. Specific initiator proteins are required to recognise the initiation point on DNA.

The initiator protein along with DNA directed RNA polymerase initiates the synthesis of RNA primer for the formation of new DNA chain.

2. *Unwinding of DNA.* The unwinding proteins bind to the nicked strand of the duplex and open up a loop or buble, separating the two strands of DNA duplex.

3. *RNA priming.* The DNA directed RNA polymerase now synthesizes the primer strands of RNA *(RNA primer)*. The priming RNA strands are complementary to the two strands of DNA and are formed of 50 to 100 nucleotides.

4. *Formation of DNA on RNA primers.* The new strands of

DNA are formed in the 5'-3' direction from the 3'-5' template DNA by the addition of deoxyribonucleotides to the 3' end of RNA primer. The addition is affected by DNA polymerase III in the presence of ATP. The unwinding proteins separate the duplex strands ahead of the replication fork.

The leading strand of DNA synthesized in 5'-3' direction. The following strand of DNA is formed in its opposite direction. The DNA strands are synthesized in short segments, consisting of 1,000 to 2,000 nucleotides. These segments are called *Okazaki fragments.*

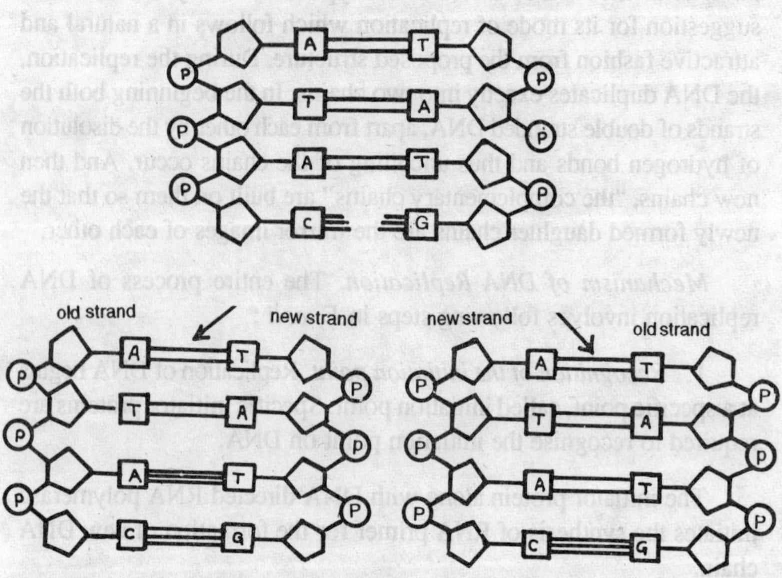

Fig. 6.17. Process of DNA duplication.

5. *Excision of RNA primers.* Once a small segment of an Okazaki fragment has been formed, the nucleotides of RNA primer are removed from the 5' end one by one by the action of 5'-3' exonuclease activity of DNA polymerase-I.

6. *Joining of Okazaki fragments.* The gaps left between Okazaki fragments are filled with complementary deoxyribonucleotide residues by DNA polymerase-I. Finally the adjacent 5' and 3' ends are joined by *DNA-ligase.*

In eukaryotic cells, the mechanism of DNA replication is expected to be more complex than in prokaryotes. The replication of eukaryotic DNA begins at multiple points of origin. Their RNA primer is formed of about 10 nucleotides and the Okazaki fragments are much shorter (about 100 to 150 nucleotides).

Enzymes for DNA synthesis. Different prokaryotic and eukaryotic cells have been formed to contain three kinds of enzymes that act on DNA namely nucleases, polymerases and ligases.

1. *Nuclease Enzymes*. These enzymes attack on the DNA molecules directly. There are two type of nuclease enzyme :

(a) *Exonuclease enzyme*. The exonuclease begins its activity either at a free 3'-OH end of a polynucleotide and progressively cleave the bands as the 3'-OH side of the phosphodiester back bone or its begins at a free 5'-P end and digests the polynucleotides in a 5'-3', direction. In both the cases the enzyme travels along the chain in a stepwise manner.

(b) *Endonuclease enzyme*. Endonuclease enzyme also attacks one of the two sides of phosphodiester linkages, but they react only with those bonds that occur within the interior of a polynucleotide chain. If the polynucleotide chain is single stranded (e.g., viral DNA) such an attack will obviousy cut the chain into two pieces. If however, the polynucleotide strand is a member of a DNA helix (e.g., prokaryotic and eukaryotic DNA), a single endonucleolytic cut will create a nick in the helix; the helix remain in one piece but it now possesses a gap that contains two free ends, which can serve as substrates for exonucleases. A nicked double helix suffers a localized disrupted of its secondary structure. The molecules becomes free to bend or rotate around its intact strand and the two broken ends are free to dangle. The increased molecular motion in the region of the nick will very likely interrupt hydrogen bonding between bases in the vicinity of the nick, thus effecting a limited "unraveling" of the helix.

2. *DNA Polymerase*. For many years, *Korenberg polymerase* was believed to be the enzyme responsible for the replication of the *E.*

coli chromosomes. But recently *Delucia* and *Cairus* isolated a true replicase from *E.coli* and named as *DNA Polymerease*. The polymerase catalyses the formation of a polymer. The synthesis of a new complementary chain becomes possible in the presence of polymerase enzyme. There are three categories of DNA polymerase enzymes:

DNA polymerease I. Discovered by *Korenberg* in 1955 and has been studied in *E. coli*. It is believed to take part in the repair of DNA. DNA polymerase I is a single polypeptide chain with a molecular weight of about 109,000. There may be as many as 400 molecule of DNA polymerase I in *E.coli*.

DNA polymerase II. It is also made up of a single polypeptide chain having a molecular weight of about 90,000. The number of DNA polymerase II molecule is about 40 in *E. coli*. This enzyme can elongate Okazaki segments even in the absence of polymerase I.

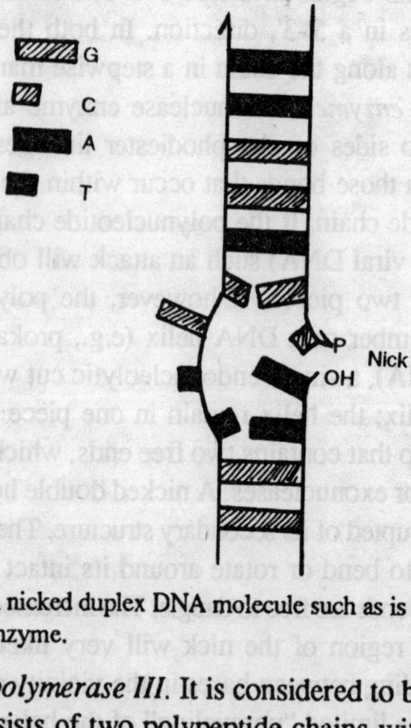

Fig. 6.18. A nicked duplex DNA molecule such as is formed by an endonuclease enzyme.

DNA polymerase III. It is considered to be a true replicase of *E. coli*. It consists of two polypeptide chains with molecular weight of 140,000 and 40,000. It is most effective and can polymersize about 15000 nucleotides per minute in *E. coli*.

3. *DNA Ligases.* DNA ligase enzymes are capable of catalyzing phosphodiester bond formation between free 3'-OH and free 5'-P groups of a nick of DNA which is created by endonuclease enzyme, thereby restoring an intact DNA duplex. Many DNA ligases have already been discovered. The ligase enzyme from *E.coli* requires the presence of oxidized nicotinamide adenine dinucleotide (NAD^+) a cofactor, whereas, the ligase enzyme specified by T_4 bacteriophage requires ATP to bring about the joining reaction.

RNA primers in DNA replication. Short segments of RNA are first synthesized before actual DNA replication starts. These RNA segments are later excised and filled with DNA with the held of enzyme DNA poymerase I. It was found that if RNA synthesis is stopped by a drug rifampicin, DNA replication stops thus proving the role of RNA primers in DNA replication. There are several other evidences which established this point.

Types of Replication

Three different types of replication of DNA have been discovered:

(1) Semiconservative
(2) Conservative
(3) Dispersive

(1) *Semiconservative.* Here the replication of DNA involves the unwinding of the double helix, but no rupture of the separated polyuclotide strands. Thus the primay DNA is causevned during the replication process. According to this mechanism, the unwinding of the original strands would be accompanied or followed by the guided synthesis of the complementary strands of each of the two sparated polyuclotide strand. The linearly arsembly, catalyzed by enzymes, of the nucleoties of the complementary sequence would be guided by their paining with the Watson-Crick partanres on the original strands. In this manner, the proper nucleotide sequence would be automatically assumed. The DNA helix would thereby regenerated each of the strand of the original helical duplex uniting with its freshly synthesized complement to form a total of two new double stranded structures. The semi-convervative

model thus predicts that entire polynucleotide strands of the parantal DNA will be passed on to the progeny in intact form. Each molecule of progeny DNA will receive either one half or none of its nucleotides from the parental DNA.

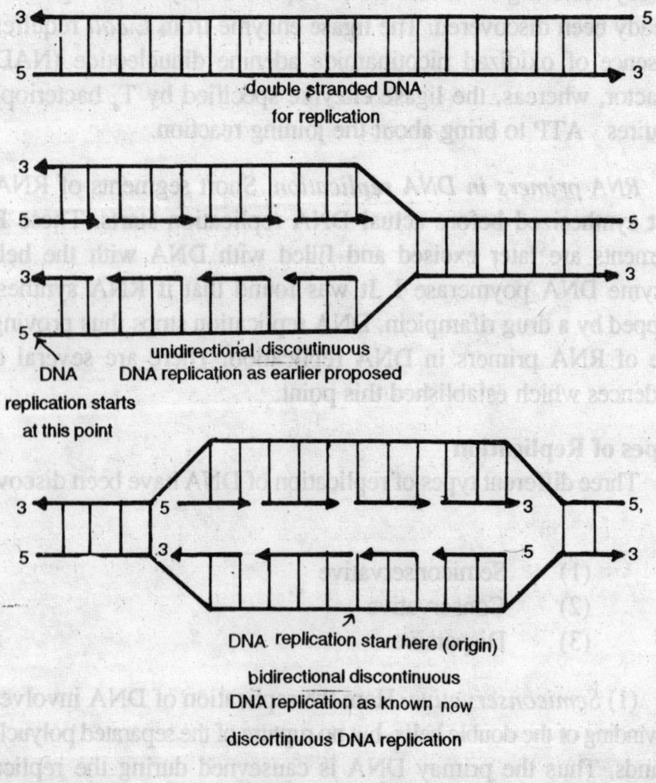

Fig. 6.19. Discontinuous DNA replication.

(2) *Conservative.* According to this mechanism both the primary and secondary structure of the parental DNA are conserved. Thus the replication process would involve neither the rupture of the individual strands nor the unwinding of the helical duplex. The details of these propose that the sequences of hydrogen bonded base pairs, rather than individual bases, serves the genetic determinant. According to this picture, particular base pairs might preferentially bind specific nucleotides and determine their sequential position in the daughter polynueleotide.

The convervative model allows for no redistribution of parental DNA among the progeny. The original DNA is transferred as an intact unit or not at all.

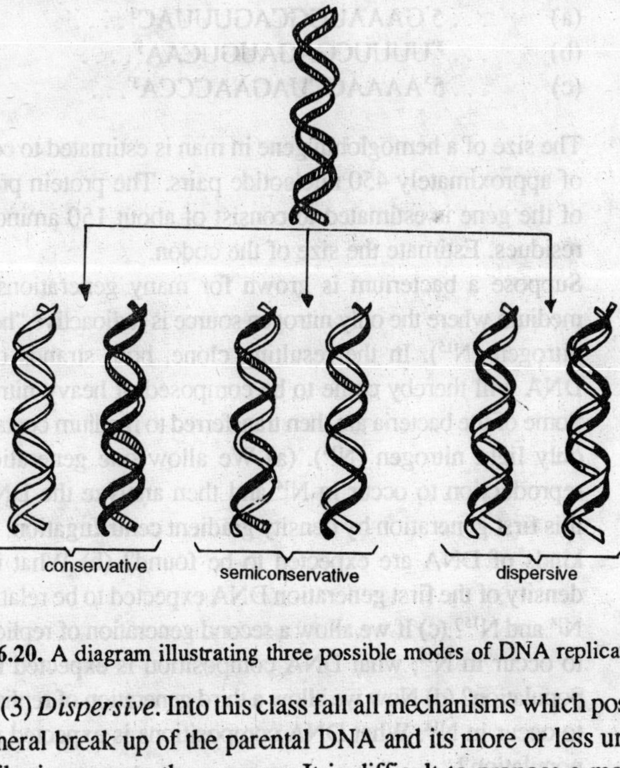

Fig. 6.20. A diagram illustrating three possible modes of DNA replication.

(3) *Dispersive*. Into this class fall all mechanisms which postulate a general break up of the parental DNA and its more or less uniform distribution among the progeny. It is difficult to propose a model of this type which can possibly allow the preservation of nucleotide sequence. Moreover, the existing experimental evidence is uniformly inconsistent with this kind of scheme.

REVIEW QUESTIONS

DNA and Protein Synthesis

1. In a system containing 70% U and 30% C and an enzyme that links ribotides at random into a synthetic mRNA, determine the relative frequencies with which all possible triplet condons would be expected to be formed.

2. Using the information in convert the following mRNA segments into their polypeptide equivalents

(a) . . . 5'GAAAUGGCAGUUUAC3'. . . .
(b) . . . 3'UUUUCGAGAUGUCAA5'. . . .
(c) . . . 5'AAAACCUAGAACCCA3'. . . .

3. The size of a hemoglobin gene in man is estimated to consist of approximately 450 nucleotide pairs. The protein product of the gene is estimated to consist of about 150 amino acid residues. Estimate the size of the codon.

4. Suppose a bacterium is grown for many generations in a medium where the only nitrogen source is radioactive "heavy" nitrogen (N^{15}). In the resulting clone, both strands of the DNA will thereby come to be composed of heavy nitrogen. Some of the bacteria are then transferred to medium containing only light nitrogen (N^{14}). (a) We allow one generation of reproduction to occur in N^{14} and then analyze the DNA of this first generation by density gradient centrifugation. What kinds of DNA are expected to be found? (b) What is the density of the first generation DNA expected to be relative to N^{14} and N^{15}? (c) If we allow a second generation of replication to occur in N^{14}, what DNA composition is expected in the population? (d) Now we allow a third generation of replication to occur in N^{14}. What DNA compositions is expected in the population?

5. Given a single strand of DNA . . . 3'TACCGAAGTAC5'.... construct (a) the complementary DNA chain, (b) the mRNA chain which would be made from this strand.

6. Acridine dyes can apparently cause a mutation in the bacteriophage T4 by the addition or subtraction (deletion) of a base in the DNA chain. A number of such mutants have been found in the rII region of T4 to be single base addition type (+) or deletion type (-) mutants. A normal or wild type rII region produces a normal lytic period (small plaque size) in the host bacterium *E. coli* strain B. Phage T4 mutant in the rII region rapidly lyses strain B, producing a larger plaque.

Several multiple mutant strains of T4 have been developed. Suppose the mutant sites are close together and also in a region of the chromosome which is not essential to the normal functioning of the protein specified by this gene. Determine the lytic phenotypes (large plaque or small plaque) produced by the following rII single base mutations in *E. coli* B assuming a triplet codon: (a) (+), (b) (+)(-), (c) (+)(+), (d) (-)(-), (c) (+)(-)(+), (f) (+)(+)(+), (g) (-)(-)(+), (h) (-)(-)(-), (i) (+)(+)(+)(+), (j) (+)(-)(+)(+)(+).

7. In polynucleotides containing uracil and cytosine produced from a mixture of ribotides where uracil is in excess, more serine is incorporated into polypeptides than proline. However, when cytosine is in excess more proline is incorporated than serine. (a) Without reference to list the codons containing uracil and cytosine which could possibly be coding for serine and for proline. (b) Using determine the percentage of the various amino acids expected to be incorporated into polypeptides by a synthetic polynucleotide in which uracil constitutes 60% and cytosine 40% of the U-C mixture. (c) How much more phenylalanine is expected to be incorporated into protein than proline?
8. If the DNA of an *E. coli* has 4.2 x 10^6 nucleotide pairs in its DNA, and if an average cistron contains 1500 nucleotide paris, how many cistrons does it possess?
9. The following experiment was performed: a short pulse of radioactive isotopes of the 20 amino acids is given to rabbit reticulocytes as they are synthesizing hemoglobin. The introduction of radioactive label occurs when some hemoglobin molecules are partly completed but unlabeled. Shortly after the pulse, completely finished hemoglobin molecules are isolated and analyzed for the location of the radioactive label. Where would you expect to find the label, and why?
10. The DNA of phage *lambda* has 1.2 x 10^5 nucleotides. How many proteins of molecular weight 40,000 could be coded by this DNA ? Assume a molecular weight of 100 for the average amino acid.

11. In phage T4, deletion 1589 lacks part of the A and B cistrons of the rII region and shows no A but partial B activity in *E. coli* strain B. When a single defect (deletion or addition) is induced in the A cistron by acridine, the B cistron activity is suppressed. What hypothesis is supported by these observations?

12. The frequency of spontaneous mutation at the R (plant colour) locus in maize is very high (492 per 10^6 gametes). The gene for red aleurone (Pr) is estimated to mutate in 11 out of 10^6 gametes. How many plants must be investigated on the basis of probability to find one with mutations at both loci?

13. If the mutation rate of a certain gene is directly proportional to the radiation dosage and the mutation rate of Drosophila is observed to increase from 3% at 1000r to 6% at 2000 r, what percentage of mutations would be expected at 3500r?

14. Assuming no intensity effect is operative, which individual would carry fewer mutation: an individual who receives 25 roentgens in 5 hours or an individual who receives only 0.5 roentgens per year for his normal lifetime (60 years)? In terms of percentage, how many more mutations would be expected in the individual with the higher total dosage?

15. The "dotted" gene in maize (Dt) is a "mutator" gene influencing the rate at which the gene for colourless aleurone (a) mutates to its dominant allele (A) for coloured aleurone. An average of 7.2 coloured dots (mutations) per kernel was observed when the seed parent was dt\dt. a\a and the pollen parent was Dt/Dt, a/a. An average of 22.2 dots per kernel was observed in the reciprocal cross. How can these results be explained?

16. A strain of *Drosophila* called Muller-5 contains the dominant sex-linked mutation for bar eye (B) and the apricot allele of the sex-linked while eye locus (w^{av}) together with an inversion to prevent crossing over between them. Homozygous bar-apricot females are crossed to X-irradiated wild type males. Each F_1 female is placed in a separate vial together with one or two unirradiated wild type males. Upon inspection of the F_2 what criterion can be used for scoring induced sex-linked lethals?

17. Single wild type female *Drosophila* possessing attached-X chromosomes (XXY) is mated to an irradiated wild type male. (a) In which generation can a sex-linked mutation be detected? (b) What criterion is used to detect a sex-linked viable mutation? (c) What criterion is used to detect a sex-linked lethal mutation? (d) What percentage death loss is anticipated among the F_1 zygotes due to causes other than lethal mutations?

18. A method is known for rendering chromosome 2 of *Drosophila* homozygous so that all recessive mutants induced in that chromosome by irradiation can be detected. This method uses a balanced lethal system involving three genes on the second chromosome: curly wings *(Cy)*. plum eye colour (pm) and lobe eye shape *(L)*. *A (Cy + Ll + Pm +)* female fly is mated to an X-rated wild type male. A curly, lobed F_1 male is then backcrossed to a curly, plum, lobed female. From the backcross progeny, select and intercross curly lobe flies. Predict the phenotypic ratios among the offspring when (a) no mutant was induced. (b) a viable recessive mutant was induced, (c) a semilethal or subvital was induced, (d) a lethal was induced.

19. Assume that the rate of induced mutations is directly proportional to the radiation dosage. Further suppose that 372 individuals out of 6000 incur a mutation at 2000 r, and that 610 out of 5000 individuals incur a mutation at 4000 r. Estimate the spontaneous mutation rate.

20. If the production of chromosomal rearrangements such as inversions reciprocal translocations requires two hits on chromosomes within the same nucleus, then one might expect the number of such rearrangements to increase more nearly as the square of the radiation dosage rather than in direct proportion to the dosage. If 100 r produces 2.5 single hit mutations, what percentage of rearrangements would be expected at 150 r (neglecting background mutations)?

21. The number of sex-linked lethals in *Drosophila* as detected by the *CIB* and Muller-5 techniques (Problems 9.5 and 9.25) increases in direct proportion to the amount of radiation at low dosage levels. However, at high dosage levels the amount of detectable lethal mutations fall below the linear expectations. How can this phenomenon be explained?

ANSWER TO REVIEW QUESTIONS

1. 3U = 0.343, 2U + 1C = 0.441, 1U + 2C = 0.189, 3C = 0.027.
2. (a) -glu - met - ala - val - tyr - (b) - phe - ala - arg - cys - asn - (c) - lys - thr - (nonsence), chain terminates prematurely.
3. Approximately 3 nucleotides code for each amino acid.
4. (a) All hybrid molecules with one strand containing N^{15} and the other strand containing N^{14} (b) intermediate between N^{15} and N^{14} (c) 1/2 hybrid DNA molecules : 1/2 light DNA molecules (both strands N^{14}) (d) 25% hybrid DNA : 75% light DNA.
5. (a) $^{5'}$ATGGCTCATG$^{3'}$ (b) $^{5'}$AUGGCUCAUG$^{3'}$.
6. (a), (c), (d), (e) , (g), (i) = large; (b), (f), (h), (j) =small.
7. (a) serine: UUC, UCU, CUU; proline: CCU, CUC, UCC. (b) phenylalanine = 0.36, leucine = 0.24, proline = 0.16, serine = 0.24 (c) 21/4.
8. 2870 cistrons.
9. Most molecules would be labeled more in the COOH end because synthesis is unidirectional starting at the NH_2 end.
10. 50 proteins, assuming only one chain of the DNA is transcribed into mRNA.
11. Codons of the rII region are read unidirectionally beginning at the A end.
12. 5412 x 10.^{12}
13. 10-1/2%.

14. 20% more mutations in the individual receiving 0.5 r/yr.

15. Seed parent contributes two sets of chromosomes to triploid endosperm; one *Dt* gene gives 7.2 mutations/kernel, two *Dt* genes increase mutations to 22.2/kernel.

16. Any vial which does not contain wild type male progeny is the result of a sex-linked lethal induced by X-ray treatment.

17. (a) F_1 (b) The appearance of the mutant trait in all F_1 males in a vail. (c) The absence of F_1 males in any vial. (d) 50% superfemales (XXX) are usually inviable: all zygotes without an X chromosome die.

18. (a) 2/3 curly, lobe: 1/3 wild type (b) 2/3 curly, lobe : 1/3 recessive mutant (c) greater than 2/3 curly, lobe : less than 1/3 wild type (d) all curly, lobe.

19. 0.2%.

20. 0.14%.

21. A single ionization ("hit") of a genetic element may destroy the functioning of a vital gene and result in death. Multiple hits at high dosage levels score as a single lethal event.

7

Ribonucleic Acid (RNA)

Ribonucleic acid usually occurs in nature as long unbranched polymeric molecule formed of single chain. RNA is generally present in the cytoplasm as well as in nucleolus. *Casperson* 1950 and *Brachet* in 1955 suggested that there is an association between RNA and protein synthesis. A large number of studies have been undertaken to characterise physically the nature and structure of RNA and their biochemical properties and by now we have acquired a lot of knowledge about the various types of RNA and their molecular structures but still there are various gaps to be filled to make a complete picture.

Some plant viruses (e.g., TMV, turnip yellow mosaic viruses, wound tumour viruses, etc.), animal viruses (e.g., influenza viruses, foot and mouth viruses; rous sarcoma viruses, poliomyelitis viruses reoviruses, etc.) and bacteriophages (e.g., MS2, etc.) contain ribonucleoic acid (RNA) as their genetic material. That is why in these cases it is called the *Genetic RNA*.

TABLE 7.1

Genetic RNA in Different Viruses

Type of Viruses	*Name of Viruses*	*Type of RNA*
1. Plant viruses	1. TMV	1. Single stranded
	2. Wound tumour	2. Double stranded
2. Aimal viruses	3. Influenza virus	3. Single stranded
	4. Poliomyelitis	4. Single stranded
	5. Reo virus	5. Double stranded
3. Bacteriophages	6. MS_2	6. Single stranded
	7. F_2	7. Single stranded
	8. r_{17}	8. Single stranded

Molecular Weight of RNA

The molecular weight of RNA may be determined by light scattering measurments, by sedimation velociy measurements, by viscosity measurements, by equillibrium ultracentrifingation or by polyacrylamide gel electrophoresis in formawide.

Bacteria, *Actinomycetes,* blue green algae and higher plant chloroplasts all have rRNA's of molecular weight 1.1 X 10^6(23S) and 0.55 X 10^6(16S) whereas the corresponding values for higher plants, ferns, algae, fungi and some protozoa are 1.3 X 10^6 (26S) and 0.7 X 10^6(18S). The 18S components is common to all animals, but the large 28S component has evolved from 1.4 X 10^6 daltons in sea urchins to 1.75 X 10^6 daltons is mammals. In mitochondria (mammalian), the rRNA's are somewhat smaller's the 21S species is 0.56 X 10^6 daltons and the 12S species 0.36 X 10^6 daltons. Whilst the ribosomal RNAs form most organisms posses broadly similar base compositions with G plus C content between 50 and 60 per cent, there are some exceptions to this, e.g. *Drosophila* (40%) and *Tetrahymena* (43%).

Primary Structure of RNA

RNA usually occur in nature as long unbranched polymeric molecule made up of single chain or helix of polynucleotide strand : Ribnucleic acid is composed of pentose sugars, phosphoric acid and nitrogenous bases. The pentose sugar of RNA is ribose sugar. The sugar forms the back bone along with the phosphoric acid just in a manner studied in DNA.

Phosphoric acid is also one of the constituent of the RNA which along with the pentose sugar forms the back bone, on which the nitrogenous bases are linearly arranged, as shown in the Fig. On the bases of phosphate or phosphoric acid, the chemical constituents of RNA can be written in two ways :

Nitrogenous Base + Ribose Sugar + Phosphate = Nucleotides

Nitrogenous Base + Ribose Sugar = Nucleoside

The nitrogenous bases of ribonucleic acid are of two types :

(i) Purine
(ii) Pyrimidine.

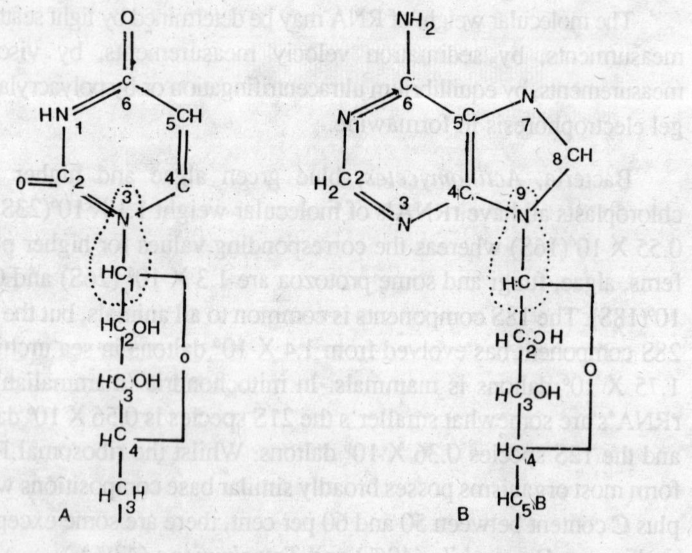

Fig. 7.1. (A) Glycosidic bond between (sugar) C_1 and (Base) N_3 (for Pyrimidine base) (B) Glycosidic bond between (Sugar) C_1 and (Base) N_9 (for Purine base).

Purine bases comprise mainly adenine (A) and guanine (G) while pyrimidine bases comprise cytosine (C) and Uracil (U).

These bases are joined to the pentoses by N-C *glycosidic bonds.* For the purine bases the glyclosidic bond is between the C_1 position of the pentose sugar and the N_9 position of the base. For the pyrimidines, the linkage joins the C_1 and N_3 positions in figure given below:

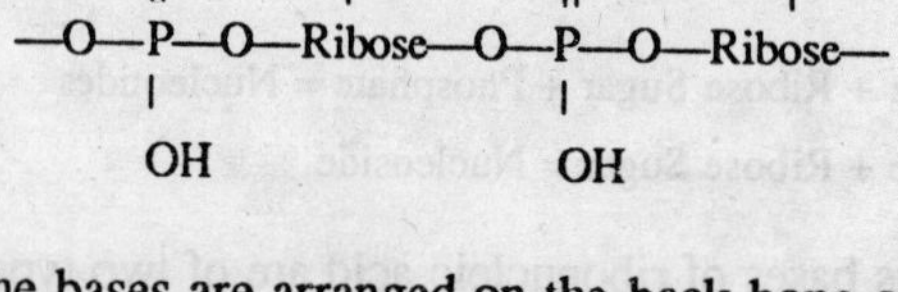

The bases are arranged on the back bone of the chain which is composed of alternating sugar and phosphate. The bases are attached to the ribose sugar in the same way, the bases in DNA are attached.

The arrangement of the bases on the back bone is shown in the following diagram :

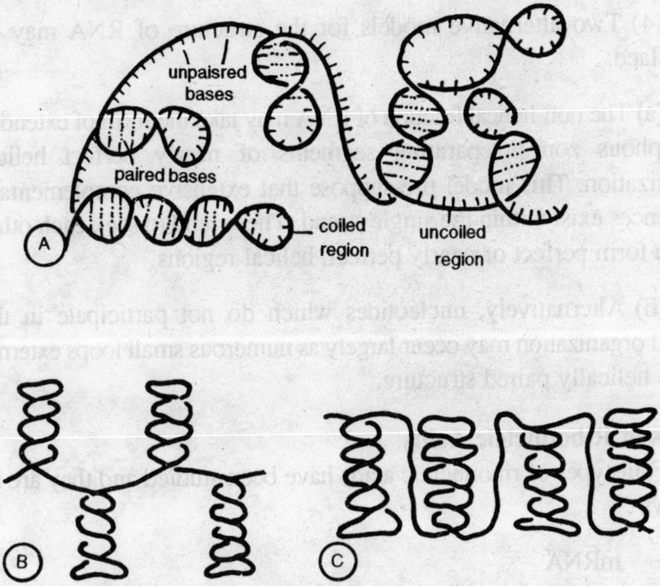

Fig. 7.2. Showing the Secondary structure of RNA.

Secondary Structure of Ribonucleic Acid

Generalization as to the secondary structure of RNA is probably less justified than in the case of DNA, as the available information is much less extensive. The available information does permit the following conclusions :

(1) There is no evidence for the existence of more than one polyribonucleotide strand in the natural RNAs.

(2) Hence the helical regions of natural RNA arise from the hair pin like bending of the single strand back upon itself.

(3) The helical regions are stabilized by A—U and C—G interactions similar to those which stabilize the confirmation of DNA. However, the irregular base composition of RNA does not allow the

incorporation of the entire molecule into double perfect helix of this type, and the helical zone co-exist with the random segments.

(4) Two alternative models for the structure of RNA may be postulaed:

(a) The non-helical fraction of RNA may take the form of extended amorphous zones separating segments of nearly perfect helical organization. This model pre-suppose that extensive complementary sequences exist within the single strand. These tend to seek each other out to form perfect or nearly perfect, helical regions.

(b) Alternatively, nucleotides which do not participate in the helical organization may occur largely as numerous small loops external to the helically paired structure.

Types of Ribonucleic Acids

Four types of ribonucleic acids have been studied and they are as follows :

(1) mRNA

(2) tRNA

(3) rRNA

(4) vRNA

1. Messenger RNA (mRNA)

Studies by *Huxley* and later by *Volkin* and *Astrachan* with T_2 bacteriophage infected (bacterial virus, *E. coli*) in 1956 led to the first existence for the unique kind of the RNA, later called messenger RNA, *Jacob* and *Monod* (1961) who first proposed this name, believed messeger to be necessary to convey the message of the genetic code from the DNA of the nucleus to the cytoplasm. This messenger is know as mRNA and it acts as a template for the translation of the DNA code into a specific protein. Thus it helps in determining the structure of protein.

mRNA molecules exist in a large variety of lengths, depending on the length of the polypeptide chain for which they code. Its molecular weight and sedimention value vary considerably. Generally the life span of mRNA is relatively short, although in some eukaryotic cells there is evidence of some degree of stability.

Further mRNA is always complementry to a portion of strand of DNA, with the exception that uracil replaces thymine. It should therefore hybridize with the DNA from which it was transcribed.

It is now an established fact that when the code has been transcribed from DNA on to mRNA, the later leaves the nucleus passes through the nuclear membrane into the cytoplasm. Here it moves to the ribosomes, the site of protein synthesis. mRNA molecules attaches reversibly to the surface of ribosome always binding to the smaller sub-units. The magnesium cation, Mg^{++} is involved in the formation of a complex between mRNA and ribosomes.

Types of mRNA

On the basis of size and the coding two types of mRNA has been described :

(a) *Monocistronic mRNA*. Mostly the mRNA carries the codons of single cistron (i.e., codes for one complete protein molecule) of the DNA. Such mRNA molecule is called monocistronic mRNA. For example, for the synthesis of a polypeptide chain of 300 to 500 amino acid residues, a monocistronic mRNA of *E. coli* contains 900 to 1500 nucleotides in its molecule.

(b) *Polygenic or polycistronic mRNA*: Sometimes a mRNA molecule carries the codes from several adjacent DNA cistrons and become much longer in size. This type of mRNA is called polygenic or polycistronic mRNA. For example, for the metabolism of the histidine protein the cell synthesizes about 10 specific enzymatic proteins and a mRNA in this case may carry codons for all the 10 enzymes.

Informosomes

In eukaryons mRNA is associated with protein forming ribonucleoprotein complexes. Some of these complexes remain free in the cytoplasm without being attached to the polyribosomes. These are named as informosomes by *Spirin*. These are very stable and can remain in the cell cytoplasm for several days. The stability is attributed by the presence of protein sheath around mRNA molecule. In informosomes proteins and mRNA are found in the ratio of 4:1. These occur in the form of particles of 80S or less.

Since informosomes are very stable, these are used by the cell is protein synthesis only when the is delay in translation. In embryonic cells, genetic apparatus starts functioning only late during organogenesis. Therefore, in early stages of cleavage when there is no rRNA, informosomes govern the process of protein synthesis.

Stability. The cell does not contain large quantities of mRNA. This is because mRNA, unlike other RNAs is constantly undergoing breakdown. It is broken down to its constituent ribonucleotides by ribonucleases. In bacteria, mRNA may be so short-lived that while one end is translating protein the other end may be undergoing breakdown. In *E. coli* the average half-life of some mRNA is about two minutes.

Eukaryotic cells, on the other hand, contain metabolically stabale mRNA. Mammalian reticulocytes are immature RBC which have lost their nuclei but have retained ribosomes. These cells can synthesize haemoglobin for many hours and even days by utilizing the mRNA transcribed by the nuclei when they were present.

2. Transfer Ribonucleic Acid

In the literature, t-RNA molecule have been variously termed tRNA, sRNA for soluble RNA, or adaptor RNA. It forms about 10-15% of the total RNA in most of the existence. Each tRNA contains approximate 80 nucleotides. The soluble or t-RNA forms large varieties of different cells, is homogenous in its molecular size having a molecular weight of about 25000.

The nucleotide sequence of tRNA was first worked out by *Holley et al.* (1965) for yeast alanine tRNA. Since then the sequence of about 75 different tRNAs have been established. According to the Holley's leaf model the tRNA is a polynucleotide chain which is folded upon itself to form 5 arms. As a result of the folding the 3' and 5' ends of the chain come near each other. An arm consists of a stem and a loop.

Bases are arranged in pairs in arms. There is no base pairing in the loops.

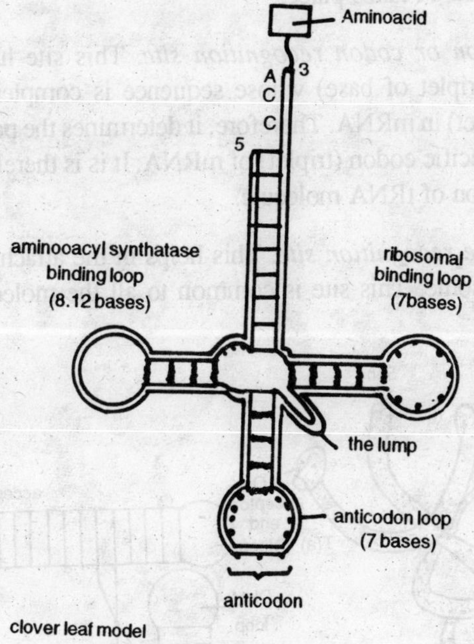

Fig. 7.3. Clover leaf model of tRNA.

The 3' end of the polynucleotide chain ends in CCA base sequence. This represents site for the attachment of activated amino acid. The end of the chain terminates with guanine base. The bent in the chain of each tRNA molecule contains a definite sequence of three nitrogenous bases which constitute the anticodon. It recognizes the codon on mRNA.

Four different regions or special sites can be recognized in the molecule of tRNA. These are :

(i) *Amino acid attachment site*. It occurs at the 3' end of the tRNA chain and has-OH group of ribose molecule free to pair. This-OH group combines with the specific amino acid in the presence of ATP forming amino-acyl tRNA. It is common to all the tRNA molecules.

(ii) *Recognition site*. It contains a specific base sequence which dictates the attachment of correct mino acid to the tRNA molecule. It matches with the amino acid activating enzyme through which attachment of amino acid to tRNA takes place.

(iii) *Anticodon or codon recognition site.* This site has three unpaired based (triplet of base) whose sequence is complementary with a codon (triplet) in mRNA. Therefore, it determines the pairing of tRNA with the specific codon (triplet) of mRNA. It is is therefore, the most specific region of tRNA molecule.

(iv) *Ribosome recognition site.* This helps in the attachment of tRNA to the ribosome. This site is common to all the molecules of tRNA.

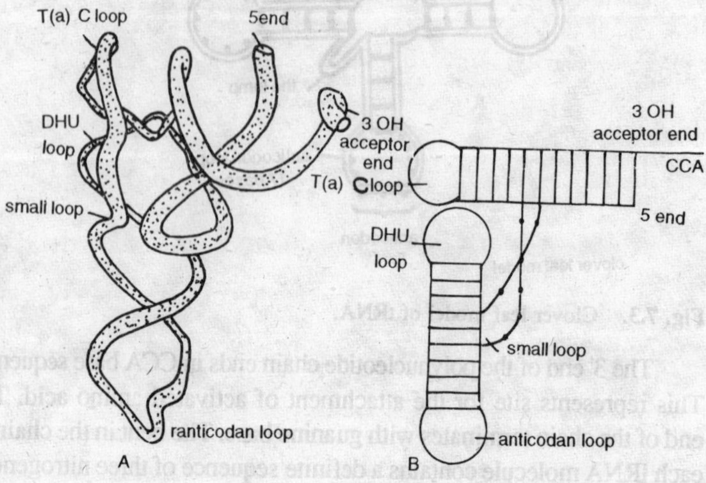

Fig. 7.4. Three dimensional structure of yeast phenylalanine tRNA.

Three dimensional structure of tRNA. In order to understand the structure-function relationship of tRNA, its three dimensional structure (TDS) should be known. With the availability of tRNA in crystalline form in 1968, TDS of tRNA could be studied through X-ray crystallography as was done for DNA double helix. About a dozen models for TDS were proposed, but the latest and most acceptable model was proposed in 1973 by *S.H. Kim*. He suggested that TDS of tRNA takes the shape of letter L with a thickness 20 A°. This can be easily derived from two dimensional clover leaf model. The extra arm which varies in different tRNA molecules can be extended without distoring TDS conformation. The CCA stem projects out and can take different orientations.

Unusual bases. In addition to the usual bases A,U,G and C, tRNA contain a number of unusual bases, and in this respect differs from mRNA and rRNA. The unusual bases of tRNA account for 5-20% of the total RNA of the cell. Most of the unusual bases are formed by methylation (addition of $-CH_3$ or methyl group to the usual bases), e.g., cytosine and guanine on methylation yield methylcytosine and methylguanine, respectively. Precursor tRNA molecules transcribed on the DNA template contain the usual bases. These are then modified to unusual bases. The unusual bases are important because they protect the tRNA molecule against degradation by RNAse. This protection is necessary because RNA is found floating freely in the cell.

Some of the unusual bases of tRNA are methylguanine (GMe), dimethylguanine (GMe_2), methylcytosine (Me), ribothymine (T) pseudouridine (Ψ), dihydrouridine (DHU_2, H_2U, UH_2), inosine (I) and methylinosine (IMe, MeI).

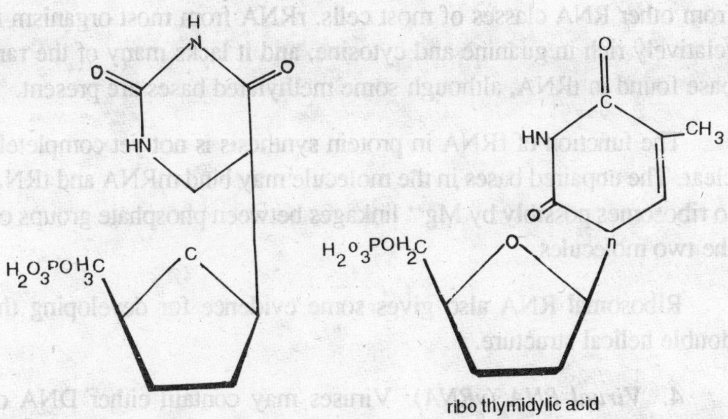

Fig. 7.5. Showing the structural formula of Pseudo-Uridylic acid and Ribo-thymidylic acid.

3. Ribsomal Ribonucleic Acid rRNA

Ribosomes are small particles of about 200A° to 250A° in diameter, and are generally present in the cytoplasm either lying free in the cytoplasm or attached to the outer membrane of endoplasmic reticulum or nuclear membrane. These particles are ribosomes containing about 60% ribonucleic acid or RNA which is ribosomal RNA or rRNA and 40% protein. It is generally agreed that cytoplasmic ribosomes or their subunits are made in the nucleus and are subsequently transferred to the cytoplasm.

However, the mode of their passages across the nuclear membrane is still a matter of controversy.

Even from the morphological evidence, it would appear that the reticular or microvesicular spheres of the nucleus could be engaged in either the synthesis of rRNA, or its assembly, or both.

The finding that hybridization of rRNA takes place with the chromatin associated with the nuleolar region suggests that rRNA is synthesized on the part of the chromosome traversing the nucleolus (the nucleolar organizer) or, in some cases, on DNA detached from the chromosome and associated with the nucleolus.

Ribosomal RNA differs in size and base contents from tRNA and from other RNA classes of most cells. rRNA from most organism is relatively rich in guanine and cytosine, and it lacks many of the rare base found in tRNA, although some methylated bases are present.

The function of rRNA in protein synthesis is not yet completely clear. The unpaired bases in the molecule may bind mRNA and tRNA to ribosomes possibly by Mg^{++} linkages between phosphate groups on the two molecules.

Ribosomal RNA also gives some evidence for developing the double helical structure.

4. *Virual RNA (vRNA):* Viruses may contain either DNA or RNA. However, each particular virus generally contains nucleic acid of only one type. Some examples of RNA viruses are influenza, turnip yellow mosaic, tomato bushy stunt, tobacco necrosis, and the particularly well studied tobacco mosaic virus (TMV).

TABLE 7.2

Comparison of the Different Types of RNA

	Ribosomal RNA (rRNA)	*Messenger RNA (mRNA)*	*Transfer RNA (tRNA)*
1. Percentage of total RNA of cell	80%	3 to 5%	10 to 20%
2. Sedimentation Coefficient	28S, 18S, 5.8S, 5S 23S, 16S, 5S	8S	3.8S
3. Number of nucleotides	5S RNA:120 nucleotides 16-18S RNA: 1,600 to 2,500 nucleotides 23-28S RNA: 3,200 to 5,500 nucleotides	*E. coli:* 900 to 1,500 nucleotides	73 to 93 nucleotides.
4. Molecular weight	23S RNA: 1.1×10^6 30S RNA: 0.55×10^6	500,000	25,000-30,000
5. Unusual bases	Small amount of methylated bases.	Small amount of unusual bases. 1 per 30-40	High content unusual bases. (*E.coli*: 1 per 100-150 nucleotides).
6. Site of synthesis	Derived from nucleolar DNA	Synthesized in nucleus on DNA template.	Synthesized in nucleus on DNA template.
7. Beginning of Synthesis	Synthesis begins at gastrulation, and increases as development proceds.	Some mRNA is found in the ovum. New mRNA is synthesized during early cleavage.	tRNA synthesis occurs at the end of cleavage stage.
8. Base relationship of DNA	No obvious base relationship to rDNA. rRNA is from only small section of DNA.	mRNA shows base relationship to DNA. It is formed from all sections of DNA.	Same as in rRNA.
9. Function	Unpaired bases may bind mRNA and tRNA to ribosomes.	Conveys genetic information from DNA of chromosomes where it takes part in protein synthesis.	Adaptor for attaching amino acids to mRNA template.

TABLE 7.3

Comparison between DNA and RNA

Deoxyribonucleic acid	Ribonucleic acid
1. DNA is the usual genetic material	1. RNA is the genetic material of some viruses.
2. DNA is uaually double-stranded. (In certain viruses DNA is single stranded, $\phi\chi$ 174).	2. Most cellar RNA is single stranded. (Some viruses e.g., reovirus, have double stranded RNA).
3. The pentose sugar is deoxyribose	3. The pentose sugar is ribose.
4. The common organic bases are adenine, guanine, cytosine and thymine.	4. The common organic bases are adenine, guanine, cytosine and uracil.
5. Base pairing: adenine pairs with thymine and guanine with cytosine	5. Adenine pairs with uracil and guanine with cytosine.
6. Pairing of bases is throughout the length of the molecule.	6. Pairing of bases is only in the helical region.
7. There are fewer uncommon bases.	7. There are more uncommon bases.
8. DNA is only of one type.	8. There are three types of RNA, messenger RNA, ribosomal RNA and transfer RNA.
9. Most of the DNA is found in the chromosomes. Some DNA is also found in the cytoplasm e.g. in mitochondria and chloroplasts.	9. Messenger RNA is formed on the chromosomes, and is found in the nucleolus and cytoplasm. rRNA and tRNA are also formed on the chromosomes, and are found in cytoplasm.
10. Denaturation (melting) is partially reversible only undertcertain conditions of slow cooling (renaturation).	10. Complete and practically instantaneous reversibility of the process of melting.
11. Sharp, narrow temperature interval of transition in melting	11. Board temperature interval of transition in melting.

12. DNA on replication forms DNA, and on transcription forms RNA.	12. Usually RNA does not replicate or transcribe. (In certain viruses RNA can synthesize on RNA chain).
13. Genetic messages are usually encoded in DNA	13. The usual function of RNA is translating messages encoded in DNA into proteins.
14. DNA consists of a large number of nucleotides, upto 4.3 million.	14. RNA consists of fewer nucleotides, upto, 12,000.

8

Protein Synthesis

The proteins are highly organized and complex macromolecules. They have significant role in the structural and functional organisation of the cells. The structural proteins constitute various cellular components (such as plasma membrane, nuclear membrane, organelles etc.) as well as some extracellular parts such as cuticle, fibres etc. The functional proteins are enzymes and some hormones which control almost all the metabolic, regulatory, sensory and reproductive activities of the cells. They also control the expression of phenotypic traits.

Chemically, the proteins are the polymers of *amino acids* and at a primary level these can be considered as polypeptide chains. The polymer in a protein consists of amino acids linked with the help of peptide bonds (-CO -NH). The peptide bond consists of a convalent bond between carboxy group (COOH) of one amino acid and the amino group (NH_2) of another amino acid.

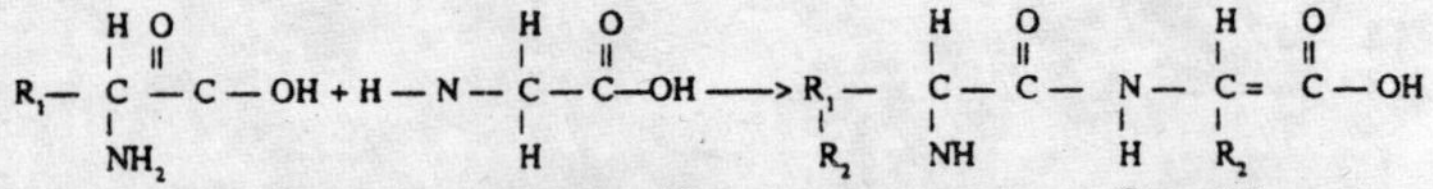

(I Aminoacid) (II Aminoacid) (Dipeptide)

Only 20 amino acids are biologically important and they participate in the protein synthesis forming polypeptide chains.

Deoxy ribonucleic acid (DNA) initiates the mechanism of protein synthesis since it carries the genetic message for protein synthesis. The information is transcribes or copied over messanger ribonucleic acid (mRNA) from one DNA strand. The mRNA translates the coded information by bringing amino acids in a particular sequence during protein synthesis through the agency of transfer ribonucleic acid (tRNA). It occurs over the ribosomes. The first step is called *transcription* which involves copying of genetic information from DNA into mRNA. The second step is called *translation* which involves the translation of information encoded on mRNA with newly synthesized protein molecules.

Flow of Genetic Informations

The protein synthesis is under direct control of DNA in most of the cases. Information of a polypeptide chain is stored in mRNA chain. The relationship between DNA, RNA and protein is called *central dogma*. From time to time central dogma has been variously proposed.

1. *One way flow of information. Crick* in 1958 suggested that DNA transmits the genetic informations by transcribing mRNA. This mRNA translates the nucleotide sequence into the sequence of amino acid.

2. *Circular flow of information.* It was suggested by *Barry Commoner* in 1968. The DNA transcribes mRNA and mRNA translates information into protein. Further proteins directs the synthesis of mRNA and this mRNA dictates DNA synthesis.

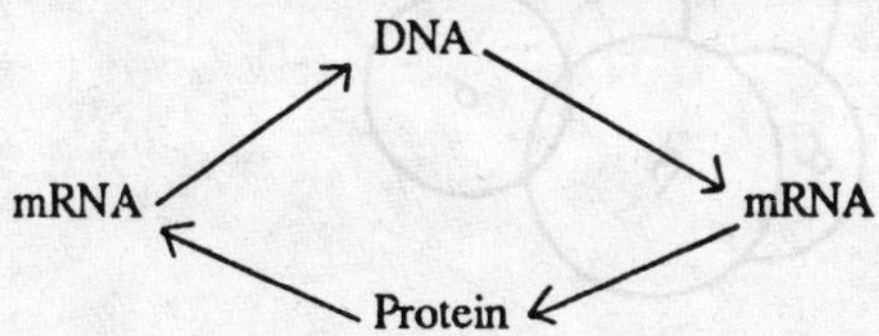

3. *Inverse flow of information. Temin* (1970) has described the presence of RNA dependent DNA polymerase enzyme, which controls the synthesis of DNA from single stranded RNA. The synthesis of DNA on RNA template in tumor virus was desmostrated simultaneously by *Temin* and *Baltimore*. This discovery was considered significant for understanding of cancer. Therefore, *Temin, Baltimore* and *Dulbecco* were awarded Nobel Prize in medicine for 1975.

DNA $\xrightarrow{\text{transcription}}$ mRNA $\xrightarrow{\text{translation}}$ Proteins

MECHANISM OF PROTEIN SYNTHESIS IN PROKARYOTES

Transcription of mRNA from DNA

The copy of a complementary mRNA strand on DNA strand is called transcription. RNA polymerase is the main enzyme involved in the transcription. All types of RNA (mRNA, t-RNA and rRNA) are synthesized with the help of this enzyme. In eukaryotic system three types of RNA polymerases are known to be present in the nucleus. In bacterial system only one type of RNA polymerase is known which is responsible for synthesis of all types of RNA. Besides RNA polymerase, transcription requires a template activated precursors and a divalent metal ion (Mg^{++} or Mn^{++}).

As earlier state that in bacterial system a single RNA polymerase is responsible for synthesis of all kinds of RNAs. It has been purified and its structure and function is now known. It consists of four polypeptide chains. There are two alpha (α) chains, one beta (β),

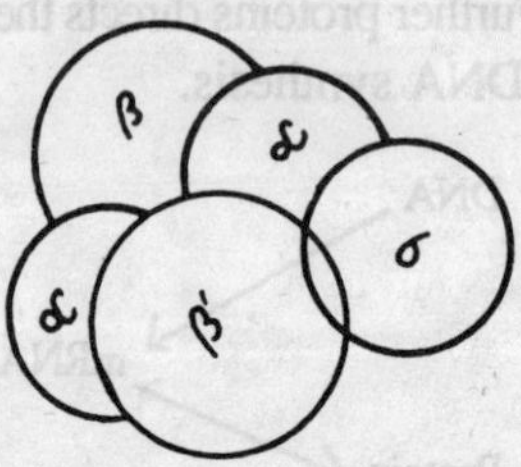

Fig. 8.1. A model of the structure of prokaryotic RNA polymerase showing association of five polypeptides (α_2 ββ σ).

one beta desh (β) and one sigma (σ) chains. The first-four form core enzyme and sigma factor units to form holoenzyme. The molecular weight of αchains is 40,000 daltons, that of β chain 1,55,000 daltons, β'chain 1,65,000 daltons and σ chain is 95,000 daltons. The attachment of σ factor is not very firm so that core enzyme (α_2ββ') can be easily isolated. Function of different polypetide chains is known β and β' help RNA polymerase in unwinding the DNA molecule for the process of transcription. The σ factor helps in recognition of start signals on DNA molecule and directs the enzyme in selecting the initiation sites. Once the mRNA synthesis is initiated, the σ factor dissociates and core enzyme brings about elongation of mRNA.

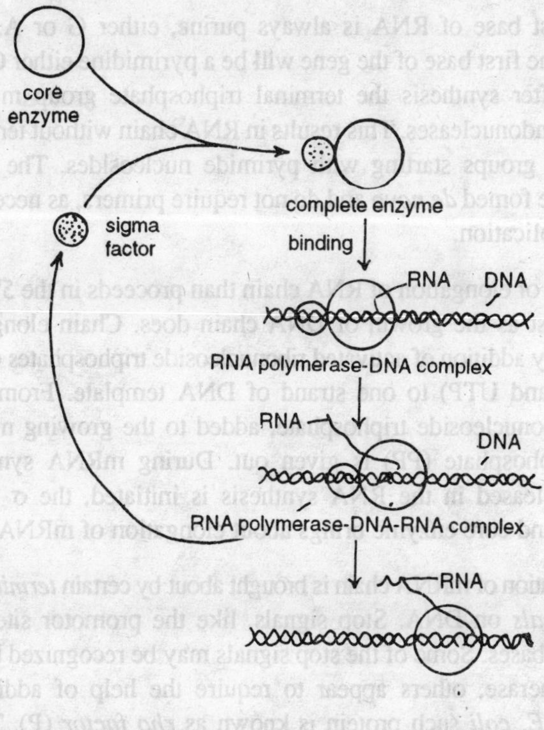

Fig. 8.2. Role of sigma factor and core enzyme of RNA polymerase during transcription.

For transcription entire chromosome or DNA molecule does not unwind but a local denaturation of the DNA helix takes place. Thus only those bases are exposed that constitute the gene to be transcribed. After unwinding only one of the two strands is copied. For synthesis of mRNA on DNA template, there are transcription units, each having a starting point and a termination signal. The start signals within DNA molecule are called *promotor sites*. These promotor sites are not identical to each other, even though all attract the same RNA polymerase.

RNA polymerase initially binds to a promotor site for RNA synthesis. It than diffuses to a binding site consisting of a sequence of six bases, which never differs by more than two bases from the following sequence 5' TATAAT. It is called *Pribnow box* and is available in almost all promoters.

The first base of RNA is always purine, either G or A. This means that the first base of the gene will be a pyrimidine either C or T in DNA. After synthesis the terminal triphosphate group may be cleaved by endonucleases. This results in RNA chain without terminal triphosphate groups starting with pyrimide nucleosides. The RNA chains can be fomed *de novo* and do not require primers, as necessary for DNA replication.

Growth or elongation of RNA chain than proceeds in the 5'—>3' direction, just as the growth of DNA chain does. Chain elongation takes place by addition of activated ribonucleoside triphosphates (ATP, GTP, CTP and UTP) to one strand of DNA template. From each activated ribonucleoside triphosphate, added to the growing mRNA chain, pyrophosphate (PP) is given out. During mRNA synthesis energy is released in the RNA synthesis is initiated, the σ factor dissociates and core enzyme brings about elongation of mRNA.

Termination of mRNA chain is brought about by certain *termination* or *stop signals* on DNA. Stop signals, like the promotor site, is a sequence of bases. Some of the stop signals may be recognized by the RNA polymerase, others appear to require the help of additional proteins. In *E. coli* such protein is known as *rho factor* (P). The P factor is discovered by *Roberts* (1962). It has a molecular weight of 2,00,000 daltons. It is made up of four polypeptide chains, each having a molecular weight of 50,000 daltons. It is' however, not certain whether it also, or exclusively binds to DNA.

Translation in Prokaryotes

During translation the geneic informations present in mRNA directs the order of specific amino acids to form protein molecule or polypeptide chain. This step is much more complex than transcription and involves the following steps:

1. *Activation of amino acids*. The 20 amino acids used in protein synthesis are activated by adenosine triphosphate (ATP). The activation is mediated by specific activating enzyme known as *amino acyl RNA synthetases*. As a result of this reaction between amino acid and ATP, a complex is formed. This complex is known as amino acyl-AMP-enzyme or amino acyl AMP complex is formed. Pyrophosphates (PP) are released. It should be realized that amino acyl RNA synthetases are specific with respect to amino acids. Therefore, 20 such enzymes are required for 20 amino acids.

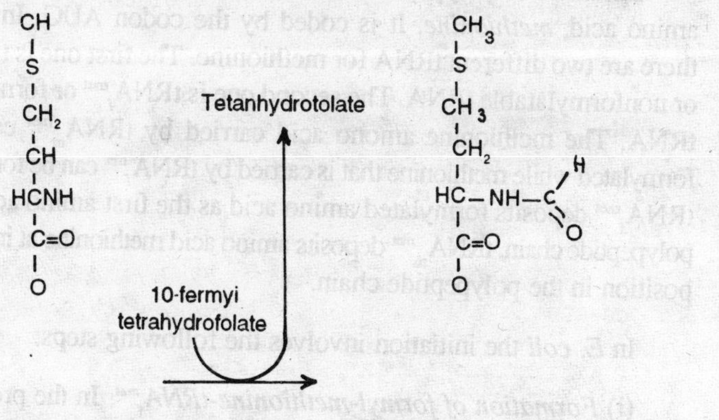

Fig. 8.3. The initiator N-formyl-methionyl-tRNA (1Met-tRNA$_1$) is generated by formylation of methionyl-tRNA$_1$ using formyl-tetrahydrofolate as cofactor.

$$AA_1 + ATP + E_1 \xrightarrow{\text{Enzyme}} (AA_1 - AMP)E_1 + PP$$

Amino acid — Enzyme — Amino acy AMP enzyme complex

2. *Attachment of activated amino acid to tRNA*. The complex formed in first step above, reacts with a particular tRNA and transfer the amino acid to tRNA with the release of AMP and enzyme amino acyl RNA synthetase.

$$(AA_1\text{-}AMP)E_1 + tRNA_1 \longrightarrow AA_1\ tRNA_1 + AMP + E_1$$

For each amino acid, there is a specific tRNA. As there are 20 different amino acids so 20 different tRNAs are there. There are cases where more than one kind of tRNA may be available for the same amino acid, but no case is known where same tRNA can recognize two or more different amino acids.

3. *Initiation of polypeptide chain.* Initiation requires the ribosome subunits, mRNA, guanosine triphosphate (GTP) as energy source, amino acid t-RNA complex and initiation factors (IF).

Initiation of polypeptide chain is always brought about by the amino acid, *methionine.* It is coded by the codon AUG. In *E. coli*, there are two different tRNA for methionine. The first one is $tRNA_m^{met}$ or nonformylatable tRNA. The second one is $tRNA_f^{met}$ or formylatable tRNA. The methionine amono acid carried by $tRNA_m^{met}$ cannot be formylated while methionine that is carried by $tRNA_f^{met}$ can be formylated. $tRNA_f^{met}$ deposits formylated amino acid as the first amino acid of the polypeptide chain. $tRNA_m^{met}$ deposits amino acid methionine at intercalary position in the polypeptide chain.

In *E. coli* the initiation involves the following steps:

(i) *Formation of formyl-methionine-$tRNA_f^{met}$.* In the presence of transformylase enzyme and formyl-tetrahydrofolic acid this takes place.

met-$tRNA_f^{met}$ + formate — f-met-$tRNA_f^{met}$

(ii) *Binding of 30 'S' subunit of ribosoe with mRNA.* mRNA becomes attached to the 30 'S' subunit to form mRNA-30 'S' complex. This process requires the initian factor 3(IF-3). Only one molecule of IF-3 binds to each 30 'S' subunit. (There are three 'F in prokaryotes namely IF-1, IF-2 and IF-3. These factors are found in 30 'S' subunit of ribosomes. Their molecular weights are 9,200, 80,000 and 23,000 daltons respectively. These factors perform different functions). This 30 'S' subunit attaches at the 5' end of m-RNA carrying AUG codon.

IF-3-30 'S' + mRNA —> IF-3-30 'S'-mRNA

Besides binding 30 'S' subunit with mRNA, the IF-3 is also necessary for (a) affecting 30 'S' subunit conformation and (b) prevents reassociation of 30 'S' and 50 'S' subunits of ribosome.

(iii) *Binding of f-met-tRNA$_f^{met}$ with IF-3-30'S'-mRNA complex.* IF-2 and IF-1 are necessary for this step. Besides GTP is also necessary. IF-2 may either attached to f-met-tRNA$_f^{met}$ to form a binary complex or may directly bind to IF-3-30 'S'-mRNA complex. In the first case IF-2-f met-tRNA$_f^{met}$ may locate or recognize AUG codon on the IF-3-30 'S' mRNA complex and associate with it. In the second case IF-2 recognizes fmet-tRNA$_f^{met}$ and later associates to form the initiation complex.

IF-3-30 'S' mRNA + fmet-tRNA$_f^{met}$ + IF-2 —

IF-3-30 'S'-mRNA-IF-2fmet-tRNAf^{met},

The mRNA has many AUG codons. Only one of these functions as an initiation codon. What factor determines that which AUG will function as the initiation codon? It is probably a specific sequence of nucleotides near AUG which is required for it to function as an initiation codon.

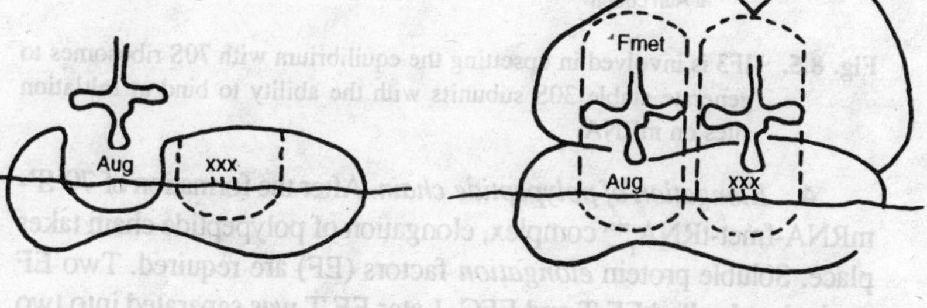

Fig. 8.4. Only tMet-tRNA can be used for initiation by 30S subunits; only other aminoacyl-tRNAs can be used for elongation by 70S ribosomes.

(iv) *Attachment of 50 'S' subunit of ribosome with above complex.* The complex formed in the above step associates with 50 'S' subunit of ribosome, to form a complete ribosome. This action depends on IF-2 activity. Energy is supplied by GTP. The IF-2 and IF-3 are now released.

IF3-30'S'-mRNA - IF-2-fmet-tRNA$_f^{met}$ + GTP —

70'S'-mRNA-fmet-tRNA$_f^{met}$ + GDP + P$_i$ + IF-2 + IF-3

The bacterial ribosome has two binding sites for tRNA. The first one is *A-site* or *amino acyl site* or *acceptor site* and the other site is *P-site* or *peptidyl site* or *polymerisation site*. Thus each ribosome accomodates two codons or six nucleotides at a time. Both the sites are located in 50'S' subunit of ribosome.

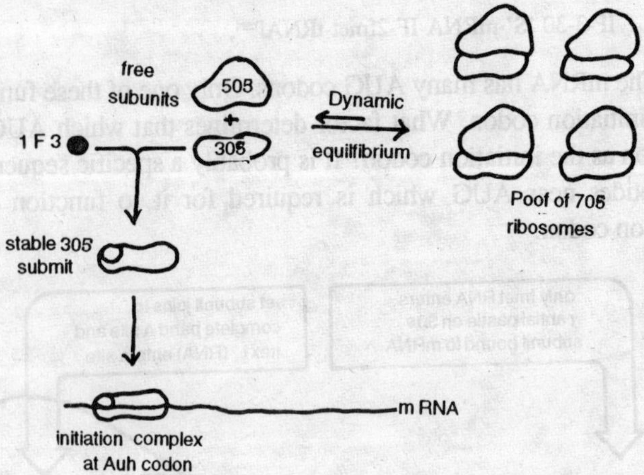

Fig. 8.5. IF3 is involved in upsetting the equilibrium with 70S ribosomes to generate stable 30S subunits with the ability to bind at Inltlation sites on mRNA.

4. *Elongation of polypeptide chain.* After the formation of 70'S'-mRNA-fmet-tRNA$_f^{met}$ complex, elongation of polypeptide chain takes place. Soluble protein *elongation* factors (EF) are required. Two EF are isolated called EF-T and EFG. Later EF-T was separated into two components, EF-Tu (unstable when heated) and EF-Ts (stable when heated). EF-Tu and EF-Ts are required for binding the amino acyl tRNA to the ribosomes. EF-G is involved in the translocation of mRNA.

The elognation process takes place in following steps :

(i) *Binding of AA_1tRNA, to the A-site of ribosome.* EF-Tu and EF-Ts are required for binding the amino acyl tRNA (AA_1tRNA_1) to A-site of ribosome. EF-Tu forms a ternary complex with AA_1 -$tRNA_1$ and GTP.

EF-Tu + GTP ——> EF-Tu-GTP

(binary complex)

EF-Tu-GTP + AA_1-$tRNA_1$ ——> AA_1-$tRNA_1$-EF-Tu-GTP.

(Ternary complex)

The P-site of ribosome being already occupied by f-met-$tRNA_f^{met}$ at the initial stage. Transfer of AA_1-$tRNA_1$ from ternary complex to A-site of ribosome now takes place.

AA_1-$tRNA_1$-EF-Tu-GTP + Ribosome ——>

AA_1-$tRNA_1$ Ribosome + EF - Tu - GDP + Pi

EF-Ts displaces GDP and associates with Tu and then GTP can again associate with EF-Tu to start another cycle for binding of amino acyl tRNA.

EF-Tu-GDP + EF-Ts ——> EFTu-EF-Ts + GDP

EF-Ts - EF-Tu + GTP ——> EF-Tu - GTP

(Binary complex)

(ii) *Peptide bond formation.* Bond formation occurs by the transfer of the first amino acid i.e. f-met-$tRNA_f^{met}$ from its tRNA at the P-site to the amino acyl tRNA (AA_1 – $tRNA_1$) just brought to the A-site. The peptide bond is formed between the carboxyl group of first amino acid (f-met) and the amino group of second one (AA_1). The reaction is catalysed by *peptidyl transferase*. This enzyme is the integral part of the 50'S' subunit of ribosome. After the formation of a peptide bond, the tRNA at P-site is deacylated and the tRNA at A-site now carries the polypeptide.

(iii) *Translocation*. The movement of ribosome relative to mRNA is called translocation. During translocation the ribosome moves along mRNA, a codon at a time, from 5' to 3' direction (5'——>3'). The peptidyl tRNA present A-site now translocated to P-site. During this process, the deacylated tRNA is liberated out from P-site. The translocation of peptidyl tRNA from A-site to P-site requires EF-G and GTP. Now the A-site of ribosome is occupied by another AA-tRNA-EF-T_u-GTP complex and the cycle continues till the ribosome comes across the termination codon i.e. UGA or UAA or UAG.

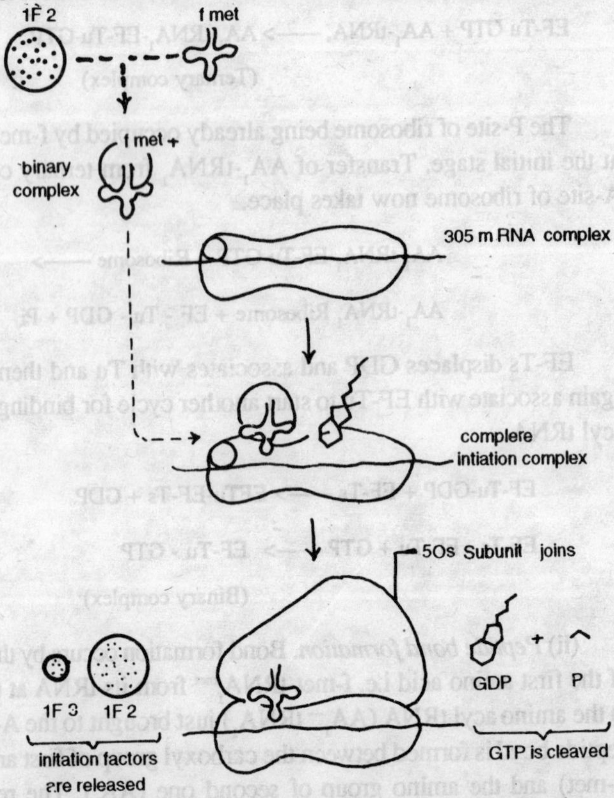

Fig. 8.6. Bacterial initiation factors are needed for 30S subunits to bind mRNA and for fMet-$tRNA_1$ to bind to the 30S-mRNA complex; they are released with cleavage of GTP when the 50S subunit arrives.

There exists a great ambiguity about relative movement of ribosome and mRNA. According to a view during translocation, the energy is

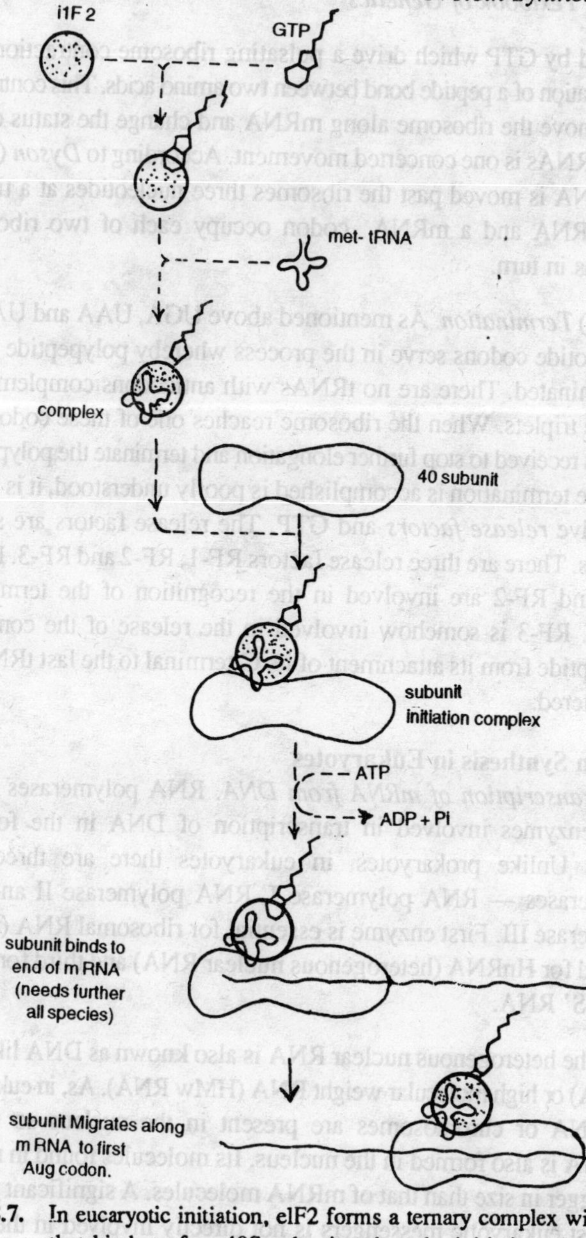

Fig. 8.7. In eucaryotic initiation, elF2 forms a ternary complex with $tRNA_1$ that binds to free 40S subunits; then the initiation plex binds to mRNA at its 5' end, from where it must migrates to the AUG initiation codon.

provided by GTP which drive a pulsating ribosome contraction after the formation of a peptide bond between two amino acids. This contraction would move the ribosome along mRNA and change the status of two bound tRNAs is one concerted movement. According to *Dyson* (1978) the mRNA is moved past the ribsomes three nucleotides at a time so that a tRNA and a mRNA codon occupy each of two ribosomal positions in turn.

(iv) *Termination.* As mentioned above UGA, UAA and UAG are trinucleotide codons serve in the process whereby polypeptide chains are terminated. There are no tRNAs with anticodons complementary to these triplets. When the ribosome reaches one of these codons, the signel is received to stop further elongation and terminate the polypeptide. How the termination is accomplished is poorly understood, it is known to involve *release factors* and GTP. The release factors are soluble proteins. There are three release factors RF-1, RF-2 and RF-3. Factors RF-1 and RF-2 are involved in the recognition of the termination codons. RF-3 is somehow involved in the release of the completed polypeptide from its attachment of its C-terminal to the last tRNA that had entered.

Protein Synthesis in Eukaryotes

Transcription of mRNA from DNA. RNA polymerases are the main enzymes involved in transcription of DNA in the forms of RNAs. Unlike prokaryotes, in eukaryotes there are three RNA polymerases — RNA polymerase I, RNA polymerase II and RNA polymerase III. First enzyme is essential for ribosomal RNA (rRNA), Second for HnRNA (heterogenous nuclear RNA) and third for t-RNA and 5'S' RNA.

The heterogenous nuclear RNA is also known as DNA like RNA (dRNA) or high molecular weight RNA (HMw RNA). As, in eukaryotes, the DNA or chromosomes are present in the nucleus so that the HnRNA is also formed in the nucleus. Its molecules found in nucleus, are bigger in size than that of mRNA molecules. A significant position of most eukaryotic messengers is not directly involved in the coding process. In addition to non-coding nucleotides, the m-RNAs have special modifications at their 5'-end and 3'-end that are not found in

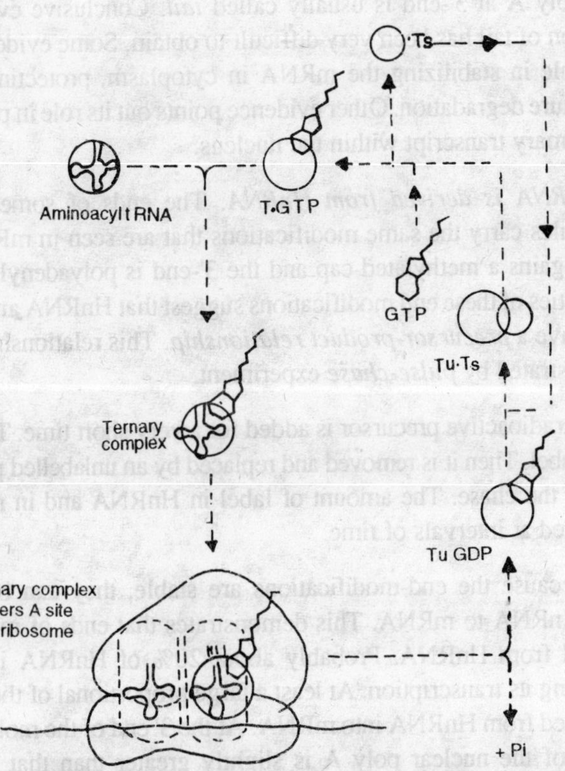

Fig. 8.8. EF-Tu. GTP places aminoacyl-tRNA on the ribosome, is released as EF-Tu. GDP, and then is regenerated as EF-Tu. GTP by the action of EF-Ts.

prokaryotic mRNAs. The 5'-end of mRNA has a cap on it. This cap is formed after the original 5' end has been synthesized. The cap on 5' end consists of 7 methyl guanine. There is good evidence that the 7 methyl guanine cap at the 5' end plays an important role in translation. It could be shown that removal of cap bads to loss of translational activity due to loss of the formation of mRNA-ribosome complex. The base sequence of the 3'end of most mRNA is very unusual. It is composed exclusively of adenine. These adenosine residues, which make up the last 50-200 nucleotides, are referred to as *poly A*. The poly A is also added to mRNA molecule after it has been synthesized.

The poly A at 3'-end is usually called *tail.* Conclusive evidence of function of tail has been very difficult to obtain. Some evidence point to a role in stabilizing the mRNA in cytoplasm, protecting it from premature degradation. Other evidence points out its role in processing the primary transcript within the nucleus.

mRNA is derived from HnRNA. The ends of some HnRNA molecules carry the same modifications that are seen in mRNA. The 5'-end gains a methylated cap and the 3'-end is polyadenylated. The similarties of these end modifications suggest that HnRNA and mRNA may have a *precursor-product relationship.* This relationship can be demonstrated by *pulse-chase* experiment.

A radioactive precursor is added for a very short time. This is the pulse label. Then it is removed and replaced by an unlabelled precursor. This is the chase. The amount of label in HnRNA and in mRNA is measured at intervals of time.

Because the end-modifications are stable, they can be chased from HnRNA to mRNA. This demonstrates that ends of mRNA are derived from HnRNA. Probably about 20% of HnRNA is capped following its transcription. At least a large proportional of the cap can be chased from HnRNA into mRNA. At the 3' end of the molecule the length of the nuclear poly A is slightly greater than that found in mRNA. A few bases are removed at the time of passage from nucleus to the cytoplasm.

In eukaryotes the termination of RNA polymerase II activity during transcription may take place through the termination sites. Obviously the 3' end of mRNA will be generated due to post-transcriptional cleavage of poly *A*. This cleavage is achieved by *snurp* (small nuclear RNA-protein complex). How the snurp helps in post-transcriptional cleavage is unknown.

Translation. This step involves translation of the language of nucleic acid into the language of proteins translation occurs in following steps :

(i) *Activation of amino acids.* This step in eukaryotes is similar to that of prokaryotic activation.

(ii) *Transfer of amino acid to tRNA*. This step in eukaryotes is also similar to that of prokaryotic transfer.

(iii) *Initiation of polypeptide chain*. This step in eukaryotes is similar to that in prokaryotes. In eukaryotes there are more initiation factors as compared to prokaryotes where three factors are present. They are named by putting prefix e to signify eukaryotic origin. These factors are eIF-1, eIF-2, eIF-3, eIF-4A, eIF-4B, eIF-4C, eIF-4D, eIF-5 and eIF-6.

In eukaryotes the methionine formylation does not occur. Following steps are involved in the initiation.

(a) *Formation of met-tRNA-elongation factor 2-GTP ternary complex*. This complex is formed in two steps — in the first GTP binds to eIF-2 forming binary complex

GTP + EIF-2 ——> GTP -eIF-2

In the second step above complex associates with met-tRNA

GTP-eIF-2 + met-tRNA ——> met-tRNA - eIF-2 - GTP.

In mammals eIF-2 has three subunits eIF-2 a-which binds to GTP eIF-2, g-which binds to met-tRNA and eIF-2 b which may be a recycling factor.

(b) *Binding of above complex with 40'S' ribosomal subunit*. The above complex associates with 40'S' subunit of ribosome to form 40 's' initiation complex.

met-tRNA-eIF-2-GTP + 40'S'——> 40'S'-met-tRNA-eIF-2-GTP

(c) *Association of mRNA with 40'S' initiation complex*. For this reaction eIF-3 is necessary. The binding of mRNA is assisted by other factors like eIF-1, eIF-4A, eIF-4B and ATP.

40'S'-met-tRNA-eIF-2-GTP+mRNA ——> mRNA-40'S-met-tRNA-eIF-2-GTP.

(d) *Association of above complex with 60'S' ribosomal subunit*. For this reaction eIF-5 is necessary. It releases eIF-2. The 40'S' and 60'S' subunits joining reaction depends on eIF-4C and GTP is hydrolysed to GDP. The above complex moves on mRNA strand towards 3'-end in the search of initiation codon i.e. AUG.

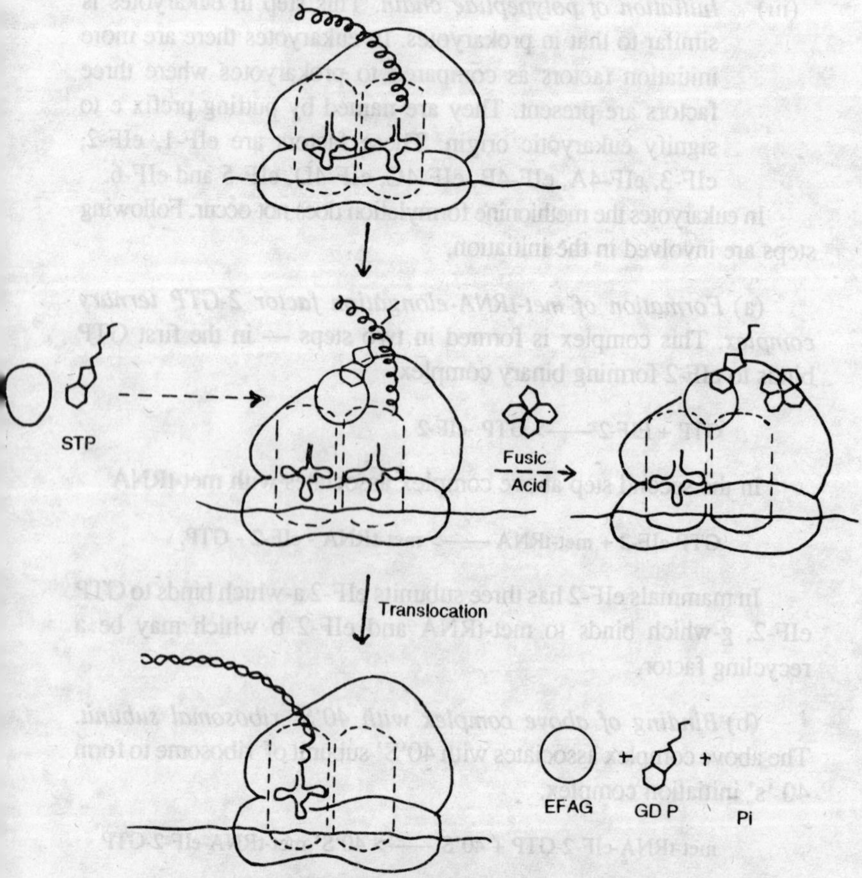

Fig. 8.9. After peptide bond formation, EF-G and GTP are needed for translocation, but then dissociate from the ribosome (unless fusidic acid is present to stabilize the post-traslocation state).

mRNA-40'S'-met tRNA-eIF-2-GTP + 60'S' ——> mRNA-80'S'-met-tRNA + eIF-2 + eIF - 3+ GDP + Pi.

The eIF-3 remains present in 40'S' ribosomal subunit, eIF-2 and eIF-3 inhibit the association of 40'S' and 60'S' ribosomal subunits.

(iv) *Elongation of polypeptide chain.* This step in eukaryotes is similar to that of prokaryotes. In eukaryotes the elongation factor is called eEF-2. This factor is needed for translocation.

(v) *Termination of polypeptide chain.* Only one release factor is known called eRF-1. GTP is necessary for binding of eRF-1 to the ribosome.

9

Genetic Code

Genetic code is a sort of relationship between the nucleotide sequence in mRNA and amino acid sequence in protein molecule. The word "code" has several meanings. Out of them one is "a system of signals communicating." A pertinent example is the Morse code. In this chapter too we are dealing with the code, for messenger RNA having a means for communicating message that directs the combination of amino acids in a specific sequence. *George Gamow* first proposed the coding scheme in 1954.

Several theories were proposed to explain the mechanism by which a particular sequence of nitrogenous bases in DNA by trnascribing complementary bases in mRNA determines the position of specific amino acid in the protein molecule. The theory which is widely accepted till now was proposed by *F.H.C. Crick.* The theory holds the existence of a genetic code and its smallest unit which codes for one amino acid is known as *codon*. Thus, we see that the code in DNA is copied in the form of mRNA. But since little was known about the base sequences in DNA, the genetic code seemed a remote goal never to be attained. But in few years the genetic code become known as a result of ingenious experimentation mainly in the laboratories of *M.W. Nirenberg, S. Ochoa* and *H.G. Khorana.*

The Genetic Code is a Triplet Code

DNA contains four kinds of nucleotides (of A,T, G and C), and proteins are synthesized from 20 different types of amino acids. A basic problem regarding the genetic code was; how many bases of DNA specify one amino acid? In a *singlet code* each base or letter would specify one amino acid. Only 4 of the 20 types of amino acids would be coded unambiguously by a singlet code. In a two-letter or *doublet code* two bases would specify one amino acid. Here 16 (4 x 4) of the 20 amino acids can be specified, but there would be ambiguous determination of a number of amino acids. *A triplet* or three-letters code was first suggested by the physicist *Gamow* in 1954. According to the triplet code three letters or bases specify one amino acid. Thus 64 (4 x 4) distinct triplets of purine and/or pyrimidine bases determine the 20 amino acids. These triplets have been called *codons.* Since there are 64 codons and only 20 amino acids it is obvious that there are many more codons than there are amino acids, i.e. the code is degenerate. Experimental evidence shows that the code is a triplet one and that 61 of the 64 codons code for individual amino acids during protein synthesis.

singlet code
A
G
C
T

doublet code			
AA	AG	AC	AT
GA	GG	GC	GT
CA	CG	CC	CT
TA	TG	TC	TT

triplet code			
AAA	GAA	CAA	TAA
AAG	GAG	CAG	TAG
AAC	GAC	CAC	TAC
AAT	GAT	CAT	TAT
AGA	GGA	CGA	TGA
AGG	GGG	CGG	TGG
AGC	GGC	CGC	TGC
AGT	GGT	CGT	TGT
ACA	GCA	CCA	TCA
ACG	GCG	CCG	TCG
ACC	GCC	CCC	TCC
ACT	GCT	CCT	TCT
ATA	GTA	CTA	TTA
ATG	GTG	CTG	TTG
ATC	GTC	CTC	TTC
ATT	GTT	CTT	TTT

Fig. 9.1. Singlet code, a doublet code and a triplet code.

Properties of Genetic Code

Following properties of the genetic code have been studied and proved expermentally:

(i) *Genetic code is non-overlapping* : When *Gamow* first proposed the genetic code, he had suggested it to be an overlapping code. No doubt *Crick* had suggested a non overlapping code but the direct experimental evidence for a non-overlapping triplet code was given by *Khorana* and his co-workers in 1965.

On the surface, it appears wasteful for 64 triplets to stand for 20 amino acid, so attempts were made to hypothesize a triplet code in which the code was overlapping and so to reduce the number of codons to about 20. But a lot of shortcommings are there in this proposal. Let us examine the case of lysine, whose codons appear to be AGG and AAA, and that of phenylalanine, whose codons appear to be UUU and UUC. Since the code letters of these two amino acids are mutually exclusive, lysine could not follow phenylalanine, or *vice versa,* in any protein; this would constitute a forbidden combinations of amino acids in proteins. Any two or three or so on amino acid combinations are possible. Therefore it can be concluded that the genetic code is non overlapping.

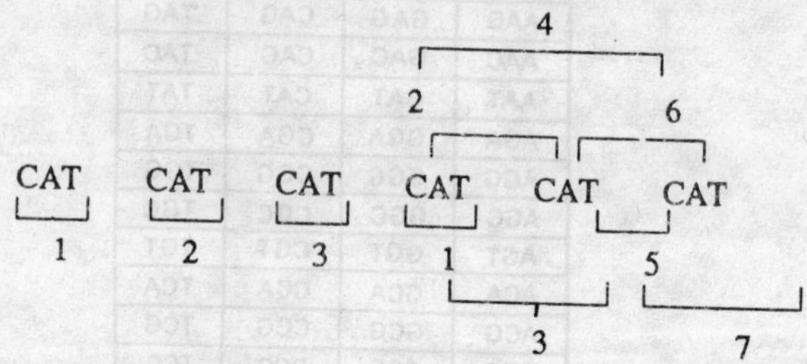

Non-overlapping Overlapping

(ii) *Genetic code is degenerate:* The code contains many synonyms, in that almost all amino acids are represented by more than one codon. For example, the three amino acids-arginine, serine and leucine-each has six synonymous codons. However, for many of the synonym codons specifying the same amino acid the first two bases of the triplet are constant, whereas the third can vary; for example, all codons starting with CC specify proline (CCU, CCC, CCA and CCG) and all codons starting with AC specify threonine. This flexibility in the third nucleotide of a codon may well help to minimize the consequences of errors.

(iii) *Genetic code is commaless* : A commaless code means that no punctuations are needed between any two words. In other words, we can say that after one amino acid is coded, the second amino acid will be automatically coded by the next three letters and that no letters are wasted for telling that one amino acid has been coded and that now second should be coded.

(iv) *Universality of the Code* : The code is based on work on the bacterium *E. coli* but it is valid for other organisms. This important characteristic of the genetic code is called its *universality*. Consider any codon. That codes for the same amino acid from the smallest organism to the largest, plant or animal. Thus, UUU codes for phenylalanine and GUC for valine in all living organisms, from *Amoeba* to ape, bacteria to the banyan tree and from cabbage to kings. The genetic code which was first developed in the bacteria about 3 billion (300 crore) years ago has not undergone any change and has been preserved in its almost original form in the course of evolution. In other words, the code is a conservative one.

The best evidence for the universality of the code comes from studies in which amino acid-charged tRNA from *E. coli* was placed with mRNA and ribosomes derived from rabbit reticulocytes. The result was the production of rabbit haemoglobin. Obviously, the bacterial tRNA was able to recognize perfectly the mammalian mRNA for haemoglobin and could work perfectly with mammalian ribosomes. Studies on amino acid sequence of homologous proteins from widely divergent and unrelated species (cytochrome—C from man, chicken, several bacteria and yeast etc.) show great uniformity although genetic divergence took place hundreds of millions of years ago. Similarly, studies with synthetic messenger RNAs and binding studies with tRNAs also point towards a universal code, though slight variations occasionally occur. Organisms with high G/C content in the DNA appear geared for more common reading of codons ending in C or G. Their proteins also are relatively rich in amino acids coded by G- and C-rich codons, for example Ala, Arg and Gly and relatively poor in

Ile, Lys, Phe and Tyr. Certain mutant microorganisms show exceptions to the absolute universality of the code. There are mutants that have acquired ability to read nonsense triplets. However, this accessory ability is only partial and nonsense triplets are nonsense much of the time. Some natural differences may exist among organisms. Initiation of polypeptide chains is encoded differently in bacteria and their bacteriophages (viruses) as compared to higher organisms. Thus, tRNA for cysteine binds well with mRNA triplet UGA (nonsense in bacteria). Biological significance of these observations remains to be seen as there is more to biology than the 64 triplets.

(v) *The code has polarity:* If a gene is to specify the same protein repeatedly it is essential that the code must be read between fixed start and end points. These points are the initiation and the termination codons, respectively. It is also essential that the code must be read in a fixed direction. In other words, the code must have polarity. It is obvious that if the code is read in opposite directions it would specify two diferent proteins, since the codons would have reversed base sequences.

For example :

	Leu	Ile	Val	Ser →
Codon -	UUG	AUG	GUG	UCG
	← Val	Leu	Leu	Ala.

(vi) *Initiation codons :* The starting amino acid in the synthesis of most protein chains is *methionine* (eukaryotes) or *N-formyl methionine* (prokaryotes). Methionyl or N-formyl methionyl-tRNA specifically binds to initiation sites containing the AUG codon. This codon is therefore called the *initiation codon.* Less often, GUG also serves as the initiation codon in bacterial protein synthesis.

(vii) *Non-sense codons:* The UAA (also called *ochre),* UAG (also called *amber* because in investigator who studied the properties of this codon belonged to the Bernstein family, and Bernstein means "amber" in German) and UDA codons do not specify any amino acid so, are called *non-sense codons.* They are also called *termination codons,* because, these codons are used by the cell to signal the natural end of translation of a particular polypetide. However, their inclusion in any mRNA results in the abrupt termination of the message at the point of their location even though the polypeptide chain has not been completed.

Deciphering or Cracking of Genetic Code

We may here briefly descibe the techniques which ultimately resulted in determining the code words for each of the 20 proteinogenic amino acids. Most of our current knowledge of the general nature of the genetic code and of the nucleotide composition and sequences of the various codons has been obtained from the following five main types of experimental approaches:

(1) *The first method* is analysis of the primary structures of the wild type and mutant proteins. In each of the mutant proteins so far analysed a mutant amino acid replaces a particular wild type amino acid. But the data so far accumulated on such amino acid replacements is very meagre and sets restrictions on the genetic code.

(2) *The second method* for establishing the genetic code consists of determining the sequence of nucleotides in the mRNA and also of amino acids in the protein molecule synthesized under its influence. The sequence analysis for proteins is now a standard biochemical technique but unfortunately the determination of the base sequence of large nucleic acid molecules (like mRNA) is yet to be achieved. Biochemists, therefore, adopted indirect methods to crack the genetic code. A cell-free amino acid incorporating system and the polymerising enzymes were known which could polymerise single mononucleoside triphosphates or their mixtures to produce long nucleotide chains of random base sequence. Initial break-through in this direction was achieved by *Nirenberg* and *Mathei* (1961) who adopted the following technique. Using the enzyme *polynucleotide phosphorylase* they prepared a syntheitc polyribonucleotide containing only one kind of base and added it to a cell-free amino acid-incorporating system from *E. coli*. In this way they prepared, for example, a poly uracil (UUU...UUU) to which they added different amino acids labelled with C^{14}. They found that this poly-U system specifically stimulated the formation of a

polypeptide which contained only the amino acid phenylalanine. The conclusion was drawn that the code word for the amino acid phenylalanine was a sequence of three uracil nucleotides (UUU) in mRNA. This was the first demonstration that synthetic polyribonucleotides could act as a messenger for polypeptide synthesis. Later on, *Nirenberg* and *Ochoa* became deeply involved in this problem. They prepared poly-G, poly-C and poly-A polynucleotides from the riboside diphosphates GDP, CDP and ADP, and employing the same technique. They observed that these homoribonucleotides stimulated the incorporation of amino acids glycine, proline and lysine to the corresponding homopeptides respectively. *Crick* has already shown that the code was triplet in nature. It was, therefore, concluded that the base triplet GGG, CCC and AAA coded respectively for the amino acids glycine, proline and lysine.

Later on, mixtures of two or more ribonucleotide diphosphates were used by *Nirenberg* and with the help of the enzyme mentioned above polyribonucleotides were prepared. Thus, using UDP and CDP in the ratio of 3:1 he obtained a polynucleotide which contained the following triplets in the descending order of frequency: UUU, UUC, UCU and CUU. The triplets containing 2Cs and 1U were the least frequent. With such a poly-UC system *Nirenberg* obtained a polypeptide containing the amino acids phenyalanine and serine in the ratio 3:1. He concluded that the code word for serine contains 2Us and 1C. This method, however, did not give the exact sequence of the three bases. *Ochoa* also used this technique and determined the code words for most of the amino acids. The exact sequence of the three bases for any amino acid was not known because all these polynucleotides had random base sequence. The assigning of the exact triplet for every amino acid was not possible at this stage.

(3) *The Third method.* The determination of the exact base sequence of mRNA triplets which códed for each amino acid was made possible by a later discovery of *Nirenberg* that each trinucleotide (triplet) caused a specific aminoacyl-tRNA to bind to the ribosome and to the specific base sequence of synthetic mRNA. With the help of this elegant technique and using trinucleotides of known sequences. *Nirenberg* and *Lader* (1965) observed that amino acyl-tRNA could be bound to

ribosomes in the presence of specific trinucleotide phosphates (triplet). In the absence of a specific triplet the amino acyl-tRNA could not bind to the ribosome and therefore, passed through the nitrocellulose filter while the tertiary complex containing amino acyl-tRNA-ribosome-mRNA was retained on the filter. Using this technique, *Nirenberg* and his associates tested 64 triplets for binding of the 20 amino acyl-tRNA's. They succeded in assigning 50 out of the 64 codons with their exact base sequences with which each of the amino acyl-tRNAs formed a union. This method was based on the assumption that binding required the same code words as did protein synthesis. This method, however, did not prove to be very satisfactory as binding stimulations observed in a number of cases were very poor and artifactual. Moreover, several authentic codons did not bind the amino acyl-tRNA.

(4) *The fourth method- Khorana's direct method,* of confirming the genetic code is the most direct method and was devised by the Nobel Prize winner *Dr. H.G. Khorana.* This method consists *in vitro* chemical synthesis of short segments of DNA of known base sequence with the help of *DNA polymerase.* From this synthetic DNA a polyribonucleotide (RNA) of strictly defined base sequence is transcribed under the catalytic influence of *RNA polymerase.* A polypeptide is then synthesized under the direction of RNA as represented in the following schematic diagram :

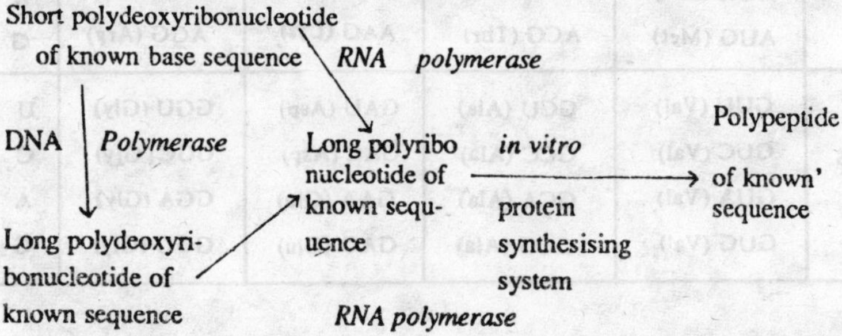

The amino acid sequence of the polypeptide so formed is then determined and correlated with the base sequence of DNA and RNA. These studies of *Khorana* and his co-workers with chemically defined messengers proved very conclusively that (1) the base sequence in DNA specifies the sequence of amino acids in proteins; (2) the information contained in DNA is conveyed through RNA and (3) the genetic code is triplet and non-overlapping in nature. The results obtained from these polypeptide synthesis in combination with the results obtained from the binding technique of *Nirenberg* and co-workers led to the deciphering and establishment of the genetic code.

FIRST LETTER	SECOND LETTER: U	C	A	G	THIRD LETTER
U	UUU (Phe)	UCU (Ser)	UAU (Tyr)	UGU (Cys)	U
	UUC (Phe)	UCC (Ser)	UAC (Tyr)	UGC (Cys)	C
	UUA (Leu)	UCA (Ser)	UAA (Ochre)	UGA ?	A
	UUG (Leu)	UCG (Ser)	UAG (Amber)	UGG (Tryp)	G
C	CUU (Leu)	CCU (Pro)	CAU (His)	CGU (Arg)	U
	CUC (Leu)	CCC (Pro)	CAC (His)	CGC (Arg)	C
	CUA (Leu)	CCA (Pro)	CAA (GluN)	CGA (Arg)	A
	CUG (Leu)	CCG (Pro)	CAG (GluN)	CGG (Arg)	G
A	AUU (Ileu)	ACU (Thr)	AAU (AspN)	AGU (Ser)	U
	AUC (Ileu)	ACC (Thr)	AAC (AspN)	AGC (Ser)	C
	AUA (Ileu)	ACA (Thr)	AAA (Lys)	AGA (Arg)	A
	AUG (Met)	ACG (Thr)	AAG (Lys)	AGG (Arg)	G
G	GUU (Val)	GCU (Ala)	GAU (Asp)	GGU (Gly)	U
	GUC (Val)	GCC (Ala)	GAC (Asp)	GGC (Gly)	C
	GUA (Val)	GCA (Ala)	GAA (Glu)	GGA (Gly)	A
	GUG (Val)	GCG (Ala)	GAG (Glu)	GGG (Gly)	G

Fig. 9.2. The triplet genetic code.

By this technique, *Khorana* not only worked out and confirmed the exact sequence of code words for all the amino acids but also clearified the roles of the 5'and 3' terminals of mRNA molecule. He showed that the translation of mRNA proceeded from its 5' hydroxyl end towards its 3'-hydroxyl end. His investigations also led to the artificial production of small segments of DNA molecule, thereby paving the way for the synthesis of artificial genes which could function in a living cell. In 1970, a Japanese student of *Khorana, Dr. Miss Otsuka,* succeeded in chemically linking six nucleotides by attaching an extra phosphoric acid artifically to a nucleotide in addition to the one it already had. By this process *Otsuka* and her co-workers at the Osaka University synthesized alanine tRNA. This synthetic RNA had exactly the same sequence of nucleotides as the one that occurs in natural cells. The same year Dr. *K.L. Agarwal,* another student of *Khorana,* succeeded in making an artificial gene. This gene, however, failed to function in a living cell because it did not possesed the necessary "start" and "stop" signals for its functioning. He reported that enough advance had been made towards incorporating these control signals in a synthetic gene. It appears that scientists are very close to synthesizing artificial gene which, when introduced into a cell, could function like a natural gene and alter the heredity of the organism. This has opened the wide but virgin field of genetic engineering with the possibility of finding a cure for genetic disorders of metabolism and of producing organisms to precise specification by introducing artificial genes into the body. But this also carries the risk of artifiical genes producing undesirable traits and monstrosities in the organisms in which they are introduced.

(5) *Fifth Method.* The fifth and final approach in this context involves inference from the results of purely genetical experiments. These included studies on mutant proteins due to the action of certain mutagens in reversing particular kinds of mutation; others involve incorporating two or more mutations in the same DNA strand and observing the effect on the phenotype. Such observations give us confidence in the genetic code as deduced from *in vitro* experiments. Moreover, extraction of tRNAs from cells and studies of the nucleotide sequence found in certain regions of these isolated tRNAs all converge to give single, consistent picture of the code.

Wobble Hypothesis

It is sometimes argued that the third base of a codon is not very important and that specificity of a codon is particularly determined by the first two bases. It has been shown, for instance that the same tRNA can recognize more than one codon differing only at the third position. This pairing is not very stable and is allowed due to wobbling in base pairing at the third position. *Crick,* in 1965 proposed a hypothesis called *Wobble hypothesis* to explain this phenomenon. He discovered that if U is present at first position of anticodon, it can pair with either A or G at the third position of codon. Similar is the case with U and I (Inosine, modified base in tRNA) found in anticodon. The pairing relationships between first base of anticodon and third base of codon are given in Table. 9.1.

TABLE 9.1
Wobble Pairing

Anticodon (first base)	*Codon (third base)*
U	A,G
C	G
A	U
G	U,C
I (inosine, resembles G)	U,C,A

This kind of wobbling allows economy of the number of tRNA molecules, since several codons meant for same amino acid are recognized by same tRNA. For instance, anticodon IGC can recognize codons GCU, GCC and GCA (coding for alanine); IGA can recognize UCU, UUC and UCA (coding for serine) and IAC can recognize GUU, GUC and GUA (coding for valine).

Contribution of *Khorana*

India born biochemst *Dr. Khorana* has devised an ingenious technique for artificially synthesizing mRNA with repeated sequences of known nucleotides. For this valuable contribution he was awarded Nobel Prize in 1970.

By using synthetic DNA, *Khorana* and his co-workers prepared chains of polyribonucleotides with known repeating sequences of two or three nucleotides as follows:

(i) Poly CUC UCU CUC UCU....

(ii) Poly CUA CUA CUA CUA....

In first case CUA and UCU are two codons arranged alternately in the polynucleotide chain. This dictates the formation of a polypeptide chain having two amino acids (leucine and serine) arranged alternatively. The second case is an example of *copolymers chains*. The polynucleotide chain is formed of repeated linkage of codon CUA. This dictates the formation of a polypeptide chain consisting of only one amino acid-leucine.

10

Sex-linked Inheritance

There are two types of chromosomes in the living organisms—*autosomes* and *sex chromosomes*. Autosomes are the chromosomes bearing genes for somatic organs and sex chromosomes carry *sex linked genes*. These genes are usually inherinted alongwith X chromosomes. In most animals the Y chromosome is usually genetically *inert or empty* i.e., it does not carry genes, and if at all it bears, the number is very small. Genes present an Y chromosomes are called *holandric genes*. Such genes produce their significant effect that too in male sex only. X chromosome is a quite long chromosome and bear a number of sex linked genes. In human beings there are two X chromosomes in females and one X in male. The males receive their X chromosome from their mother whereas daughters or females receive one X from mother and one from father.

Discovery of Sex-linkage

Although inheritance of certain characters and their association with sex was well-known for many years. One of the earliest known instance of linkage of a character to sex is bleeding diseases—*haemophilia* which appeared only in males in the royal family of Spain. But in beginning it was regarded a hereditary defect. The actual explanation was offered by *Thomas H. Morgan*in 1910 while working on *Drosophila melanogaster*. He named this phenomenon as sex-linkage.

Types of Sex-linkage

The mechanism of sex-linkage is slightly different from the ordinary linkage of genes because the sex-linked genes are present on the sex chromosomes which do not posses homologous parts. Depending upon whether the sex-linked genes are present on X or on Y-chromosome, the sex-linked inheritance can be of the following types:-

1. *Digenic:* The sex-linked genes are present on the non-homologous part of X-chromosome and are passed on from male parent to male grandsons through daughters of F_1 generation. Naturally male sex is heterogametic.

2. *Diandric:*In this case the 2X—chromosomes of female behave as if these are completely homologous and female passes its one X to female and other to male.

3. *Hologenic*: Character is directly passed from female to female.

4. *Holandric* : Genes which are in the non-homologous part of Y-chromosome are passed on directly from father to son.

Besides above, there are sex-influenced traits in which characters (secondary sexual) become changes due to the type of sex, phenotypically without any change in chromosomes. Similarly there are sex-linked lethals which cause death to individuals.

Sex Linkage in Drosophila

Inheritance was prosided by the work of Margan on Drosophila. In this fruit fly the females is homogametic *(XX)* and the male heterogametic *(XY)*.White eye colour in this fly is a sex linked factor and is recessive to the normal red eye colour. Following crosses between white eyed and red eyed *Drosophila* will justify the sex linkage in the fruit fly.

(i) *Red eyed female X white eyes male:* When a white eyed male is mated to a true breeding red eyed female, all the offsprings have red eyes. The females of the F_1 however, carry the white eye gene, while the males do not, because they receive the X chromosome from their mother. When the red eyed male and red eyed female individuals of F_1 are intercrossed, the F_2 progeny is found to include : exclusively red eyed female population and a male population with 50% red eyed individuals and 50% white eyed individuals. Thus, F_2 generation included red eyed and white eyed individuals in the ratio of 3:1

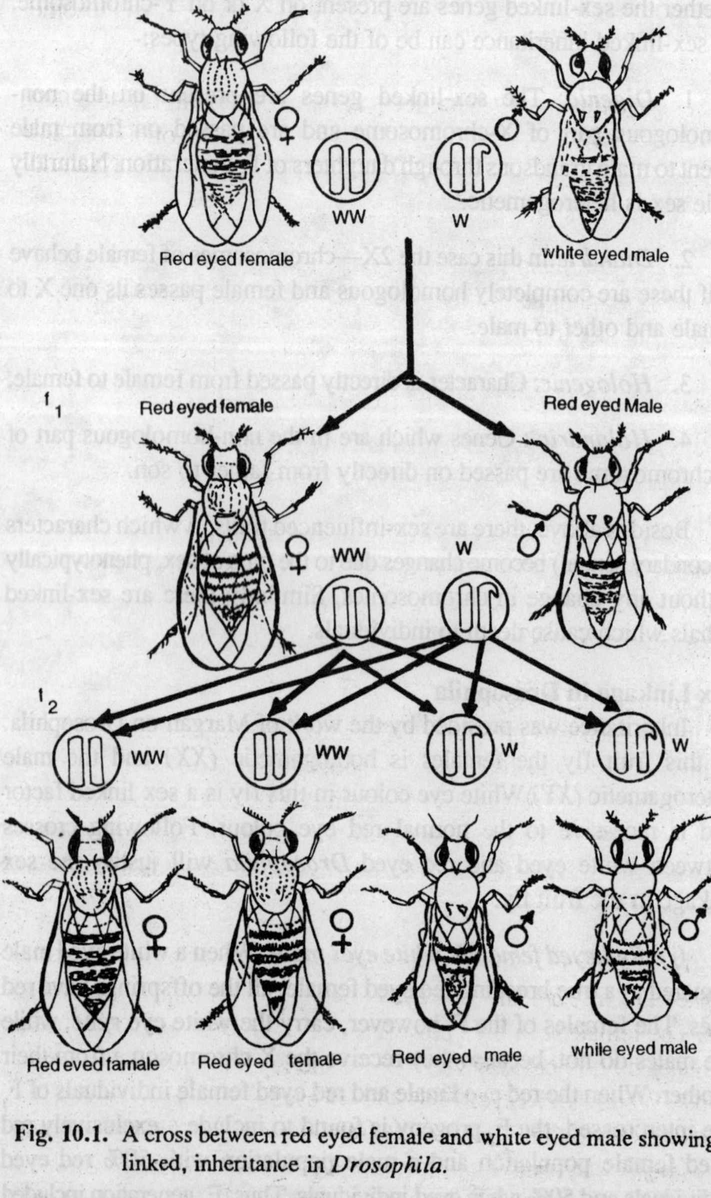

Fig. 10.1. A cross between red eyed female and white eyed male showing sex linked, inheritance in *Drosophila*.

(ii) *White eyed female X red eyed male :* When a white eyed female *Drosophila* is crossed with a red eyed male *Drosophila*, all the female individuals in the F_1 generation are red eyed and all the male individuals are white eyed. When these red eyed female individuals and white eyed male individuals of F_1 are intercrossed, the female population of F_2 generation is found to include 50% red eyed and 50% white eyed flies. Similarly, the male population of F_1 includes 50%, red eyed and 50% white eyed flies.

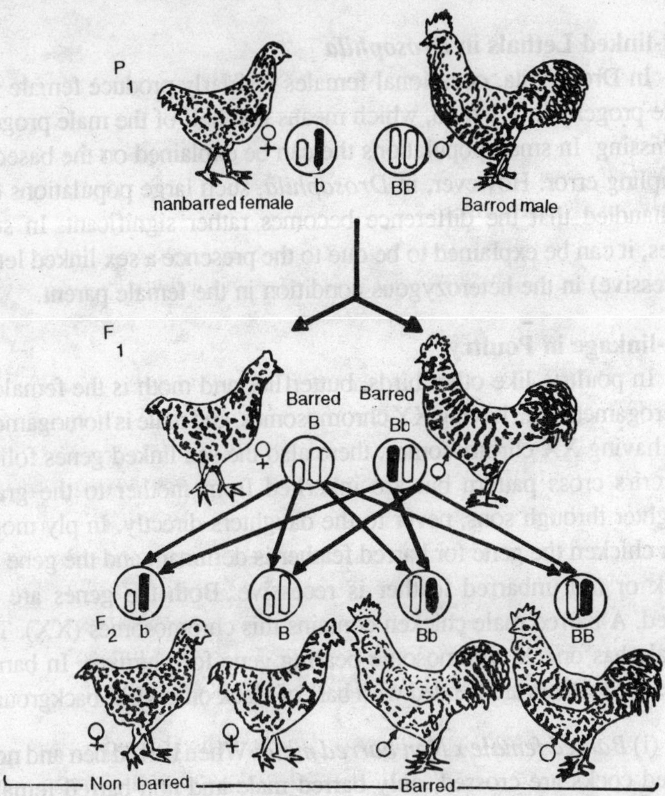

Fig 10.2. A cross between a barred hen and a non-barred cock showing sex linked inheritance.

From the above experiments it is evident that white eye colour is a sex-linked character following a crisscross inheritance. The male transmits sex-linked characters through his daughters to grandsons and never through his sons. This is because the character is contained in X chromosome of the male. From female, both her daughters and sons may receive that character. Y chromosome of male carries no normal allele for white. When a white eyed female is crossed with a red eyed male, the gene goes both to sons daughters. *Morgan* concluded that the gene for white is contained in X chromosome of male. At least 140 other genes in *Drosophila melanogaster* are reported to be sex-linked.

Sex-linked Lethals in *Drosophila*

In Drosophila, occasional females regularly produce female and male progeny in 2:1 ratio, which means that half of the male progeny is missing. In small populations the can be explained on the based of sampling error. However, in *Drosophila,* such large populations can be handled that the difference becomes rather significant. In such cases, it can be explained to be due to the presence a sex linked lethal (recessive) in the heterozygous condition in the female parent.

Sex-linkage in Poultry

In poultry, like other birds, butterflies and moth is the female is heterogametic i.e., having XY chromosomes, and male is homogametic i.e., having XX chromosomes, there also the sex linked genes follow the crirs cross pattern but are inherited from mother to the grand daughter through sons, never to the daughters directly. In ply mouth rock chicken the gene for barred feather is dominant and the gene for black or red unbarred feather is recessive. Both the genes are X-linked. A barred male chicken contains this chromosomes (XX). The female has one X chromosome bearing gene for plumage. In barred cases the feathers are banded with bars of black on a white background.

(i) *Barred female x Non barred male:* When barred hen and non-barred cocks are crossed, only barred male and non-barred females are obtained in F_1 generation. When these are intercrossed and F_2 progeny is caused, barred and manbarred individuals appear in 1:1 ratio among male as well as female populations.

(ii) Non barred female x Barred male :

When barred cock with two genes for the character is crossed with unbarred female having single recrossing gene for the character, only barred males and females are produced in F_1. These F_1 barred males and females when inbred produce in F_2, a hemizygon non-barred female, a hemizygon barred female, a hetenzyges barred male and a homozygous barren female.

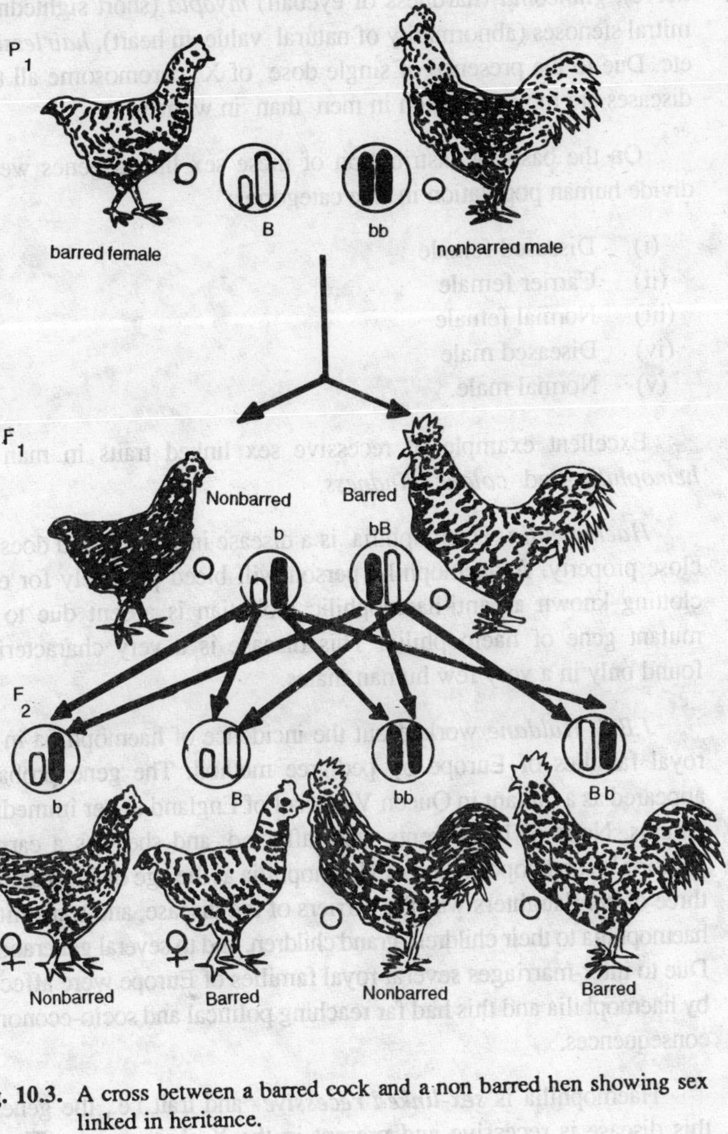

Fig. 10.3. A cross between a barred cock and a non barred hen showing sex linked in heritance.

Sex-linkage in Human Beings

R.R Gates has reported about 45 sex-linked genes in human are: *haemophilia, colour blindness, opticatrophy* (degeneration of optic nerve), *glancoma* (hardness of eyeball) *myopia* (short sightedness), mitral stenoses (abnormality of natural value in heart), *hairlessness.* etc. Due to the presence of single dose of X chromosome all these diseases are move common in men than in women.

On the basis of distribution of these sex linked genes we can divide human population in five categories:

(i) Diseased female
(ii) Carrier female
(iii) Normal female
(iv) Diseased male
(v) Normal male.

Excellent example of recessive sex linked traits in man are *hemophilia* and *colour blindness.*

Haemophilia: Haemophilia is a disease in when blood does not close properly. A haemophilic person will bleed profusely for even clotting known as anti-haemophilic globutian is absent due to the mutant gene of haemophilia. This disease is a very characteristic found only in a very few human males.

J.B.S. Haldane worked out the incidence of haemophilia in the royal families of Europe by pedigree method. The gene probably appeared as a mutant in Queen Victoria I of England or her immediate parents. None of her parents were affected, and she was a carrier. Victoria's son heopold died of Haemophilia at the age of 31 years and three of her daughters was the carriers of the disease, and transmitted haemophilia to their children, grand children, and to several generations. Due to inter-marriages several royal families of Europe were affected by haemophilia and this had far reaching political and socio-economic consequences.

Haemophilia is *sex-linked recessive* and trait i.e., the gene of this disease is recessive and present in the X chromosome. Thus a single gene produces the disease in male while two genes will produce the disease in female. A female with one same gene (heterozygous) is apparently normal but is actually carrier and transmits the gene to 50% of the sons. Haemophilics usually die at an early age and rarely survive to attain marriageable age.

Three types of haemophilia have bee studies biochemically, each affecting thee production of thromboplastin or thrombokunase.

Heomophilia A. This is the most common type of this abnormally and a large number of person suffer this which is due to reduction in Anti-Haemophilic Factor (AHF).

Haemophilia B: This is less common type than haemophilia A About 20% of three haemophilic persons suffer from this type of haemophilia. In this plasma thromboplastin component (PTC) is synthesized in lesser proportion.

Haemophilia C. In this the synthesis of Plasma Thromboplasin Anticedent (PTA) is reduced. It is due to an antosomal gene. About 1% of the haemophilic persons suffer from this type of disease.

Colour Blindness: Colour blindness is also an example of sex-linked characters. Those who suffer from colour blindness cannot distinguish between red and green and sometimes between blue and black. It is estimated that about 4 per cent of all human females are affected with the colour-blindness. Red green colour blindness is more significant that the blue black colour blindness. The red blindness is called protanopia and the green blindness *denturanopia.* X chromosome possesses a normal gene which controls the formation of colour sensitive cells in the retina. Its recessive allele fails to do its job properly and results in colour blindness.

Like haemophilia, colour blindness is also transmitted by an affected person to all his daughter and the related gene is located in his chromosome. These daughter are normal but carrier because the gene for colour-blindness is recessive and a dominant allele is present in the other X chromosome.

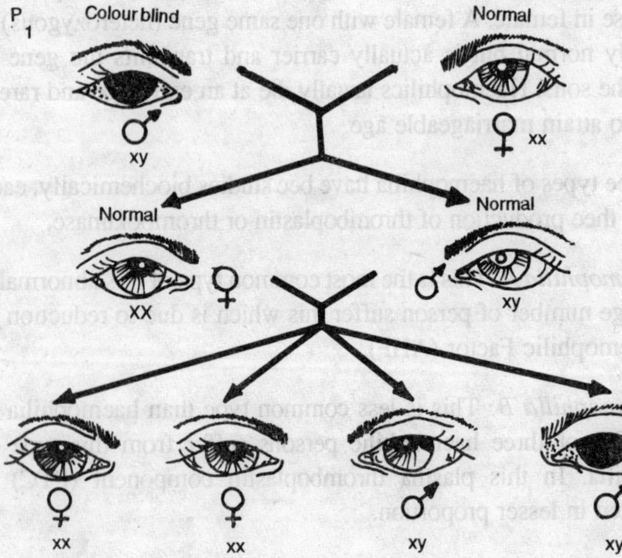

Fig. 10.4. A marriage between a red green colour blind male and a normal female showing X-linked inheritance

1. Colour blind woman and normal man

If woman (XX) is colour-blind and she marries a normal man all of her sons will be colour-blind while daughters (containing only one X for colour blindness and other X for normal colour) will be carriers which if married with colour-blind boys will produce colour-blind daughters. Thus colour-blindness is more common in man.

2. Colour blind man and normal woman

If a normal woman marries a colour-blind man all his sons will be normal while daughters will be hybrid (carriers) which if married with a normal man will produce grandson who will colour blind as shown in diagram.

Dominant Sex-linked Genes in Man

The above discussed cases of sex-linked inheritance were concerned with the sex-linked recessive genes. There are some example of dominant

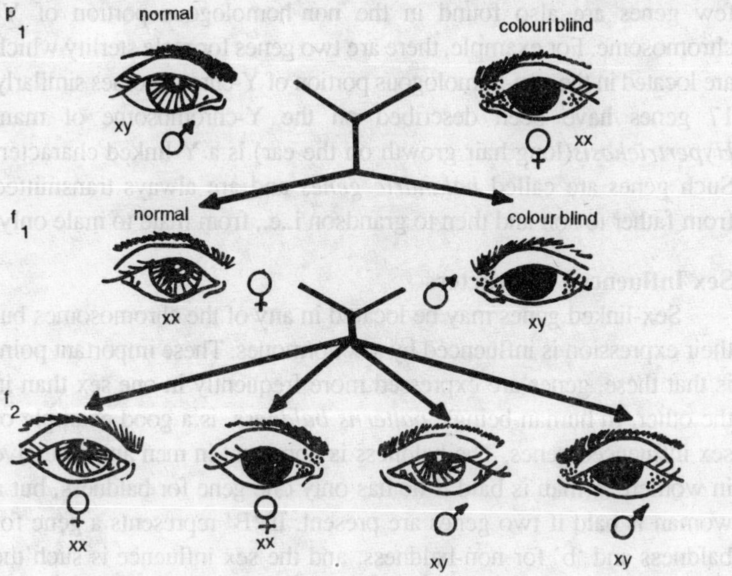

Fig. 10.5. A marriage between a red green colour blind female and a normal male showing X-linked inheritance.

sex-linked genes. One dominant sex-linked gene produces *defective enamel of teeth* and is found more frequently in women than in men because there are two X chromosomes in women and only one in male.

Fig 10.6. Hair on pinna-9 holandric character.

Inheritance Through Y-Chromosome

In *Drosophila* the sex-linked genes are found to be located in X-chromosome, and Y-chromosome has been described as empty. But few genes are also found in the non-homologous portion of Y-chromosome. For example, there are two genes for male sterlity which are located in the non-homologous portion of Y-chromosomes similarly 17 genes have been described on the Y-chromosome of man. *Hypertrichosis*(long hair growth on the ear) is a Y-linked character. Such genes are called *halandric genes* and are always transmitted from father to son and then to grandson i..e., from male to male only.

Sex Influenced Characters

Sex-linked genes may be located in any of the chromosomes but their expression is influenced by sex hormones. These important point is that these, genes are expressed more frequently in one sex than in the other. In human beings, *patterns baldness* is a good example of sex influenced genes. The baldness is dominant in men and recessive in women. A man is bald if he has only one gene for baldness, but a woman is bald if two genes are present. If 'B' represents a gene for baldness and 'b' for non-baldness, and the sex influence is such the 'B' is dominant is men and recessive women, the genotypes of various individual will be:

Genotype	*Men*	*Women*
BB	Bald	Bald
Bb	Bald	Non-bald
bb	Non-bald	Non-bald

The occurrence of baldness in may pedigences of women shows that it may be caused by a larger quality of male sex hormone in women of the genotype BB but the hormone is insufficient in women of genotype Bb.

Index fingers: The length of index finger is also a sex influenced character. If the hand is so place that the tip of the fourth finger just touches a horizontal line, it is seen that the index (second finger) is short and does not touch the line in some cases. A short index finger is dominant in man and recessive in women.

Horns of sheep: A striking example of the influence of sex hormones on phenotypes due to some antosomal genes is met with in the Inheritance of horns in sheep. In the Suffolk breed of sheep, both sexes are hornless, but in the Dorset sheep both sexes are horned. When the two breeds are crossed the F_1 males are always horned and the females are hornless. When the F_1 offspring are mated together, the F_2 ration for the males is 3 horned :1 hornless and for the females 3 hornless: 1 horned. Representing the horned condition by the autosomal gene H and the hornless condition by its allele h the results of the above breeding experiments can be explained as an example sex influenced genes as shown below

Suffolk (Hornless) X Dorest (horned)

hh HH

Hh

Hornless females and horned males

Mating of two F_1 together

	HH	Hh	Hh	hh	HH	Hh	Hh	Hh	
Males	Horned	Horned		Hornless	Horned		Hornless		Females

This again shows that in males a single gene H is able to produce horns due to the influence of male sex-hormone required to produce horns. Thus, horns, appear more frequently in male sheep than in female sheep.

Sex-limited Genes

Sex-limited genes are those genes which are present in the autosomes but whose expression in determined by the presence or absence of one of the sex-hormones and, therefore, these express themselves only in one sex. These differ from the sex-linked genes, which are actually located in the sex chromosomes. The sex-limited genes control the expression of primary and secondary sexual characters. The sex-linked genes may express themselves in either sex, though with differential frequency. The sex-limited genes on the other hand express their effects in only one sex and their action is clearly related to the sex-hormones. For example, beard-development in human beings is a sex-limited character. The genes for deep male voice and male musculature are expressed only in the presence of male sex-hormone. Similarly, genes for feminine voice, development of breast and feminine musculature are sex-limited characters. The abnormalities in hormone secretion may result in the development of breast in male or beard in female (hormone imbalance). Similarly bulls have genes or milk production which they transmit to their daughters but they a then ones are unable to express this traits.

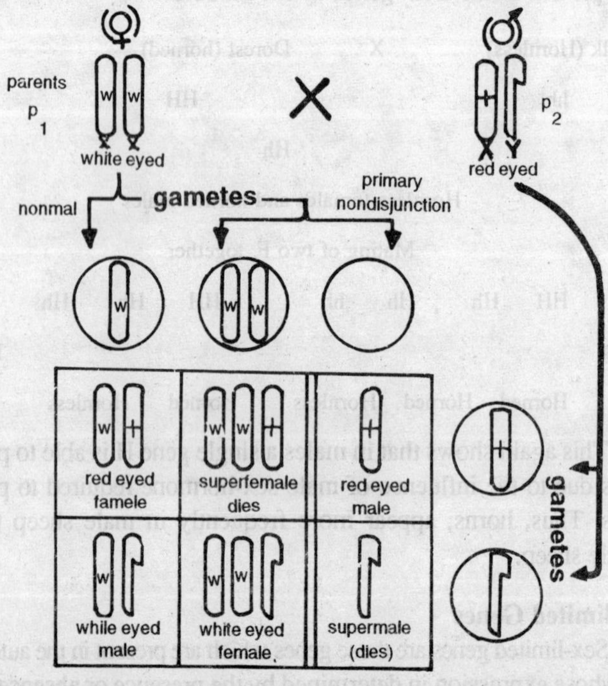

Fig. 10.7. Inheritance of eye colour in *Drosophila* due to primary non-disjunction of sex chromosome.

Non-disjunction

Non-disjunction is a phenomenon in which the two X-chromosomes fail to separate during the meiotic division preceding the formation of gametes. The credit for this discovery goes to *C. Bridge* (1930). He observed this in *Drosophila melanogaster*.

We have seen that when a white eyed females is crossed with red eyed male, all the daughters produced are red and sons white eyed. But sometimes there have been found some exceptions. One fly out of 2000 or 3000 F_1 offspring had a different eye colun, i.e., red in male and white in female. *Bridges* explained this abnormality through the following points.

1. White eyed exceptional females must carry 2X chromosomes because they are female and both must have come from mother because mother is white eyed.

2. Red eyed males will have only one X chromosome, coming from red eyed father because the male is red eyed and Y is empty.

3. Often it happens that XX chromosomes of female fail to disjoin or synapse during meiosis, the condition is termed primary non-disjunction. It results into the production of three types of ova, viz. (i) normal, with one X-chromosome (ii) with 2X chromosomes, and (iii) lacking X chromosome. When these eggs are fertilized by red eyed males, 4 types of offspring are produced.

(i) white eyed females with XXY
(ii) red eyed males with X only
(iii) superfemale with XXX and
(iv) individuals with no X chromosome.

(iii) and (iv usually die out. Thus we see that white eyed female is peculiar in having a Y chromosome and red eyed males are not peculiar in having a Y chromosome and hence these males are not fertile.

Secondary non-disjunction. Bridges crossed the white eyed females (XXY), produced due to primary non-disjunction, with a normal red eyed male and observed the result of secondary non-disjunction in their offspring. In the result 96% of females are red eyed and 4% are white eyed and among the sons 96% are white eyed and 4% are red eyed. During meiosis, the female produces four types of eggs: (i) egg with one X-chromosome normal type (X); (ii) eggs with two X-chromosomes (XX); (iii) eggs with X and Y-chromosomes (XY) (iv) eggs with an empty Y-chromosomes(Y).

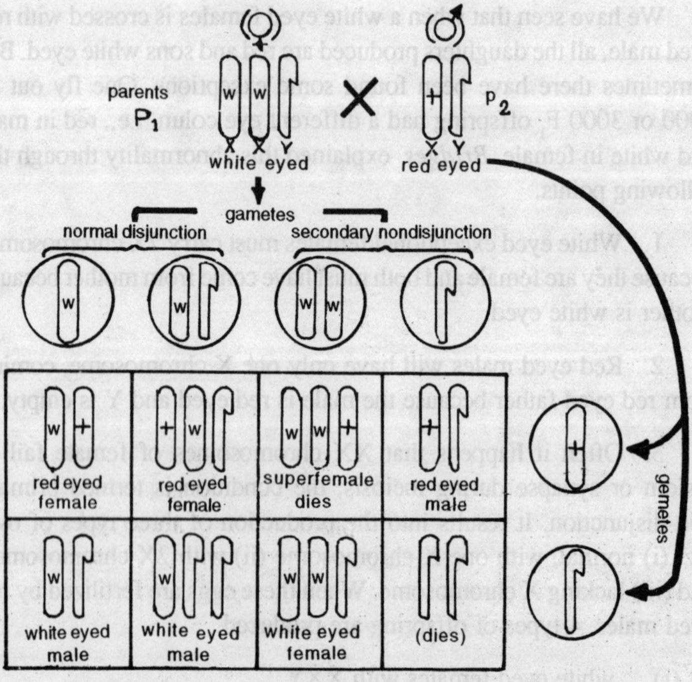

Fig. 10.8. Inheritance of eye colour is *Drosophila* due to secondary non-disjunction of sex chromosome.

Males produced two types of sperms as usual (X) and (Y). As a result of fertilization of these eggs with normal sperms, 8 types of zygotes are formed from which following types of individuals are born:

(i) Normal females with XX-red eyed
(ii) Normal males with XY -white eyed
(iii) Superfemales; with XXX chromosomes
(iv) Individuals with YY chromosomes, lacking X.
(v) Abnormal red eyed females with XXY chromosomes.
(vi) Abnormal white eyed males with XYY chromosomes.
(vii) Red eyed males with XY chromosomes.
(viii) White eyed females with XXY chromosomes.

These results are found due to secondary non-disjunction of XXY chromosomes of female, famed due to primary non-disjunction of the parent female. The XXY chromosomes during oogensis could not synapse perfectly at the time of meiotic division, this condition is known as secondary non- disjunction. *Bridges* verified these facts by cytological examination of all types of flies.

11

Determination of Sex

The problem of sex determination has been one of the most puzzling biological mysteries upto the year 1900. A number of theories were postulated at different times by way of explaining the intricacies of the phenomenon. Most of earlier views, however, are only of historical interest.

Hippocrates and other theorists believed that age or vigour of the parents was responsible for the sex determination of young. The older or more vigorous parents try to impress their sex upon the offspring.

Theory considered the ripeness of ovum at the time of fertilization as a factor pre-determining sex. According to other idea if an eggs gets fertilized shortly after its ovulation it produces female while the other egg which remains in the oviduct for sometime before fertilization produces male.

Nutritional Theory was also attributed to the determination of sex. *Schenk* demonstrated that by controlling the diets of mothers required sex may be produced. He assumed that high degree of nourishment to mother brings bout female offspring while less nutrition causes male offsprings.

An eminent genetical historian *Conway Zirkle* assumed that direction of wind at the time of coition determined the sex of the generation. Males will be produce if wind is northwide and female if it blew southwards.

Various authors thought that one ovary gives rise to male producing eggs and other ovary gives rise to female producing eggs. *Galen* asserted that right side of the body is warmer and its right ovary produces males while left side ovary produces female. It was also thought that the *semen* from right testis produced male while that from left testis produced female. Similarly healthy fathers produced sons and weak father produced daughters.

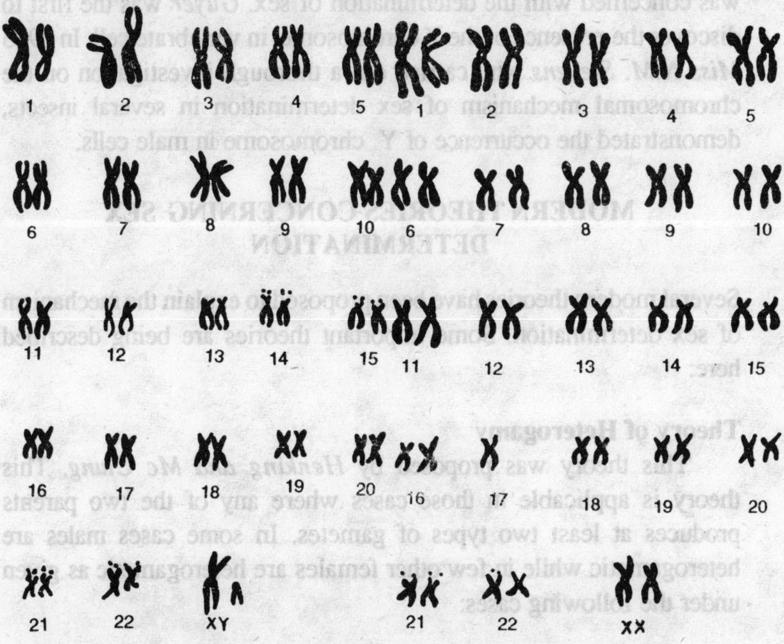

Fig. 11.1. Metaphase chromosomes of man and woman in mitosis.

Scientific investigation over a long period have revealed the fact that the sex of an animal or plant is determined by one or a pair of chromosomes. In every body cell of an organism two types of chromosomes occur namely the *autosomes* which are not concerned with sex determination but with other characters and the sex chromosomes or the *allosomes* which are concerned with the determination of sex. The sex chromosomes which form a pair normally may either be similar or dissimilar. The similar ones are known as the X chromosomes while those that are dissimilar consist of an X and a Y chromosome.

It was only during the latter part of 19th century that significant investigations which relate the chromosomes to sex were carried out. In 1891 *H. Henking*, a German Zoological demonstrated the presence of X chromosome in the male of a species of bug Pyrrhocorris. In 1901 Mc. Clung, and American Zoologist working on the spermatogenesis of the grasshopper *Xiphidium fasciatum* suggested that the X chromosome was concerned with the determination of sex. *Guyer* was the first to discover the presence of the X chromosome in vertebrate cell. In 1905 *Miss N.M. Stevens* who carried out a thorough investigation on the chromosomal mechanism of sex determination in several insects, demonstrated the occurrence of Y chromosome in male cells.

MODERN THEORIES CONCERNING SEX DETERMINATION

Several modern theories have been proposed to explain the mechanism of sex determination. Some important theories are being described here:

Theory of Heterogamy

This theory was proposed by *Henking and Mc Clung*. This theory is applicable in those cases where any of the two parents produces at least two types of gametes. In some cases males are heterogametic while in few other females are heterogametic as given under the following cases:

(a) *Heterogametic Males:* In such cases, females produce only one type of ova, each bears one X chromosome but males produce two types of sperms, 50% with X chromosome and remaining 50% with Y as without any sex chromosomes, as shown in the following cases:

(i) *XX-XO type :* In certain insects specially those of the orders Hemiptera (true bugs) and Orthoptera (grasshoppers and roaches), the female having two X chromosomes (hence referred as XX) and are thus homogametic, while the male having only one X chromosome (hence referred as XO). The presence of an unpaired X chromosome determines the masculine sex. The male lacking in one X chromosome produces two type of sperms: half with X chromosome and half without X chromosome. The sex of the offspring depends upon the sperm that fertilizes the egg (each of which carries a single X chromosome.)

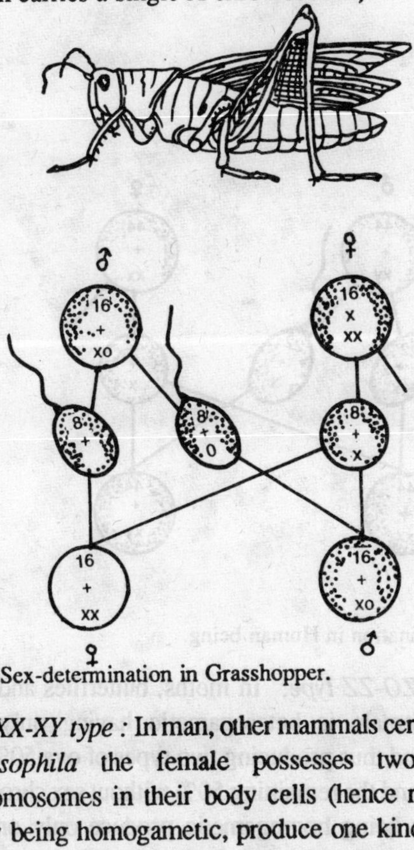

Fig. 11.2. Sex-determination in Grasshopper.

(ii) *XX-XY type :* In man, other mammals certain insects including *Drosophila* the female possesses two homomorphic X chromosomes in their body cells (hence referred as XX) and they being homogametic, produce one kind of eggs, each with one X chromosoine. The males of these organisms posses one X chromosome and one Y chromosome (hence referred as XY). The males having two heteromorphic sex chromosomes produce two kinds of sperm: half with X chromosome and half with Y chromosome. The sex of embryo depends on the kind of sperm. An egg fertilized by a X-bearing sperm, produces a female, but, if fertilized by a Y bearing sperm, a male is produced.

(b) *Heterogametic female:* In this the heterogametic nature is restricted to the female sex only resulting two types of ova as showing in the following examples.

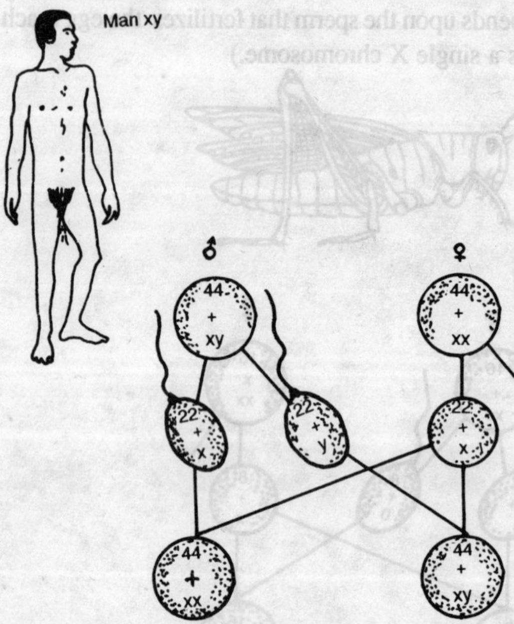

Fig. 11.3. Sex-determination in Human being.

(i) *XO-XY or ZO-ZZ type:* In moths, butterflies and domestic chicken the female is heterogametic having only one sex chromosome and thus producing two types of ova 50% with sex chromosomes and the remaining 50% without sex chromosomes, whereas males being homogametic produce only one type of sperms.

(ii) *XY-XX or ZW-ZZ type:* This type of sex determination occurs in fishes, reptiles and birds. Here the female sex has one X or Z and one Y or W chromosome. This is the heterogametic stage resulting two types of ova with X or Z and with Y or W chromosome. Males are homogametic and produce only one type of sperms.

(c) *Haploid-Diploid type:* In honey bee, wasp, ants and certain other hymenopterans males are usually haploid and females diploid. In females reduction division occurs during oogenesis where as it fails to

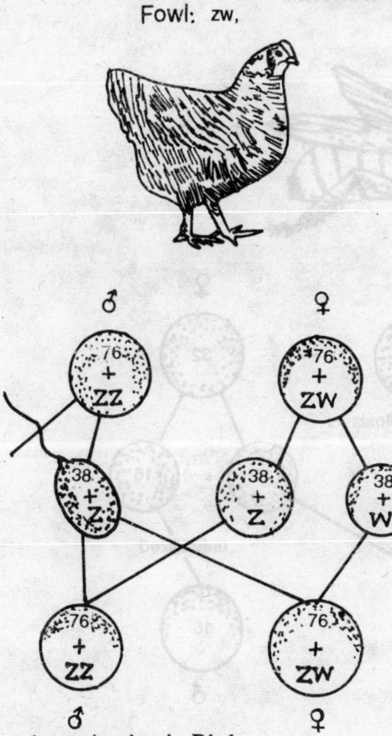

Fig. 11.4. Sex-determination in Birds.

occur during spermatogenesis. Wherever ova are fertilized the diploid zygotes develop either into worker females or into queens. But if the eggs develop parthenogenetically i.e., without fertilization—males or drones are formed.

Metabolic Theory

This theory was proposed by Dr. *Oscar Riddle* of Carnegie Institute (U.S.A.). He worked on rotifers and eggs of pigeons and doves. He formulated the view that sex is determined by the degree of metabolism. By altering the metabolic rate by altering the input of oxygen the sex of the organism can also be altered. Pigeons and doves have been known to produce two kinds of eggs in a single clutch, one types developing into males and the other into females. The male producing eggs have increased rate of oxidation, larger water contents and less protein storage. Similarly the female producing eggs have low rate of oxidation, less water contents and more protein storage.

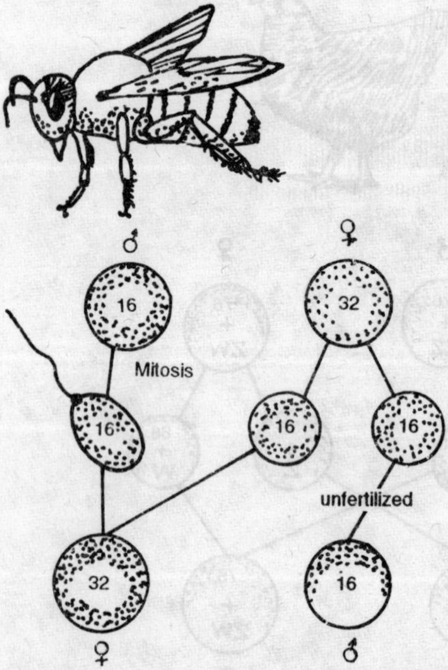

Fig. 11.5. Sex-determination in Honey bee.

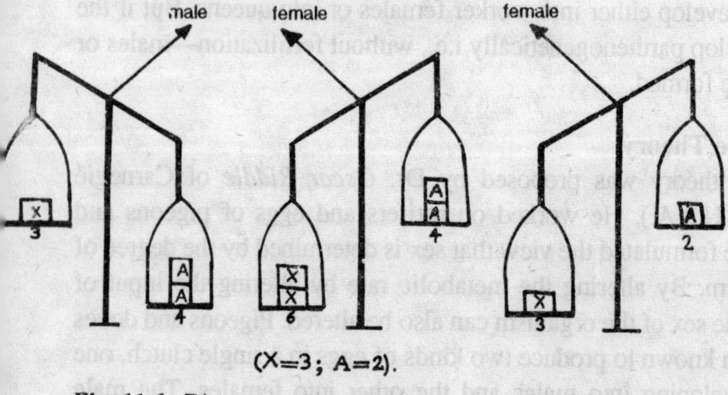

(X=3 ; A=2).

Fig. 11.6. Diagrammatic representation of ganic balance theory.

Genic Balance Theory

This theory was postulated by *Calvin Bridge*. He worked on

Drosophila melanogaster. According to him the X chromosome carry the female determining factors while autosomes the male determining factors. Y chromosome does not take part in sex-determination. Thus the formation of any sex depends upon the balance between X chromosomes and the autosomes. An imbalance between these two results in the development of an inter-sex. Actually the organism with a ratio between 0.5 and 1.0 are inter-sexes and posses a mixture of male and female characters. Super male and super female are generally weak, sterile and poorly viable.

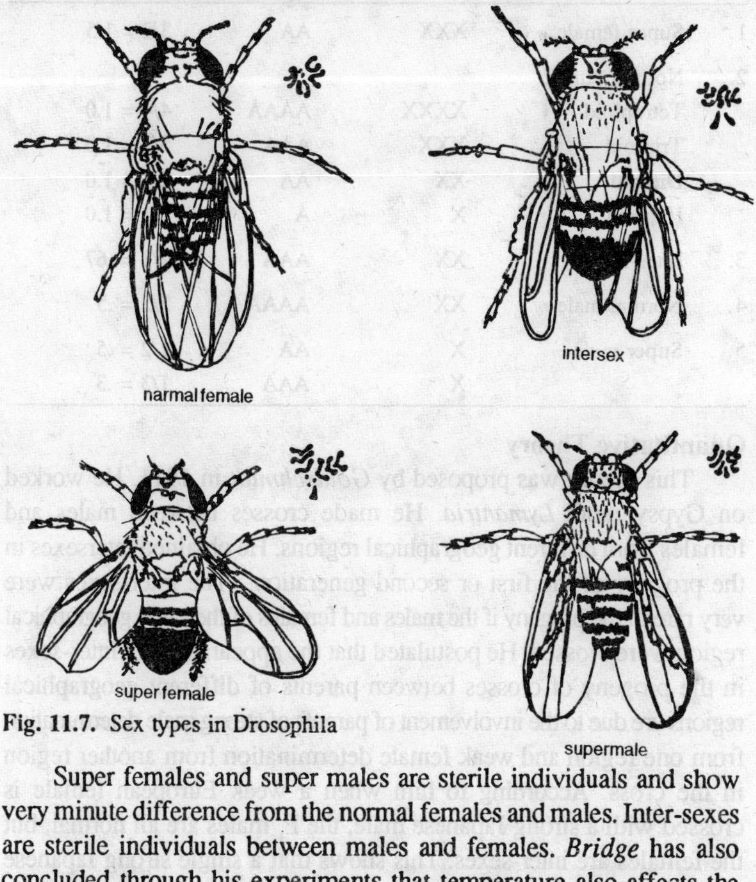

Fig. 11.7. Sex types in Drosophila

Super females and super males are sterile individuals and show very minute difference from the normal females and males. Inter-sexes are sterile individuals between males and females. *Bridge* has also concluded through his experiments that temperature also affects the sex-determination. He observed that the average sex type in inter-sexes produced by a super female at a higher temperature was shifted towards femaleness and in those produced at lower temperature was shifted towards maleness. This shows that the higher temperature has a feminising and low temperature a masculinizing effect on *Drosophila melanogaster*.

TABLE 11.1

Sexual types in Drosophila Melanogaster

S.N.	*Sex*	*X Chromosomes*	*Sets of autosomes*	*Ratio (X/A)*
1.	Super female	XXX	AA	3/2 = 1.5
2.	Normal female			
	Tetraploid	XXXX	AAAA	4/4 = 1.0
	Triploid	XXX	AAA	3/3 = 1.0
	Diploid	XX	AA	2/2 = 1.0
	Haploid	X	A	1/1 = 1.0
3.	Inter-sex	XX	AAA	2/3 = .67
4.	Normal male	XX	AAAA	2/5 = .5
5.	Super male	X	AA	1/2 = .5
		X	AAA	1/3 = .3

Quantitative Theory

This theory was proposed by *Goldschmidt* in 1911. He worked on Gypsy moth *Lymantria.* He made crosses between males and females from different geographical regions. He obtained intersexes in the progeny of the first or second generation. The inter-sexes were very rare in the progeny if the males and females of the same geographical region were crossed. He postulated that the appearances of inter-sexes in the progeny of crosses between parents of different geographical regions are due to the involvement of parents of strong male determination from one region and weak female determination from another region in the cross. According to him when a weak European female is crossed with a strong Japanese male, the F_1 males are all normal, but the females are inter-sexes. This shows that a single strong Japanese X chromosome is sufficient to overcome the effects of the female determining factors, so that the XY individuals differentiate into inter-sexes, instead of developing into females. The F_2 from the cross again results in normal sons, but half of the daughters are normal and the other half inter-sexual. In a reciprocal cross between strong Japanese female and a weak European male, all the offsprings in F_1 are normal, but in F_2 females are normal, while half of the males are inter-sexes and the other half normal.

Hormonal Theory

There are many instances where sex becomes modified by the hormones which are secretions of endocrine glands. This theory is based upon the observations of *Crew* in chicks. He found that a hen which laid fertile eggs, accidentally lost its ovary, stopped laying eggs, and developed male comb, male plumage and became a cock. It finally functioned as male and became father of two chickens. *E. Witschi* observed similar hormonal effects in salamanders.

Sex-reversal: The above case of sex-reversal was explained by assuming that as soon as the ovary was removed or destroyed, the ovarian hormones were stopped and after sometime the dormant testes present as rudiments in almost all the female birds started functioning. The male hormones were produced and resulted in the appearance of male secondary sexual characters and the formation of sperm. Cases of sex reversal have been observed in fishes, amphibians, birds and even in mammals.

The artificial removal of gonads of either sex before puberty in mammals and even in man, results in the development of secondary sexual characters of the opposite sex. The female characters can be induced in a castrated male by injecting female hormone and *vice versa.* Cases of sex reversal in man have been found due to the irregularities in the functioning of adrenal and pituitary glands.

A classical example of modification of sex by hormone action is furnished by an example of *freemartin*. In cattles there are sometimes twins. In twin one is male and the other member is female which sometimes becomes a sterile inter-sex called freemartin. This freemartin is sterile unsexed female having female type external genital organs but internal organs having male tendencies. *F.R. Lillie* explained the

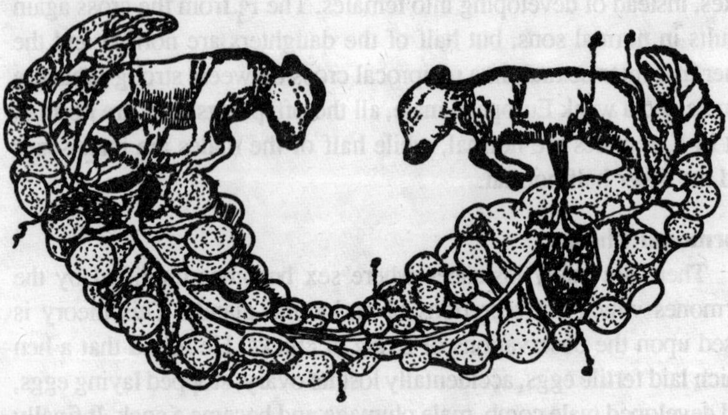

Fig. 11.8. Freemartin in cattle twins.

production of freemartin as a result of fusion of foetal membranes when twins were in uterus. Due to fusion there occurred common circulation of same blood between twins. The male animal was more rapidly developed and this sent male hormone which interfered with the differentiation of female sex organs. This male hormone inhibited the normal development of female sex with the result that it is born with ovaries which tend to from testicular tissues producing never ova.

Environmental Factors

Sometimes external environmental factors determine the sex of individuals as seen in the variety of animal species belonging to different groups.

(i) *Bonellia*: The effect of environment in the determination of sex in *Bonellia* was extensively studied by *F. Baltzer* in 1935. The adult female is about an inch long and fairly complex anatomical organization. Male is of the size of large Protozoa and has rudimetary organs. The males live as parasites in the uterus of the females. All larvae of Bonellia are genetically and cytologically similar. If a particular larva settles on the proboscis of an adult female, it becomes a male individual. On the other hand, if a larva develops in isolation (i.e., in water) it develops into a female. Further, if a incompletely developed male is detached from the proboscis of female, it becomes an intersex. Thus, it becomes evident that proboscis of adult female secretes some hormone-like substance which suppresses femaleness and initiation maleness in the larvae which are attached with it.

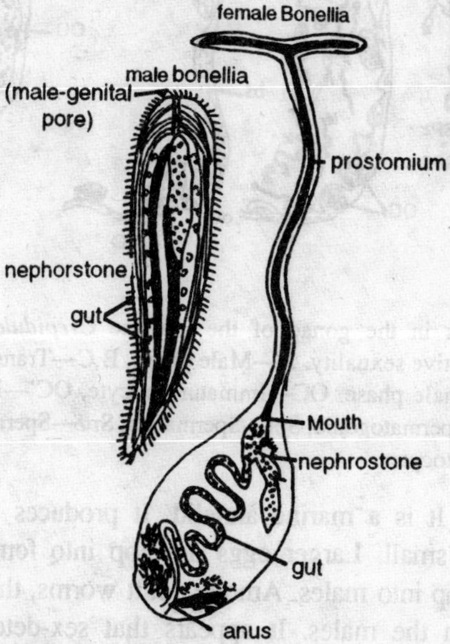

Fig. 11.9. The female and male of the echiuroid *Bonellia viridi*.

(ii) *Crepidula:* Another example of sex-determination by enviroment is a marine mollusc *Crepidula*. In this species it has been observed that young *Crepidula* remains functional male and produce only sperm but no ova. As it matures the male phase is changed by transitional phase to a functional female phase. The female phase continues for the remainder of the animal's life. It is suggested that environment somehow regulates the activity of certain genes to produce only male phenotype during young stages, while inhibits them during adult stages and simultaneously initiating other genes to produce female phenotype.

Ophryyotrocha: It is a segmented animal in which the male and female sexes are determined by the number of segments of a body has. If the number of segments is more than 20, the animal is female otherwise it is a male. By cutting the number of segments sex from female to male can easily be changed.

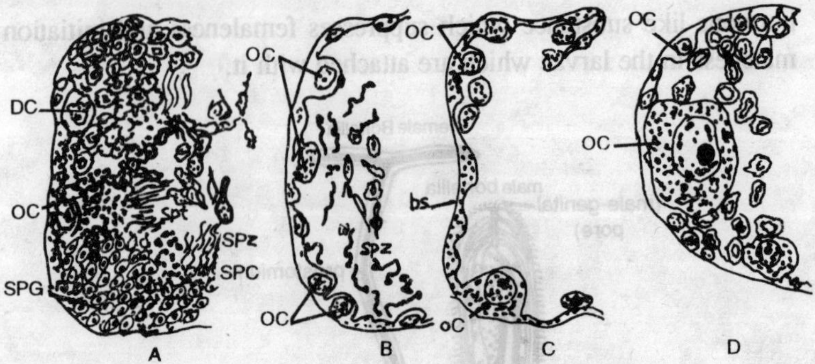

Fig. 11.10. Changes in the gonad of the mollusc *Crepidula* experiencing consecutive sexuality. A—Male phase, B,C—Transitional phases; D—Female phase. OC—Immature oocyte, OC'—Mature oocyte: Spg—Spermatogonia, Spt—Spermatids, Spz—Spermatozoa, Spc—Spermntocytes.

Dinophilus: It is a marine annelid. It produces two types of eggs—large and small. Larger eggs develop into females and the smaller on develop into males. Among adult worms, the females are much larger than the males. It appears that sex-determination in *Dinophilus* depends upon the quantity of the cyoplasm in the egg cell which can be utilized for the development of the embyo.

Echiuroid Worm. Weisz (1971) describe that the sex-determining factor in echiuroid worm is environmental CO_2 Relatively low concentration of CO_2 is sea water causes free swimming larvae or this worm to develop as females. Added respiratory CO_2 produced by an adult female or with a larva already determined as a female then undetermined larva develops as a male.

In Greek (*Gyne,* woman, *aner,* man and *morphe,* fam) the term gynandromorph refers to the appearance of both male and female characters at different halves of the body in the same organism. In insects, the secondary sexual characters are not influenced by the hormones secreted by the gonads and the differentiation occurs independently depanding upon the genetic constitution of the cell in various parts of the body. When the genotype of the cell in different parts of the body differs with respect to the sex-linked characters, a

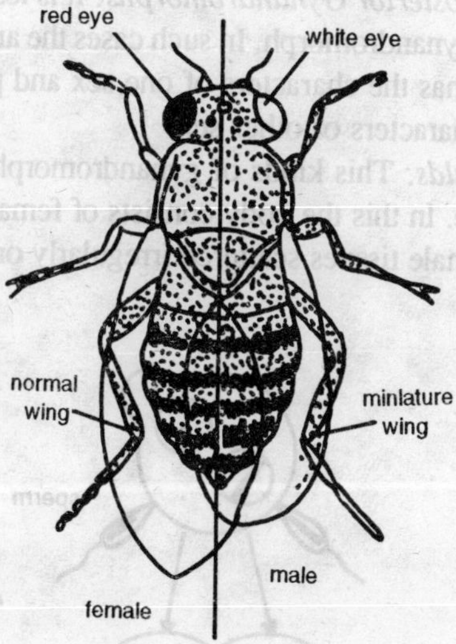

Fig. 11.11. Bilateral gynandromorpis in Drosophila.

sex mosaic results and is known as *gynandromorph.* Gynandromorphs differ from the inter-sexes. The later are genetically similar throught their bodies and usually sterile, but a gynandromorph consists of two genetically different tissues. Some of the cells of a gynandromorph are genetically male, and others are female.

Gynandromorphic cases have been studied in several species of insects like Drosophila, bees and in silk worms. In a gynandromorph of *Drosophila,* obtained by *Morgan'* the right half of the fly showed male characters viz., shorter wings, black tipped abdomen and a sex comb of the first leg. The left side showed contrasting female characters. The right eye was white and the left one red.

Kinds of Gynandromorphs

Gynandromorphs can be classified into the following types on the basis of the position of sex tissue the gynanders have:

(i) *Bilateral gynandromorphs:* If the line of division between male and female tissues of a gynander passes through the middle of the body the stage is called *bilateral gynandromorph.*

(ii) *Anterio-posterior Gynandromorphs:* It is less common than bilateral gynandromorph. In such cases the anterior region of the body has the characters of one sex and posterior region has the characters of other sex.

(iii) *Sex piebalds:* This kinds of gynandromorph is of very rare occurence. In this the body consists of female tissue having spots of male tissues scattered irregularly or the *vice versa.*

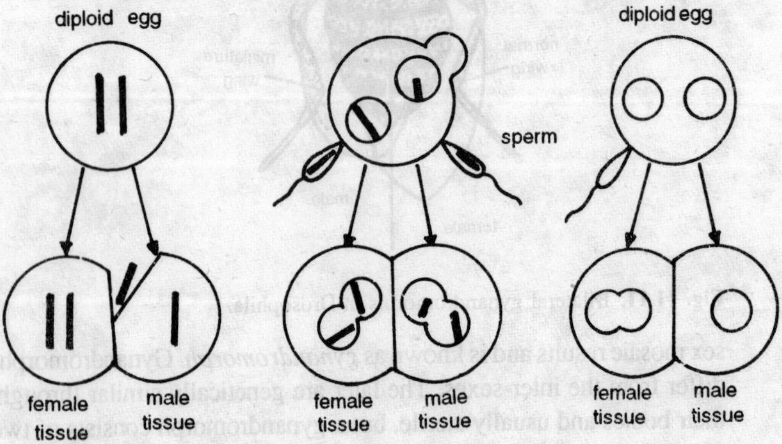

Fig. 11.12. Showing the origin of gynandromorph.

Origin of Gynandromorphs

Three different hypothesis have been offered to explain the origin of gynandromorphs.

(i) *By elimination of X-Chromosome:* Usually gynander starts its development with two X's. But in the later phase during cell division one X gets lost or disappeared in daughter cells. The two X chromosomes in the usual mitoses become divided into four X chromosomes (daughters). The one daughter receives only two X's and other daughter cell gets only one X while fourth X becomes lost due to abnormal cell division. The daughter cell receiving two X's forms the female tissue while other daughter cell having one X developes into male tissue. It occurs in *Drosophila* (Fig. 11.12).

(ii) *By retention of polar nucleus in the egg.* In silk worms female is XY. During meiosis X and Y get separated, either of one (X or Y) going to egg and polocyte. But sometimes polar body nucleus remains in the egg alongwith egg nucleus. Thus these eggs will be binucleate (XY). As a result of fertilization two sperm cells (which contain X) fuse with X and Y chromosome separately giving rise to male (XX) and female (XY) tissues respectively.

In *bees* also binucleate eggs are found (either due to fusion of polar body nucelus with the egg nucleus or parthenogenetically) producing gynanders. In such cases when fertilization takes place one sperm cell fuses with the one nucleus of egg forming *female tissue* while other nucleus will remain single producing *make tissue.*

Muller described another possibility of the gynander formation in a parasitic wasp -*Harbrobracon.* In this wasp females are heterozygous and males are either homozygous or haploid. The egg during oogenesis may come to contain polar nucleus as a result of fusion. Suppose egg nucleus is *a* and polar nucleus is *b:* then both *a* and *b* will be inside the egg as it is binucleate egg fertilized by two sperms (suppose having *b*), the fertilized nuclei will be *ab* and *bb*. The first *(ab)* will from female tissue and second *(bb)* will develop into male tissue homozygous.

Barr Body—*Barr* and *Bertram* (1949) reported a deeply stained chromatin body (i.e. a chromocentre) the nerve cells of female cat which was absent in the male. This chromatin body is called *sex chromatin* or *Barr body* after the name of its discoverer.

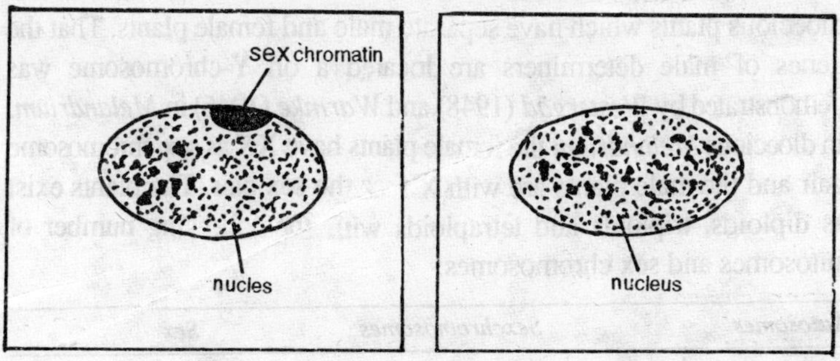

Fig. 11.13. Sex chromatin in the nucleus of female cat (A). Male cat (B) has no sex chromatin in its nucleus.

Such Barr body has also been observed in most of the body cells (e.g., skin, oral epithelium and blood cells) of man and other mammals. Human females have the Barr body in the nuclei of their body cells in higher proportion than males and are therefore, referred as *sex chromatin positive*. The human males are called *sex chromatin negative*.

Similar darkly stained sex chromatin has also been reported from the female neutrophils where it is called *drumstick body*. The relationship between sex chromatin and sex chromosomes has recently been elucidated (*Ohno, Kaplan and Kinosita 1959*). It has been demonstrated that the sex chromatin is derived only from one of the two X-chromosomes. The other X chromosomes behaves like an autosome and it not heteropy[illegible]otic at interphase. The number of corpuscles of sex chromatin at interp[illegible]se is equal to nx-1. It means that there is one Barr body less than the number of X-chromosomes. Besides its genetical significance, Barr body has a wide field of medical application and offers the possibility of relating the origin of certain congenital disease to chromosome anomalies.

Sex Determination in Plants

The mechanism of sex determination has been studied in a large number of plant materials. A detailed account of sex-determination in some plants like *Melandrium, Coccinia, Viscun* and *Aspararguns* etc., has been made.

In *Melandrium* majority of the plants are monoecious and the problem of sex-determination does not arise in them as in the case of dioecious plants which have separate male and female plants. That the genes of male determiners are located a on Y-chromosome was demonstrated by *Westergad* (1948) and *Warmke* (1946) in *Melandrium*. In dioecious *Melandrium* the female plants have XX as sex chromosome pair and the male plants are with XY as the sex pair. The plants exist as diploids, triploids and tetraploids with the following number of autosomes and sex chromosomes:

Autosomes	*Sexchromosomes*	*Sex*
2A	XX	Female
2A	XY	Male

3A	XX	Female
3A	XY	Male
3A	XXX	Female
3A	XXY	Male
3A	XXXY	Male
4A	XXXX	Female
4A	XXXXY	Male

In is clear from the above table that Y-chromosome is the sex-determiner. Even in the presence of four X-chromosomes, one Y-chromosomes expresses the male characteristics. Whenever a plant has a Y-chromosomes, irrespective of the number of X-chromosomes, the plant is a male. If it lacks a Y-chromosome, it is a female, it is a female. It seems that the Balance ratio theory of *Bridges* does not good in *Melandrium*.

In Misletoes *(Viscum)*. Several species of Viscum are dioecious, which are found in Africa, Madagascar, Europe, Asia and Taiwan. In *Viscum fischeri*, for instance, male plants have 2n=23, out of which, at meiosis, 14 chromosomes form seven bivalents and the remaining nine chromosomes form a ring (7" + 1.9). Seven bivalents disjoin normally, but the ring of nine chromosomes disjoins 4:5 thus giving rise to two types of gametes, one with 11 and the other with 12 chromosomes. The female plant has 2n=22 forming 11 bivalents (11") disjoining normally and giving only one type of gametes with 11 chromosomes each. Thus male is heterogametic due to translocations without any sex chromosomes involved. Two types of male gametes will give rise to two sexes. This phenomenon is also common in several other species of *Viscum*.

In Liverworts Sex chromosomes in *Liverworts* were first of all studied by *Allen*. The gametophytes of *Liverworts* are haploid in nature and bear sex organs. The male gametophyte in the case of *Sphaerocarpos* has seven autosomes and a Y-chromosome, while the female gametophyte has seven autosomes and an X-chromosome. The zygote produced inside the venter of the archegonium is diploid and has 14 autosomes and one X and one Y-chromosome. The zygote develops into a sporophyte. The spore mother cells undergo meiosis to produce haploid spores. Each spore has seven autosomes and one X or one Y-chromosome. The spore carrying a Y-chromosome develops into a male and the spore carrying an X-chromosome develops into a female gametophyte. The nature of the developing gametophyte depends upon the type of sex chromosome that it receives.

Sex mechanism in *Chlamydomonas* Perhaps the sex mechanism operation in *Chlamydomonas eugametos* is the best understood of all, although the male and female cells may not be distinguished morphologically. Whether a particular cell will be male (minus) of female (plus) will depend on the ratio of *cis* and *trans* isomers or crocetin dimethyl ester, the proportions of cis being always greater in female (plus). The ratio of isomers in male is just the reverse of what is found in female. In other words, if the *cis* and *trans* isomers in minus of a variety of this species is in the ratio of 25:75, in the plus strain the ratio will be reverse i.e. 75:25. The ratios of these two isomers in the four varieties of C, eugametos are as follows:

TABLE 11.2

Ratio of CIS: Trans Isomers of Crocetin Dimethylester in Different Varieties of C.Eugametos

Variety	*Male*		*Female*	
	Cis:	*Trans*	*Cis:*	*Trans*
Simplex	25	75	75	25
Synoica	35	65	65	35
Typica	15	85	85	15
Brauanii	5	95	95	5

If the hypothesis is correct that the expression of sex is dependent on the relative proportion of the above two isomers, then any two varietal males showing different levels of cis and transisomers should mate; and indeed when two plus strain of *synoica* and of *typica* or of *braunii* or for that matter any two minus strains of the above combinations differing in levels of two isomers were placed together, the sexual fusion between them took place.

In Asperagus *Asparagus* is a dioecious form. However, rarely the female flowers bear rudimentary anthers and the male flowers bear rudimentary pistils. Thus, the rare male flowers having poorly developed pistils may seeds. In one such case when the seeds obtained from a rare male flower were raised into plants; the male and female plants were found to be present in 3:1 ratio. When the male plants raised thus were used to pollinate the female flowers on female plants, only two-third of them showed segregation indicating that the sex is controlled by a single gene. In this case, maleness should be dominant over femaleness and the male plants should ordinarily be heterozygous.

In Maize *Maize* is a monoecious plant and has male as well as female flowers. The ears (female flowers) normally develop along the sides of the stalk and tassel (male flowers) at the top of the plant. The two genes Bs and Ts control the normal appearance of female and male flowers. The homozygous recessive genotype *bs bs* results in maize plant which is male, i.e. it does not develop female flowers. The homozygous recessive genotype it is brings about maize plants without male flowers and the female flowers are present in the lateral as well as in the terminal positions. The doubly homozygous individuals, i.e. *bs bs ts ts* are pistilate, the ears being present in the terminal position only.

The genotypes and the phenotypes in maize are given below:

Genotype	*Phenotype*
Bs-Ts-	Normal monoecious
bs bs ts ts	Female, ears terminal only
bs bs Ts-	Male
Bs-ts ts	Female, ears terminal as well as lateral.

It is clear that the homozygous recessive *ts ts* genotype trans forms the male flowers into female flowers. The female flowers develop in this genotype where the male flowers should have been under the influence of the dominant gene *Ts*. The homozygous recessive *bs bs* hinders the development of female flowers only in the lateral position. In the presence of genotype *ts ts* the development of terminal female flowers in place of male flowers is not affected. Variations in two genes simply converts a monoecious plant in to a dioecious plant indicating minor differences in monoecious and dioecious plants. The dioecious state is evolutionarily more advanced than the monoecious state. It aids in cross-fertilization — an advantage to be plant breeder.

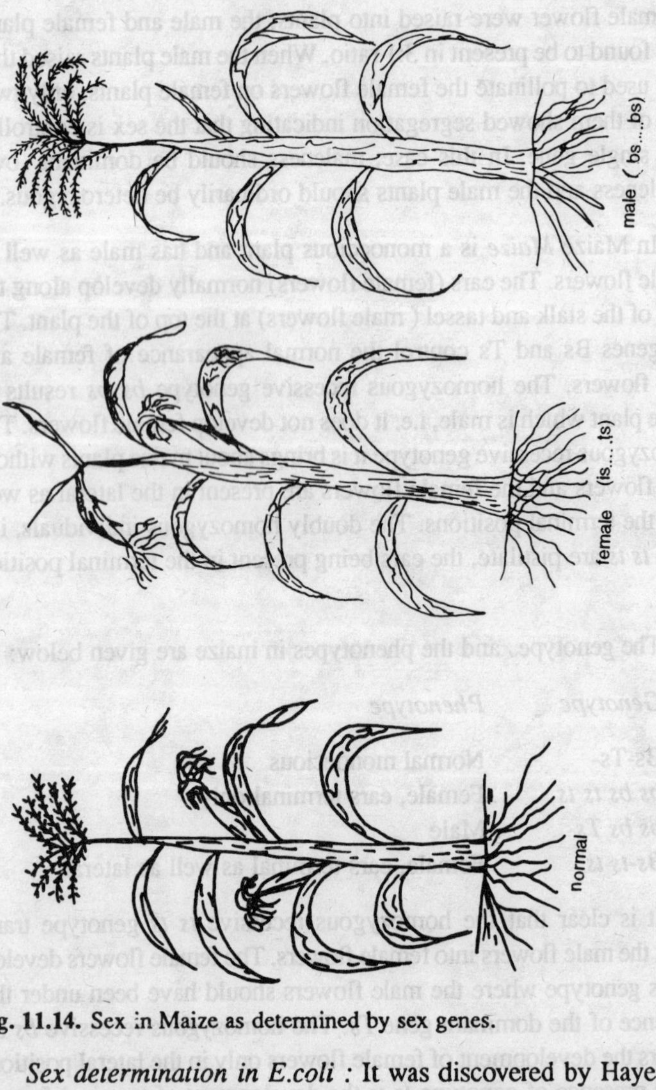

Fig. 11.14. Sex in Maize as determined by sex genes.

Sex determination in E.coli : It was discovered by Hayes that during conjugation, one *E, coli* cell behaves like a donor cell having a sex factor in the cytoplasm (F^+) or acts as the male type and the other cell is the recipient, does not possess the sex factor (F^-), or the female type. The sex factor (F) therefore, is the sex determiner.

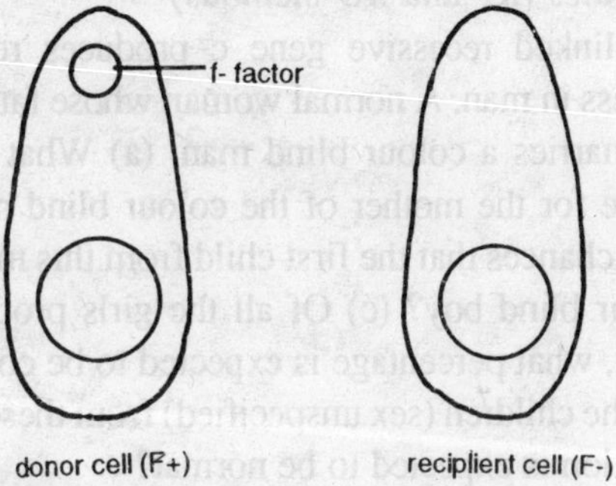

Fig. 11.15. Diagram showing sex differences in E, coli.

Sex Reversal in plants

In plants sex reversal have been found to be produced by the genetic, cytological and Environmental causes:

(i) *Genetic causes:* Jones in 1932 was able to produce dioecopis plants from the monoecious plants of corn. He discovered that the genic constitution of monocious plants is *Sk Sk Ts Ts* and when *Sk* is mutated to *sk,* the presence of homologous *(sk sk)* makes the plants male, because *Ts* is responsible for male sex and in absence of *Sk* ears do not develop. The same effect have been observed in papaya and other plants.

(ii) *Cytological causes:* In many plants cytological differences has been reported to cause the sex reversal. The examples of this type are Cannabis sative, Humulus, Rumex etc., when different kinds of inter-sexes have been observed. For example, in Cannabis, stamens have been seen on the female and carpels on the male plants. The sex determination in this plants is XY type and the irregularities in cell division cause the sex reversal in these plants.

(iii) *Environmental causes:* Light, nutrition, temperature, diseases, abnormal conditions of growth and mutatians etc., have been reported to cause sex reversal. Schaffner reported that when hemp is grown in shallow soil and diffused light, it shows sex reversal.

REVIEW QUESTIONS

Sex-determination and Sex-linked Inheritance

Heterogmetic Males (XY and XO methods)

1. A sex-linked recessive gene c produces red-green color blindness in man. A normal woman whose father was colour blind marries a colour blind man. (a) What genotypes are possible for the mother of the colour blind man? (b) What are the chances that the first child from this marriage will be a colour blind boy? (c) Of all the girls produced by these parents, what percentage is expected to be colour blind? (d) Of all the children (sex unspecified) from these parents, what proportion is expected to be normal?

2. The gene for yellow body colour y in *Drosophila* is recessive and sex-linked. Its dominant allele y^+ produces wild type body colour. What phenotypic ratios are expected from the crosses (a) yellow male x yellow female, (b) yellow female x wild type male, (c) wild type female (homozygous) x yellow male, (d) wild type (carrier) female x wild type male, (e) wild type (carrier) female x yellow male?

3. A narrow reduced eye called "bar" is dominant sex-linked condition (B) in *Drosophila,* and the full wild type eye is produced by its recessive allele B^+. A homozygous wild type female is mated to a bar-eyed male. Determine the F_1 and F_2 genotypic and phenotypic expectations.

4. Sex determination in the grasshopper is by the XO method. The somatic cells of a grasshopper are analyzed and found to contain 23 chromosomes. (a) What sex is this individual? (b) Determine the frequency with which different types of gametes (number of autosomes and sex chromosomes) can be formed in this individual. (c) What is the diploid number of the opposite sex?

5. Male house cats may be black or yellow. Females may be black, tortoise-shell pattern, or yellow. (a) If these colours

are governed by a sex-linked locus, how can these results be explained? (b) Using appropriate symbols, determine the phenotypes expected in the offspring from the cross yellow female x black male. (c) Do the same for the reciprocal cross of part (b)? (d) certain kind of mating produces females, half of which are tortoise-shell and half are black; half the males are yellow and half are black. What colours are the parental males and females in such crosses? (e) Another kind of mating produces offspring, 1/4 of which are yellow males, 1/4 yellow females, 1/4 black males and 1/4 tortoise-shell females? What colours are the parental males and females in such crosses?

6. In the plant genus Melandrium, sex determination is similar to that in man. A sex-linked gene (l) is known to be lethal when homozygous in females. When present in the hemizygous condition in males (lY), it produces blotchy patches of yellow-green colour. The homozygous or heterozygous produces normal dark green colour. From a cross between heterozygous females and yellow -green males, predict the phenotypic ratio expected in the progeny.

7. The recessive gene for white eye colour in *Drosophila* (w) is sex-linked. Another recessive sex-linked gene governing eye colour is vermilion (v), which when homozygous in females or hemizygous in males together with the autosomal gene for brown eye (bw/bw), also produces white eye. (a) What phenotypic results are expected among progeny from mating a white-eyed male of genotype (bw/bw, vw^+/y) with a white-eyed female of genotype (bw^+/bw, vw/v^+w)? Hint. See Problem 5.5. (b) What phenotypic proportions are expected in the progeny from the mating of a vermilion female heterozygous at the brown locus but not carrying the white allele with a male which is white due to the w allele but heterozygous at the brown locus and hemizygous for the vermilion allele? (c) Determine the expected phenotypic ratio in the *F1* and F_2 from the reciprocal cross of problem 5.5(a).

8. In the Rosy Gier variety of carrier pigeon, a cross was made between gray-headed females and creamy-headed males. The F_1 ratio was 1 gray-headed female: 1 gray-headed male : 1 creamy-headed male. (a) How many these results be explained? (b) Diagram this cross using appropriate symbols.

9. Chickens have an autosomal dominant gene (C) that produces a short legged phenotype called "creeper" in heterozygotes. Normal legs are produced by the recessive genotype (cc). The homozygous dominant genotype (CC) is lethal. A dominant sex-linked gene (B) produces barred plumage, the recessive allele (b) produces non-barred plumage. (a) Determine the phenotypic expectations among progeny (of both sexes) from the cross of a barred creeper female and a non-barred creeper male. (b) Determine the phenotypic ratios within each sex for part (a). (c) Two chickens were mated and produced progeny in the following proportions: 1/12 non-barred males, 1/6 non-barred creeper females, 1/12 barred males, 1/12 non-barred females, 1/6 non-barred creeper males, 1/6 barred creeper males, 1/12 barred females and 1/6 barred creeper females. What are the genotypes and phenotypes of the parents?

10. A dominant autosomal inhibitor (I-) as well as a recessive autosomal inhibitor (cc) prevents any colour from being produced in chickens. The genotypes *I-C*, *I-cc* and iicc all produces white chickens; only the genotypes *iiC*-produces coloured birds. A recessive sex-linked gene k produces slow growth of the primary wing feathers. Its dominant allele k^+ produces fast feathering. A white *(IICC)* slow feathering male is mated to a white *(iicc)* fast feathering female. What are the F_1 and F_2 phenotypic expectations?

11. The presence of feathers on the shanks of the Black Langshan breed of chickens is due to the dominant alleles at either or both of two autosomal loci. Non- feathered shanks are the result of the double recessive genotype. A dominant sex-linked gene (B) places white bars on a black bird. Its recessive

allele (b) produces non-barred (black) birds. Trihybrid barred males with feathered shanks are mated to dihybrid non-barred females with feathered shanks. Determine the F_1 phenotypic expectations.

Genetic Balance

12. In *Drosophila*, the ratio between the number of X chromosomes and the number of sets of autosomes (A) is called the " sex index." Diploid females have a sex-index (ratio X/A) = 2/2 = 1.0. Diploid males have a sex index of 1/2 = 0.5. Sex index values between 0.5 and 1.0 give rise to inter-sexes. Values over 1.0 or under 0.5 produce weak and inviable flies called superfemales (meta-females) and supermales (meta-males) respectively. Calculate the sex index and the sex phenotype in the following individuals: (a) AAX, (b) AAXXY, (c) AAAXX, (d) AAXX, (e) AAXXX, (f) AAAXXX, (g) AAY.

Haplodiploidy

13. If the diploid number of the honey bees is 16, (a) how many chromosomes will be found in the somatic cells of the drona (male), (b) how many bivalents will be seen during the process of gametogenesis in the male, (c) how many bivalents will be seen during the process of gametogenesis in the female?

14. Seven eye colours are known in the honey, bee, each produced by a recessive gene at different locus: brick (bk), chartreuse (ch), ivory (i) cream (cr), snow (s), pearl (pe) and garnet (g). Suppose that a wild type queen heterozygous at the brick locus (bk^+/bk) was to be artificially inseminated with a mixture of sperm from seven haploid drones each exhibiting a different one of the seven mutant eye colours. Further assume that the semen contribution of each male contains equal concentrations of sperm, that each sperm has an equal opportunity to enter fertilization, and that each zygote thus formed has an equal opportunity to survive. (a) What percentage of the drone offspring is expected to be brick-eyed? (b) What percentage of worker offspring is expected to be brick-eyed?

Single Gene Effects

15. In the single-celled haploid plant *Chlamydomonas*, there are two mating types, (+) and (-). There is no morphological distinction between the (+)sex and (-) sex in either the spore stage or the gamete stage (isogametes). The fusion of (+) and (-) gametes produces a 2n zygote which immediately undergoes meiosis producing four haploid spores, two of which are (+) and two are (-). (a) Could a pair of genes of sex account for the 1:1 sex ratio? (b) Does the foregoing information preclude some other form of sex-determination? Explain.

16. Sex-determination in the wasp *Bracon* is either by sex alleles or haplodiploidy. A recessive gene called "veinless" (v) is known to assort independently of the sex alleles; the dominant allele v^+ results in wild type. For each of the eight crosses listed below, determine the relative frequencies of progeny phenotypes within each of 3 categories: (1) haploid males, (2) diploid males, (3) females, (a) v/v, s^a/s^b x v^+, s^a, (b) v/v, s^a/s^b x v^+, s^c, (c) v/v, s^a/s^b x v, s^b, (d) v/v, s^a/s^b x v, s^c, (e) v/v^+, s^a/s^b x v, s^a. (f) v/v^+, s^a/s^b x v, s^c, (g) v^+/v^+, s^a/s^b x v, s^a, (h) v^+/v^+, s^a/s^b x v, s^c.

ANSWERS

1. (a) Cc or cc (b) 1/4 (c) 50% (d) 1/2.

2. (a) All offspring yellow (b) all females wild type, all males yellow (c) all offspring wild type (d) all females wild type : 1/2 yellow males (c) females and males: 1/2 wild type : 1/2 yellow.

3. F_1: B+/B bar eyed females, B^+/Y wild type males; F_2 females: $1/2B^+/B^+$ wild type : $^1/_2$ B + bar eye; F_2 males: $1/2B^+/Y$ wild type : 1/2B/Y bar eye.

4. (a) male (b) 1/2(11A + 1X) : 1/2(11A) (c) 24.

5. (a) A pair of codominant sex-linked alleles.

	Females	Males
black	C^BC^B	C^BY
tortoise-shell	C^BC^y	—
yellow	C^yC^y	C^yY

(b) All males yellow, all females tortoise-shell (c) all males black, all females tortoise-shell

(d) Tortoise-shell x female x black male (e) tortoise-shell female x yellow male.

6. 1/3 dark green females : 1/3 males with yellow-green patches : 1/3 males dark green.

7. (a) Males all white eyed; females: 1/4 vermilion : 1/4 wild type : 1/4 white : 1/4 brown (b) males and females: 3/4 vermilion : 1/4 white (c) males and females: 3/8 wild type : 3/8 vermilion : 1/8 brown : 1/8 white.

8. (a) Sex-linked gene with one allele lethal when hemizygous in females or homozygous in males.

	Male	*Female*
gray	HH	HW
cream	HH^1	—
lethal	H^1H^1	H^1W

(b) P: HW x HH^1; F_1: 1/3 HH gray male : 1/3 HH^1 cream male : 1/3 HW gray female.

9. (a) 1/6 non-bar, normal leg females : 1/6 bar, normal leg male: 1/3 non-bar, creeper female : 1/3 bar, creeper male

(b) Males: 2/3 bar, creeper : 1/3 bar, normal leg; females: 2/3 non-bar, creeper : 1/3 non-bar, normal leg (c) bar, creeper male (CcBb) X non-bar, creeper female (CcbO).

10. F_1: males: white, fast; females: white, slow; F_2 males and females: 13/32 white, fast : 13/32 white, slow : 3/32 coloured, fast : 3/32 coloured, slow.

11. Males and females: 15/32 bar, feathered : 15/32 non-bar, feathered : 1/32 bar, non-feathered : 1/32 non-bar, non-feathered.

12. (a) 0.5 male (c) 0.67 intersex (e) 1.5 superfemale.
(b) 1.0 female (d) 1.0 female (f) 1.0 female (triploid).

13. (a) 8 (b) none; meiosis cannot occur in haploid males (c) 8.

14. (a) 50% (b) 7.14%.

15. (a) Yes (b) No. A sex chromosome mechanism could be operative without a morphological difference in the chromosomes, gametes, or spores.

16.

	Haploid Males		*Diploid Males*		*Females*	
	Wild type	*Veinless*	*Wild type*	*Veinless*	*Wild type*	*Veinless*
(a)	0	all	all	0	all	0
(b)	0	all	0	0	all	0
(c)	0	all	0	all	0	all
(d)	0	all	0	0	0	all
(e)	1/2	1/2	1/2	1/2	1/2	1/2
(f)	1/2	1/2	0	0	1/2	1/2
(g)	all	0	all	0	all	0
(h)	all	0	0	0	all	0

12

Interaction of Genes

A study of inheritance of charters whether a particular gene is completely dominant or completely recessive. F_2 ratio of 3:1 in a monohybrid cross suggests that one gene is completely dominant and its allele is completely recessive. F_2 ratio of 1:2 in a monohybrid cross indicates that none of the two alleles is completely dominant, but their dominance is incomplete. Sometimes the Interaction between alleles is more complex and we get a genotypic ratio which is completely different from 3:1 or 1:2:1. Sometimes in dihybrid and polyhybrid crosses too, the expression of genes is not independent of each other and depends on the presence or absence of other gene or genes. The fact that genes are not merely separate entities producing distinct individual effects, but that they could interact with each other to give completely novel phenotypes, was discovered by Mendelism. The occurrence of such interactions suggests that the phenotype we observe is not due to individual genes but arises from complex developmental processes, although the genes themselves follow the regular laws of inheritance. The overall phenotype of an organism, therefore, is the result of epigenetic interactions of a network of gene controlled reactions. A few simple genetic interactions are discussed in the following paragraphs:

Lethal Genes

There are genes which control certain phenotypic traits, and at the same time also influence the viability of the individuals. For instance, white eyes and vestigial winged flies have lower viabilities than the wild types. There are still other genes which have no effect on the phenotype but influence the viability. This influence on viability may be of such an order that the individual may fail to survive. Such genes which cause the death of the individual carrying it are known as *lethal genes*.

In the lethal effect is dominant over its normal allele, all individuals carrying it will die and the gene cannot be transmitted to next generation. Such dominant lethal genes will, therefore, be lost in the same generation. On the other hand, recessive lethals are carried in heterozygous condition, and will express themselves only in the homozygous condition.

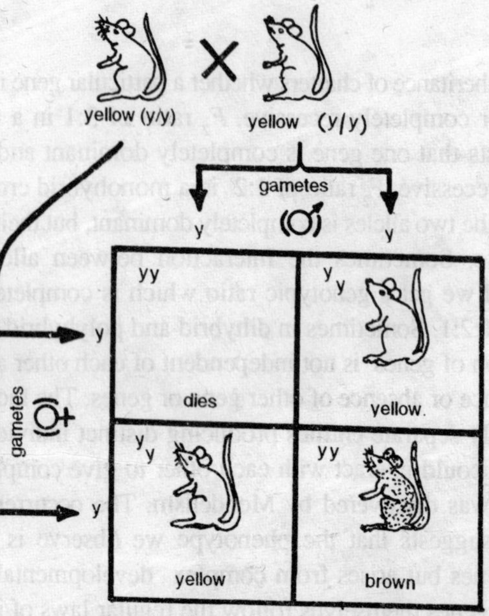

Fig. 12.1. Showing lethality in mice.

Lethality in mice: Cannot (1905), a french geneticist, studied the inheritance of mouse body colour and discovered lethality in mice. He noted that the yellow body colour is controlled by a dominant gene but the yellow mice are never true breeding. When a cross is made between two yellow mice, yellow and black offsprings are produced in a ratio of 2:1. A back cross of yellow to black revealed that yellow mice are always heterozygous for the yellow gene and that individuals homozygous for the yellow gene cannot be found. This is because yellow has a dominant phenotypic effect and in homozygous condition the gene is lethal and the mice bearing such genes die in vitro.

Lethality in Dorsophila. In *Drosophila* genes like curly wings (Cy), plum eyes (Pm) and stubble bristles (Sb) influence the viability of the flies when present in homozygous condition. For instance, an individual Cycy (heterozygous) will have curly wings and will survive. However, homozygous dominant i.e. CyCy will not survive. Homozygous recessive, *cycy* will have normal phenotype and will survive. These genes, therefore, have dual effect. They are dominant with respect to the phenotype (e.g., curly wings, plum eyes and stubble bristles) but are recessive with respect to the lethal effect.

Lethality in Chicken: Some chicken develop abnormal short and crooked legs. This stage is called *Creeper*. This is due to a dominant gene C. The normal development is due to a recessive gene C. When a cross is made between two heterozygous creepers, the creepers and normal are produced in ratio of 2:1 instead of 3:1. The lethality is deserved in dominant homozygous stage CC.

Creeper	X	Creepers
Cc		Cc
|		|
Normal	Creeper	Creeper(dies off)
cc	Cc	CC
25%	50%	25%

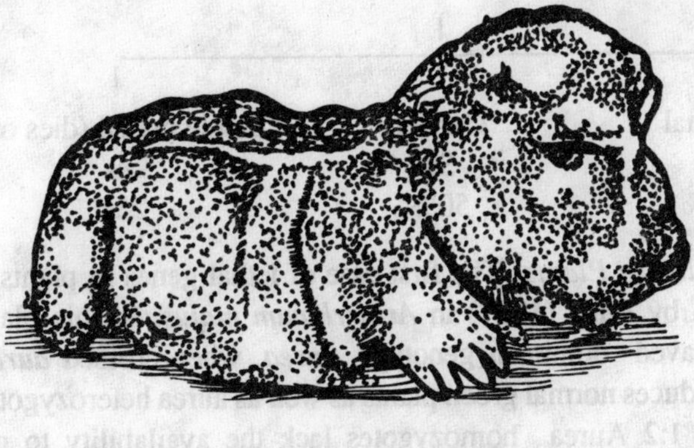

Fig. 12.2. Bulldog calf. This character is controlled by a single recessive gene in homozygous form.

Lethality in Cattle: More than 25 lethal genes have been studies in cattle. The excellent example of lethality is the *bulldog character*.

In this the head is abnormal with short muzzel like of of a bulldog. The leg is short and palate is clefted. These are produced when two dexter races are crossed. The bulldog calf dies off within few days after birth on account of the effect of homozygous recessive lethal gene.

Lethality in Human beings

Several cases of lethality and intermediate lethality have been observed. But here only two cases are being described.

(i) *Brchyphalangy:* Brchyphalangy is the stage of short fingers with two joints, the middle bone being extremely shortened and often fused with one of the other two bones of the finger. *Molir* described one case where one child was born without any fingers or toes and did not survive.

(ii) *Sickle celled anaemia:* This abnormality is due to a recessive gene and in this the shape of erythrocyte is changed from circular to shaped and the amount of hemoglobin in also reduced. Such cases produced when a cross is made between two heterozygous individuals. It results carrier and normal offspring in a ratio of 2:1 instead of 3:1. The lethality is developed in a homozygous recessive case i.e.ss.

Carrier	X	Carrier
Ss		Ss

Normal	Carrier	Sickle celled (dies off)
SS	Ss	ss
25%	50%	25%

Lethality in Plants: The first case of lethal genes in plants was discovered by *Baur* (1907) in *Aritirrhinum majus*. In this plant, a yellow leaved dominant genotype, *aurea,* when crossed *aurea X aurea,* produces normal green plants as well as aurea heterozygotes in a ratio of 1:2 Aurea homozygotes lack the availability to make chlorophyll and die as seedlings or sometimes even before germination.

2. Complementary Genes

A classical example of interaction of genes is the complementation

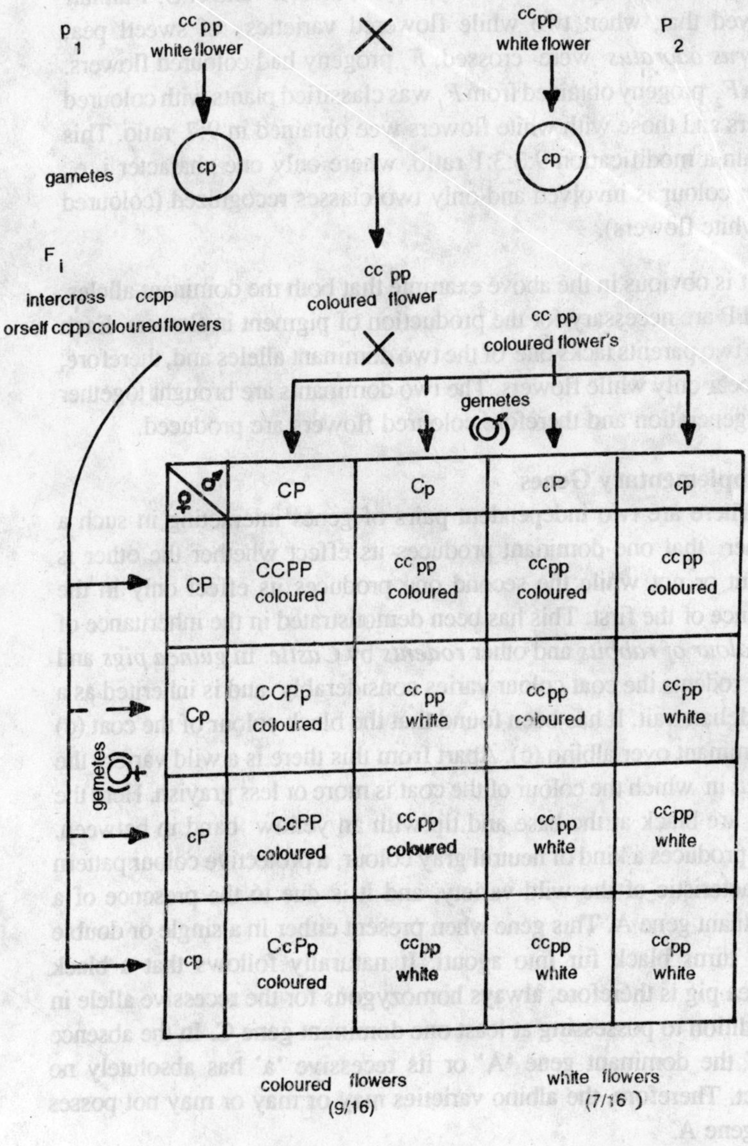

Fig. 12.3. Complementary interaction in two varieties of pea.

between two genes meaning that both genes are necessary for the production of a particular phenotype. *W Bateson* and *R.C. Punnett* observed that, when two while flowered varieties. of sweetl pea, *Lathyrus odoratus* were crossed, F_1 progeny had coloured flowers. When F_2 progeny obtained from F_1 was classified plants with coloured flowers and those with white flowers wee obtained in 9:7 ratio. This is again a modification 9:3:3:1 ratio, where only one character i. e., flower colour is involved and only two classes recognized (coloured and white flowers).

It is obvious in the above example that both the dominant alleles, C and P are necessary for the production of pigment in flowers. Each of the two parents lacks one of the two dominant alleles and, therefore, both bear only while flowers. The two dominants are brought together in F_1 generation and therefore coloured flowers are produced.

3. Supplementary Genes

There are two independent pairs of genes interacting in such a manner that one dominant produces its effect whether the other is present or not while the second one produces its effect only in the presence of the first. This has been demonstrated in the inheritance of *cot colour of rabbits* and other *rodents* by *Castle*. In *güinea pigs* and other rodents the coat colour varies considerably and is inherited as a Mendelian trait. It has been found that the black colour of the coat (c) is dominant over albino (c). Apart from this there is a wild variety the *agouti* in which the colour of the coat is more or less grayish. Here the hairs are black at the base and tip with an yellow band in between. This produces a kind of neutral gray colour, a protective colour pattern characteristic of the wild variety, and it is due to the presence of a dominant gene A. This gene when present either in a single or double dose turns black fur into agouti. It naturally follows that a black guinea pig is therefore, always homozygous for the recessive allele in a addition to possessing at least one dominant gene C. In the absence of C the dominant gene 'A' or its recessive 'a' has absolutely no effect. Therefore, the albino varieties may or may or may not posses the gene A.

The genetic constitution of the 3 different kinds of guinea pigs may be represented as follow: Pu ... 'Caa; pure albino AAcc; Agouti CcAa.

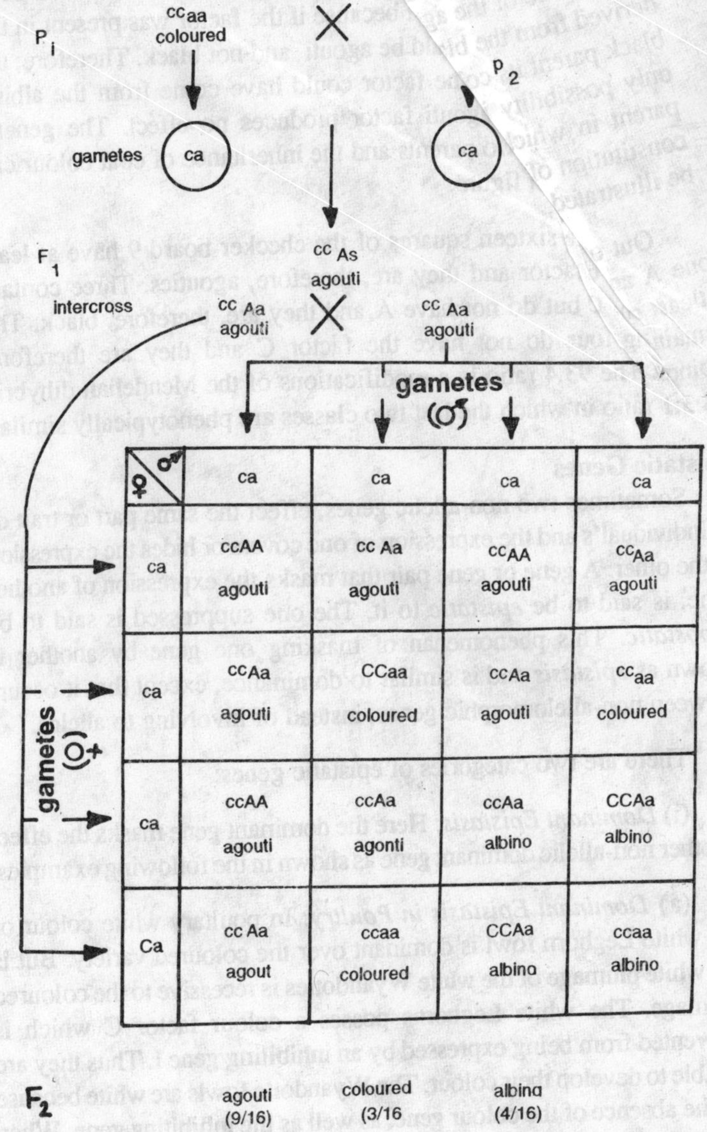

Fig. 12.4. Showing supplementary interactions in mice.

In a cross bet[...]e a breeding black guinea pig with an albino variety carr[...]ominant gene A, all the F_1 offsprings were found to be of th[...]ype. The agouti factor could not have been derived from th[...]parent because if the factor was present in the black parent its[...] would be agouti and not black. Therefore, the only possibili[...]at the factor could have come from the albino parent in wh[...]e agouti factor produces no effect. The genetic constitution [...]e two parents and the inheritance of coat colour can be illustrat[...]s in figure.

Out [...] the sixteen squares of the checker board 9 have at least one A a[...] c factor and they are, therefore, agouties. Three contain atleast one C but do not have A and they are, therefore, black. The remaining four do not have the factor C and they are therefore, albinos. The 93.4 ratio is a modifications of the Mendelian dihybrid 9:3:3:1 ratio in which the last two classes are phenotypically similar.

Epistatic Genes

Sometimes two non-allelic genes, effect the same part or trait of an individual's and the expression of one covers or hides the expression of the other. A gene or gene pair that masks the expression of another gene, is said to be *epistatic* to it. The one suppressed is said to be *hypostatic*. This phenomenan of masking one gene by another is known as *epistasis* and is similar to dominance, except that it occurs between non-allelomorphic genes, instead of involving to alleles.

There are two categories of epistatic genes:

(i) *Dominant Epistasis:* Here the dominant gene masks the effect of other non-allelic dominant gene as shown in the following examples.

(a) *Dominant Epistasis in Poultry:* In poultary white colour of the white Leghorn fowl is dominant over the coloured variety. But is the white plumage of the white Wyandottes is recessive to the coloured plumage. The white Leghorns posses a colour factor C which is prevented from being expressed by an inhibiting gene I. Thus they are unable to develop their colour. The Wyandotte fowls are white because of the absence of the colour gene, as well as the inhibiting gene. When white leghorns (CCII) and white Wyandotte (ccii) are mated, the F_1

hybrids are all white (CcIi) with small black markings. But when the white fowls of the F_1 are crossed among themselves, the offspring are in the peculiar ratio of 13/16 white and 3/16 coloured.

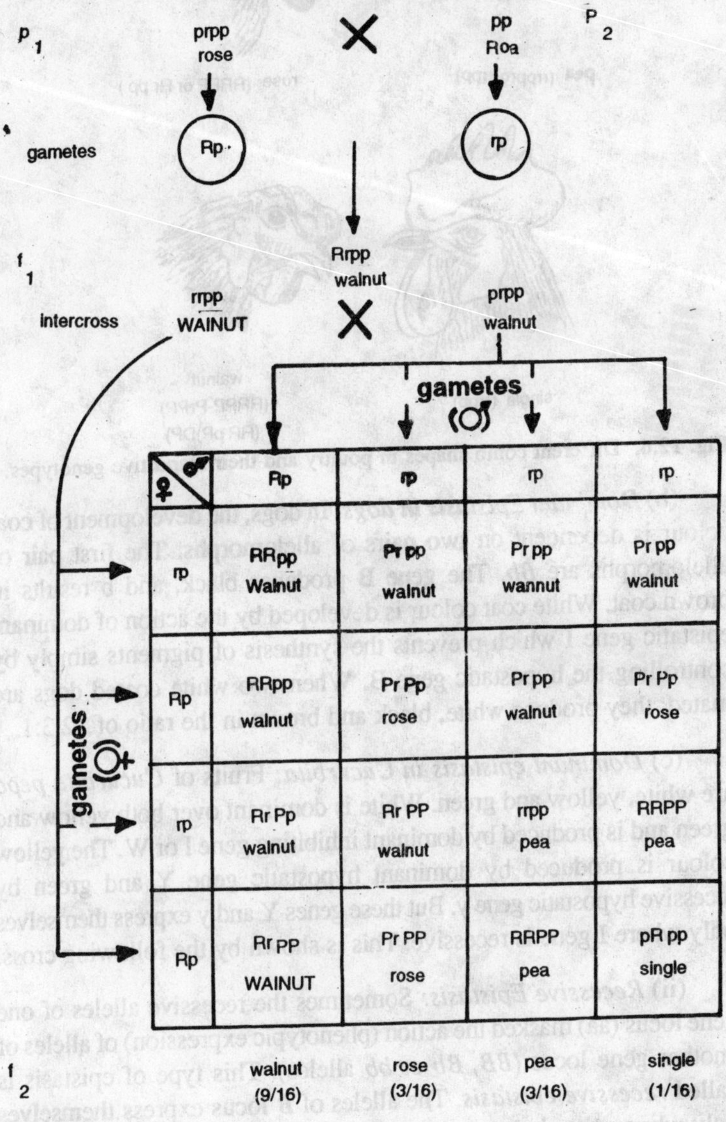

Fig. 12.5. Showing dominant epistatic genes.

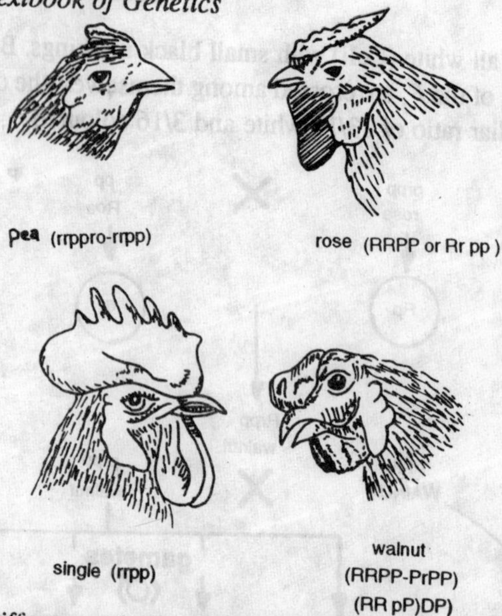

Fig. 12.6. Different comb shapes in poultry and their respective genotypes.

(b) *Dominant Epistasis in dogs:* In dogs, the development of coat colour is dependent on two pairs of allelomorphs. The first pair of allelomorphs are *Bb*. The gene B produces black, and b results in brown coat. White coat colour is developed by the action of dominant epistatic gene I which prevents the synthesis of pigments simply by controlling the hypostatic gene B. When two white coated dogs are mated, they produce white, black and brown in the ratio of 12:3:1.

(c) *Dominant epistasis in Cucurbita:* Fruits of *Cucurbita pepo* are white, yellow and green. White is dominant over both yellow and green and is produced by dominant inhibiting gene I or W. The yellow colour is produced by dominant hypostatic gene Y and green by recessive hypostatic gene y. But these genes Y and y express themselves only where I gene is recessive. This is shown by the following cross.

(ii) *Recessive Epistasis:* Sometimes the recessive alleles of one gene locus (aa) masked the action (phenotypic expression) of alleles of another gene locus *(BB, Bb* or *bb* alleles). This type of epistasis is called *recessive epistasis*. The alleles of *B* locus express themselves only when epistatic locus— *A* has dominant alleles (e.g., *AA* or *Aa).*

Due to recessive epistasis the phenotypic ratio 9:3:3:1 becomes modified into 9:3:4 ratio.

Recessive Epistasis in Mice

In mice various types of epistatic genetic interactions have been reported. The most interesting case is of recessive epistasis in coat colours. The common house mouse occurs in a number of coat colours, i.e., agouti, black and albino. The agouti colour pattern is commonly occurred on (wild type) and is characterized by colour banded hairs in which the part nearest the skin is grey, then a yellow band and finally the distal part is either black or brown. The albino mouse lacks totally in pigments and has white hairs and pink eyes.

When a homozygous agouti or grey *(BBAA)* is crossed with a homozygous albino *(bb aa)* in F_1 all agouti offsprings appear. When, the F_1 agouti are crossed among themselves in F_2 agouti, black and albino offsprings appear in the ratio of 9:3:4.

Recessive Epistasis in Man

This is seen in the inheritance of albunism in man. Sometimes an albino child is born to parents who are normal or have dark skin. Each of these parents carries a recessive gene for albunism. The child who is homozygous for the genes for albinism, is unable to develop any pigment in the skin although the might inherit genes for pigmentation also. This is because the gene for albinism is epistatic to genes for pigmentation of the skin, hair and eyes.

Deaf-Mutism Man

Sometimes a child who is deaf and mute is born to normal parents. The normal hearing trait is due to the presence of two dominant genes *(AABB)*. The abnormal condition of deaf mutism is due to the presence of either of the two recessive genes *(a* or *b)* in homozygous condition. Thus, a person with the genotype *aaBB* would be a deaf-mute because the gene *'a'* is epistatic to B for normal hearing. Similarly, *a* person with the genotype *AAbb* is also *a* deaf mute, because the gene *b* is epistatic to the gene *A* for normal hearing. This is called *duplicate recessive epistasis*. Thus, deaf and mute children are born to parents who are normal but heterozygous for the genes for normal hearing as shown in the Figure. When two deaf-mutes marry, all of their children may have normal hearing and speech as both the genes for deafness are recessive.

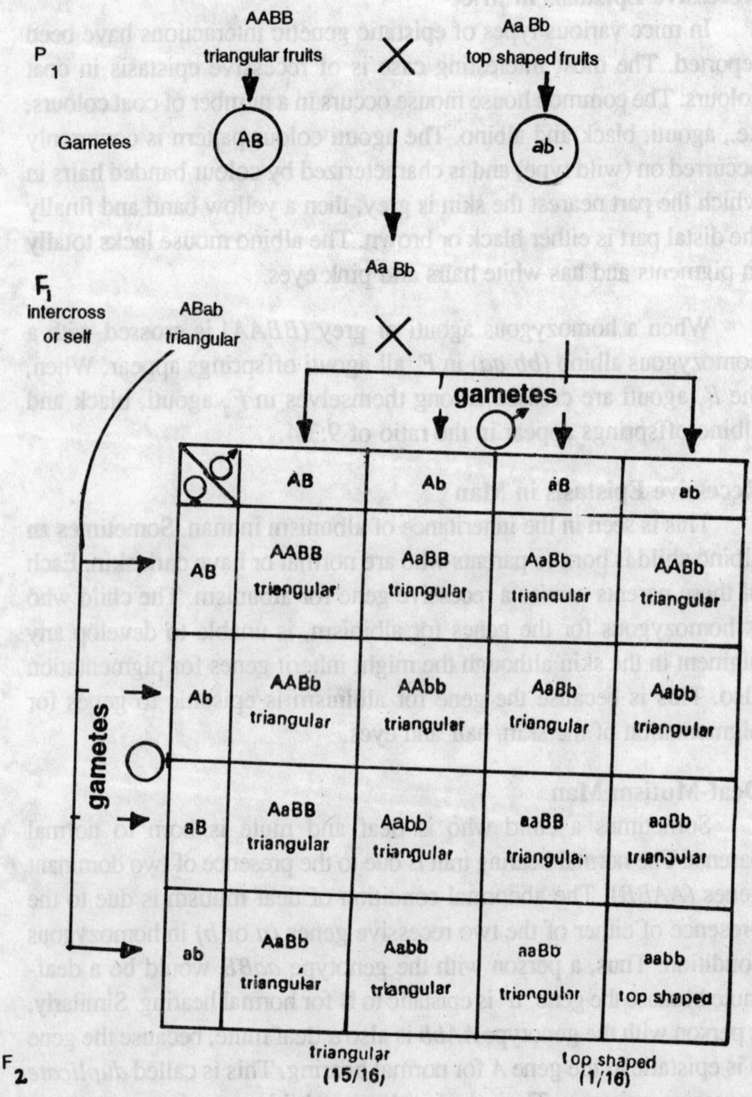

Fig. 12.7. Showing duplicating genes.

Duplicate Genes

Sometimes two different genes determine the same phenotype. Such genes are called *duplicate genes*. For example, endosperm colour in maize is determined by two genes such as Y_1 and Y_2. When both or either of the genes are dominant endosperm is yellow. When both the genes are recessive, endosperm is colourless or white. Yellow colour is dominant and white recessive. A cross between plants with yellow $(Y_1 Y_1 Y_2 Y_2)$ and white $(y_1 y_1 y_2 y_2)$ endosprms yield progeny which are heterozygous for bo'h the genes $(Y_1\ y_2\ Y_2\ y_2)$ and, therefore, have yellow endosperm. Plants of F_2 generation have either yellow or colourless endosperm and segregation yellow; white is in the ratio of 15:1. Seven out of sixteen progeny are homozygous dominant for both or either of the genes and are true breeding, producing only yellow seeded progeny on selfing. Four out of sixteen are dominant and heterozygous for only one and, therefore, on selfing segregate into three yellow-seeded: one white seeded progeny. Another four out of sixteen F_2 progeny are heterozygous for both the genes and produce 15 yellow seeded: one white seeded progeny on selfing. One of sixteen is homozygous recessive for both genes and remains true-breeding, producing only white-seeded plants. Inheritance of fruit shape in *Capsella bursa pastor* is follows the same pattern. Fruits are triangular in the presence of two $(T_1\ T_2)$ or one $(T_1$ or $T_2)$ dominant gene. Recessive homozy-gosity for both the genes $(t_1 t_1 t_2 t_2)$ results into oblong fruits.

Cumulative or Quantitative Genes

These genes are quantitative in nature and bring out different degrees of phenotypic expression according to the number of doses of determiners. The greater the genes will be, larger will be its effect. The traits influenced by duplicate factors are also of this type. A case of this type came to the notice of the famous plant breeder *Kolreuter*. He crossed a tall variety of tobacco with a dwarf variety. The F_1 were intermediate between the two parents. The melanin pigment in the human skin experimented by *Davenport* by crossing Negro white crosses in an another classical example of cumulative genes. The inheritance of quantitative characters was also explained by the Swedish plant breeder, *Nillson-Ehle*. According to his explanation one gene will have a positive effect, two genes will have a greater effect and total effect of three genes will be still greater and so on. The effect adds up with each increase in the number of genes so that it is said to be cumulative. Such genes are described as *multiple genes* or *polygenes* or *cumulative genes.*

Modifying Factors

In a number of instances a character determined by one main gene is modified or changed slightly by the other genes. Such genes are known as *modifying genes.* They cannot produce any separate effect except for a slight modification in the character produced by another gene. A typical instance of this kind was described by *Castle* with reference to the coat colour in guinea pigs. In these animals the body colour has an yellow background. The development of the pigment black or brown over this yellow background is influenced by two factors or genes called the *modifying factors* or *modifying genes.* These are the extension and restriction factors. The former whenever present, occurs alongwith the pigmentation factor and allows the development of pigmentation all over the body but when the restriction factor is present the pigmentation becomes confined to eyes and the skin of the fore and hind limbs. The yellow factor which forms the background is unaffected by the modifying factor. If *'B'* and *'b'* are taken to represent the pigmentation factors controlling clack and chocolate brown respectively, and *'E'* and *'e'* to represent the extension and restriction factors controlling the black and chocolate brown colour and *'Y'* the yellow colour unaffected by *'E'* but concealed by *'B'* then the genotype for an extended brown guinea pig will be *bbEEYY;* while that of ordinary black eyed yellow guinea pig will be *BBeeYY.*

When an extended brown guinea pig *bbEEYY* was crossed with an ordinary black eyed yellow *BBeeYY* one; all the young ones were found to be black *BbEeYY.* When the F_1 hybrids were crossed among themselves 4 phenotypic varieties namely Black, Brown, Black eyed yellow, and Brown eyed yellow appeared in the ratio 9:3:3:1 in the F_2 generation. This can be diagrammatically expressed as in figure.

Pleiotropy

The term *pleiotropy* refers to the situation in which a gene influences more than one trait. Many such instances have been discovered. In fact, all genes (whether mutant or wild-type allelic forms) may be pleiotropic, with their various effects simply not yet recognized. Even though a structural gene may have many end effects, it has only one primary function, that of producing one polypeptide (in some cases, one RNA molecule). This polypeptide may give rise to different expressions at the phenotypic level.

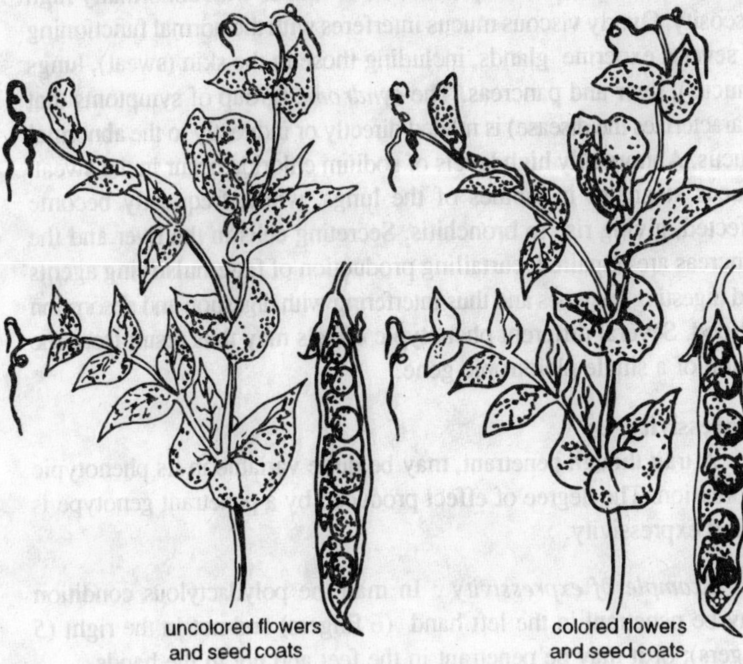

Fig. 12.8. Pleio tropy in pea plant. The gene for flower colour also affects the colour of seeds and of the leaf axils.

The *Hb*² allele provides a classic example of pleiotropy. It not only causes hemolytic anemia (in the homozygous state) but also results in increased resistance to one type of malaria, that caused by the parasite *Plasmodium falciparum.* Because the increased resistance of falciparum malaria occurs in Hb^AB Hb^bB heterozygotes, such heterozygous individuals have a selective advantage in geographical region where this type of malaria is prevalent. The sickle-cell allele also has pleiotropic effects on the development of many tissues and organs such as bones, the lungs, the kidneys, the spleen and the heart.

Cystic fibrosis is a hereditary, metabolic disorder in children that is controlled by a single autosomal recessive gene. The gene apparently specifies an enzyme that produces a unique glycoprotein. This glycoprotein results in the production of mucus with abnormally high viscosity. Overly viscous mucus interferes with the normal functioning of several exocrine glands, including those in the skin (sweat), lungs (mucus), liver and pancreas. The *syndrome* (group of symptoms that characterizes the disease) is related directly or indirectly to the abnormal mucus. Abnormally high levels of sodium chloride occur in the sweat. Mucus stagnates in tabules of the lungs, which frequently become infected, giving rise to bronchitis. Secreting cells in the liver and the pancreas are impaired, curtailing production of fat-emulsifying agents and digesting enzymes and thus interfering with digestion and absorption of food. Several different phenotypic effects may thus result from the action of a single pleiotropic gene.

Expressivity

A trait though penetrant, may be quite variable in its phenotypic expression. The degree of effect produced by a penetrant genotype is called expressivity.

Example of expressivity : In man the polydactylous condition may be penetrant in the left hand (6 fingers) and not in the right (5 fingers); or it may be penetrant in the feet and not in the hands.

Effect of environment on expressivity. The expressivity or a given gene is often influenced by environmental conditions. For instance, the expressivity of completely penetrant gene for vestigial wings in *Drosophila* is influenced by the temperature at which the fly develops, with the effect being most obvious at lower temperature.

Other examples of environmental effects on the expressivity of a gene includes such cases as the differences in the severity of symptoms of an inheritable allergy, or the differences in height of identical twins who have been raised in different home (with different diets), or who have had different medical histories (one with a serious childhood disease, the other escaping this disease)

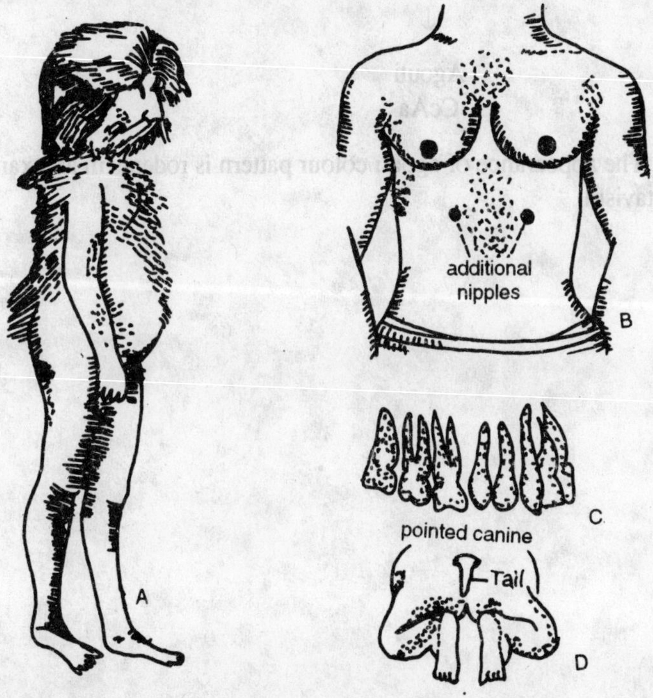

Fig. 12.9. Some human features of Atarism.

Atavism

Characters may and often remain hidden generation after generation through the effect of inhibiting or epistatic factor or some other gene interaction. Occasionally, some wild character which was present in the ancestral forms reappears in the offsprings. This was first discussed by Charles Darwin and was called *atavism.*

Atavism results on account of chance combination of genes that allows a long suppressed character to reappear. For example, homozygous gene c is albino rats does not permit the appearance of agouti pattern. But in an outcross, when gene c is replaced by its allele C, agouti colour appears in the offsprings.

Black	x	Albino
CCaaa		ccAA
Ca		cA

↓

Agouti
CcAa

The appearance of agouti colour pattern is rodents in an example of atavism.

13

Multiple Alleles

The occurrence of segregation of characters in hybrids, and the formation of heterozygote indicate that the genes are responsible for the expression of various characters. Behaviour of flower colour in pea plants shows that there is some gene responsible for the production of colour in flowers. Similarly in man the occurence of brown and blue eyes shows that there are genes responsible for the production of different eye-colours in man. The above changes (difference of colour in flower or in eye) are produced by the variants of a particular gene, responsible for that character. It has been assumed that a gene in the past has mutated to another form resulting into the formation of contrasting pair of characters, these are known as *alleles*. The alleles occupy the same position in the different chromosomes. When a particular gene has several mutated forms giving rise to several alternate states or variants, the term multiple alleles can be applied. Thus when a gene is known to have a number of alternative forms, the latter together with the original gene are called *multiple alleles*. Multiple alleles must not be confused with the polygenes which means a group of genes occupying different loci and control the same character.

Characteristics of Multiple Alleles

The important features of multiple alleles are summanized below:

(i) Multiple alleles always occupy the same locus in the chromosome.

(ii) Since all the alleles occupy the same locus, crossing over does not occur between them.

(iii) Only two members of such alleles (in contrast to many in polygenic system) are present at a time in a diploid.

(iv) The wild type allele is nearly always dominant, while the other mutant alleles in the series may show dominance or there may be an intermediate phenotypic effect.

(v) When any two of multiple alleles are crossed, the phenotype is of a mutant type and not the wild type.

(vi) There are no wild type recombinants in the double mutant heterozygote.

(vii) The gametes are hapolid containing only one chromosome of each type, hencefore are allele of the multiple *us* an each gamete.

(viii) The entire series of multiple alleles are represented by the same basic letter and the different forms are indicated by superscripts or subscripts.

Multiple Alleles in *Drosophila*

Eye colour in *Drosophila* is a well-known example of multiple alleles. Fifteen different alleles of this particular gene, located in the X-chromosome, are known. These are named after the colour that they produce, red, eosine, apricot, buff, blood, cherry, honey, ivory, pearl, coral, tinged, white, etc. As already stated that the entire senses of multiple alleles are represented by the same basic letter and the different forms are indicated by superscript or subscript. Thus, in case of eye colour of *Drosophila*, W represents the gene for red eye colour; W^{bl}, blood; W^{e}, eosine; W^{z}, apricot; w, white. In this case the original wild population had white types due to a recessive gene w. All the other forms arose as a result of different mutations in the wild population. The gene for red eye colour is represented by capital W because it is dominant over all the other alleles in the series.

Another example of multiple alleles has been taken from the shape and size of wings-in *Drosophila*. The fly has three alleles of gene Lof size of wings—(i) L^{+} for normal or wild type wings. (ii) L^{vg} for vestigial wings and (iii) L^{a} for antlered wings. The wild type gene L^{+} is dominant to Lvg and La, l while both recessive alleles (L^{vg} and L^{a}) have incomplete dominant recessive relationships, therefore, their cross gives a new mutant form as given below:

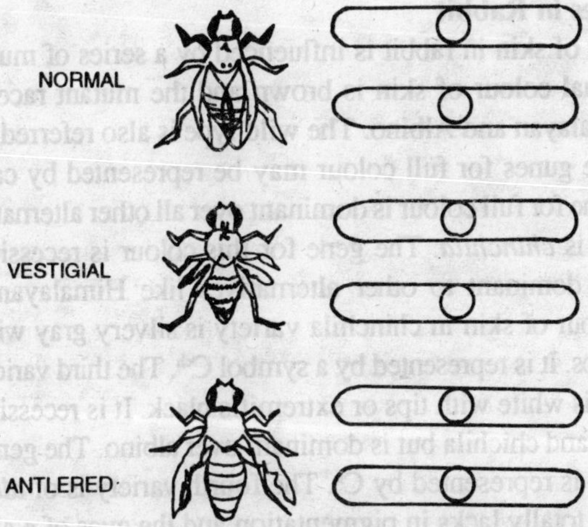

Fig. 13.1. Multiple alleles of wings in *Drosophila.*

P:	Vestigial-winged fly		Antlered-winged fly
	$L^{vg} L^{vg}$X	X	$L^{a} L^{a}$
F_1:		$L^{vg} L^{a}$	
		Intermediate	
F_2	$L^{vg} L^{vg}$,	$L^{vg} L^{a}$, $L^{vg} L^{a}$,	$L^{a} L^{a}$
	Vestigial	Intermediate	Antlered

Thus, in *Drosophila,* the gene L occurs in the three allelic forms (L^{+}, L^{vg} and L^{a}) and their possible genotype can be summarized as follows:

Phenotypes	*Genotypes*
Long-wings	$L^{+} L^{+}$, $L^{+} L^{vg}$, $L^{+} L^{a}$
Vestigial-wings	$L^{vg} L^{vg}$
Antlered-wings	$L^{a} L^{a}$
Intermediate-wings	$L^{vg} L^{a}$

Multiple Alleles in Rabbit

The colour of skin in rabbit is influenced by a series of multiple alleles. The usual colour of skin is brown and the mutant races are Chinchila, Himalayan and Albino. The wild type is also referred to as *full colour*. The gunes for full colour may be represented by capital letter C. The gene for full colour is dominant over all other alternatives. Another colour is *chinchila*. The gene for this colour is recessive to full colour but dominant to other alternatives like Himalayan and albino. The colour of skin in chinchila variety is silvery gray with or without white tips. It is represented by a symbol C^{ch}. The third variety—the *Himalayan* is white with tips or extremitis black. It is recessive to both full colour and chichila but is dominant over albino. The gene for Himalayan coat is represented by C^h. The fourth variety is of *albino*. The albino coat totally lacks in pigmentation and the eyes of a albino also remain pink due to lack of pigment in *iris* of eye. The gene for albino is represented by *c* which is recessive to all other forms.

TABLE 13.1

Various Genetypes and their Phenotypes due to Multiple Allelic Series for Skin Colour in Rabbit

Genotypes	*Phenotypes*
CC, CC^{ch}, CC^h, Cc	Coloured (wild)
$C^{ch}C^{ch}$, $C^{ch}C^h$, $C^{ch}c$	Chinchilla (light grey)
C^hC^h, C^hc	Himalayan
cc	albino

When crosses are made between rabbits with different coat colours, then it is revealed that full colour (C) is dominant to chinchilla (C^{ch}), himalayan (C^h) and albino (c). Further, the allele for chinchilla (C^{ch}) is found recessive to allele C but dominant over C^h and c. Likewise, the allele for himalayan (C^h) is found recessive to allele C and C^{ch}, but dominant over allele c and, allele c for albino coat is found to be recessive to all alleles of this multiple allelic series. These results clearly indicate that the allele C^{ch} (chinchilla), C^h (himalayan) and c (albino) are mutant or allelic forms of wild type gene C. The dominance heirarchy may be symbolized as follows :

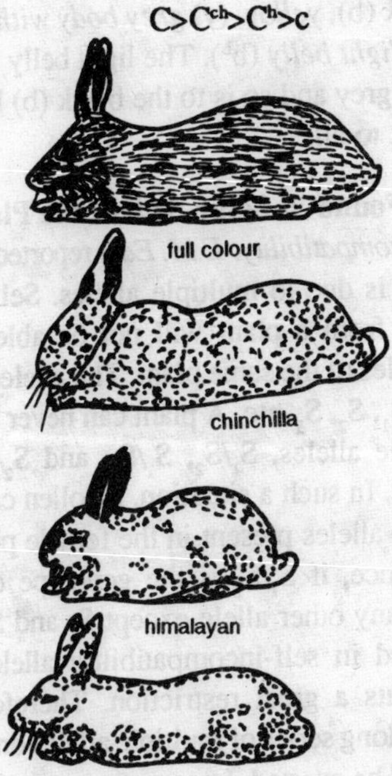

Fig. 13.2. Coat colour in rabbit.

Multiple Alleles in Mice

Coat colours of mice and cat also offer examples of multiple allelism. These are also found in various insects and plants. Blood groups in cattles are determined by as many 80 alleles.

Two series of multiple alleles are found associated with the determination of coat (skin) colour in mice. Species number 1 is at the albino-locus and comprises of four genes. The normal allele in this series in for *grey coat colour* and is represented by (+) in the figure. The other genes in the series are mutant genes—*albino (a), medium light grey* (a^m), and *extreme light* (a^e). all the three mutant genes are recessive to the normal allele.

The second series of the alleles is at the black locus and is in a different chromosome. Here the normal allele is one of the numerous genes necessary for *grey coat colour* (+). The mutant genes of the black series are: *black* (b), *yellow* (y) *grey body with light belly* (G^L), and *black body with light belly* (b^L). The light belly effect of both G^L and b^L is recessive to grey and so is to the black (b) but the yellow of the series is dominant to all.

Multiple Alleles as Found in Some Self-sterile Plants

Gametophytic Incompatibility: E.M. East reported that self-sterility in the tobacco plants is due to multiple alleles. Self sterility would mean that the pollen from a plant will not be able to bring about fertilization in the ovules of the same plant. The alleles in such a series can be designated as S_1, S_2, S_3, etc. A plant can never be homozygous. If there are only three alleles, S_1/S_2, S_1/S_2, and S_2/S_3 are the only possible combinations. In such a situation, a pollen carrying an allele different from the two alleles present in the female plant will be able to function. For instance, if S_1/S_2 is the genotype of female plant, pollen carrying S_3 or any other allele except S_1 and S_2 will function. The principle involved in self-incompatibility alleles in shown in Table below. This puts a great restriction. Therefore, a selection pressure in favour of a long series of multiple alleles on the functioning ability of pollen would be exerted. In some cases, as many as 40 such alleles are known to be present in the same series.

TABLE 13.2

Multiple Alleles for Eye Colour in *Drosophila* and Various Phenotypes due to Homozygous and Heterozygous Conditions

Allete	*Phenotype in homozygous condition*	*Heterozygotes*	
		Genotypes	*Phenotypes*
+	red	+/w	
w	white	$+/w^{ch}$	
w^{ch}	cherry	$+/w^{b}$	
w^{b}	blood	$+/w^{e}$	red eyes
w^{e}	eosin	$+/w^{apr}$	
w^{apr}	apricot	$+/w^{iv}$	
w^{iv}	ivory	$+/w^{cr}$	
w^{cr}	cream	other heterozygotes	intermediate

TABLE 13.3

Genotypes of Progenies Obtained due to Crosses Between Various Self-incompatible Types

	S_1S_2	S_1S_3	S_2S_3
S_1S_2	—	S_1S_3 S_2S_3	S_1S_3 S_2S_3
S_1S_3	S_1S_2 S_2S_3	—	S_1S_2 S_2S_3
S_2S_3	S_1S_2 S_1S_3	S_1S_2 S_1S_3	—

Multiple Alleles in Human Beings

The excellent example of multiple allelism is taken from the blood groups in human beings. Various blood groups with different combinations of the related alleles will be discussed in details under a separate heading The Human Genetics.

Multiple Allelism in Maize

In the maize plant a series of multiple alleles have been observed influencing seed colour. Here, the normal seed colour in purple, the pigment occuring in the aleuron layer of cells just beneath the seed coat. The normal gene that governs the alevion layer may be represented by the letter A. One successive mutation in this gene makes it white and a second recessive mutation changes it to light purple. The corresponding mutated genes may be expressed as *'a'* for white and *'a'* for light purple etc. These mutations are found to affect the other regious of the plant also.

Pleiotropism: We have seen that one gene is generally responsible for one character; but there are also such examples where we find that one gene, besides the production of its character, also affects the other characters of the body. Thus such genes have pleiotropic effects. Following are the examples of such cases :

1. The same gene is found to effect the flower colour alongwith the colour of seed coat and the colour of the leaf axil in pea.

2. In *Drosophila* gene responsible for the size of wing also affects the eye colour, shape of spermatheca, and position of dorsal bristles.
3. In lupine the height of plant, and colour of seed coat are affected by the gene responsible for flower colour.
4. Members of a series of multiple alleles may affect different combinations of body parts of organism. Gene R in maize determines purple (anthocyanin) colour in different parts of plants; four of these alleles have following affect :

   Rr—Coloured seeds (aleuron), coloured plants.
   Rg—Coloured seeds (aleuron), colourless plants.
   rr—Colourless seeds (aleuron), coloured plants.
   rg—Colourless seeds (aleuron), colourless plants.

Therefore, gene appears to consist at least two parts or components, one determines the aleuron colour and other the plant colour. The aleuron colour may charge (mutate) separately from plant colour component although two kinds of changes are not independent.

5. In Beetle, *Harmonia axyridis* different alleles produce black colour in different parts of wing cover. Heterozygous individuals for any two alleles are black in all those parts which are make black by either allele separately.

Theories of Multiple Alleles

Three important theories have been postulated to explaining the origin and occurance of multiple allelism. These are:

(i) *Theory of point mutation:* This theory postulates that multiple alleles are the result of mutation occurring in the same locus in different directions.

(ii) *Theory of Positional Pseudoallelism*: This theory postulates that multiple alleles result not by gene mutations at the same locus but that the genes concerned occupy different loci which are closely situated in the chromosome. These closely linked genes at different loci are referred to as *pseudoalleles,* and affect the expressions of normal genes due to position effect. For instance in *Drosophila,* the normal eye colour is red, but there are other shades like white, pink apricot etc. When apricot flies are crossed with white, the F_1 hybrids are intermediate, proving allelic nature of the two genes. But recently *Lewis* found that these hybrid flies produced some red offsprings when crossing over is greatly increased in the vicinity of white locus. It means these red flies are cross-overs. *Lewis*, therefore, concluded that apricot and white are not actually alleles but they occupy adjacent loci and are pseudoalleles. These suppress the expression of normal gene.

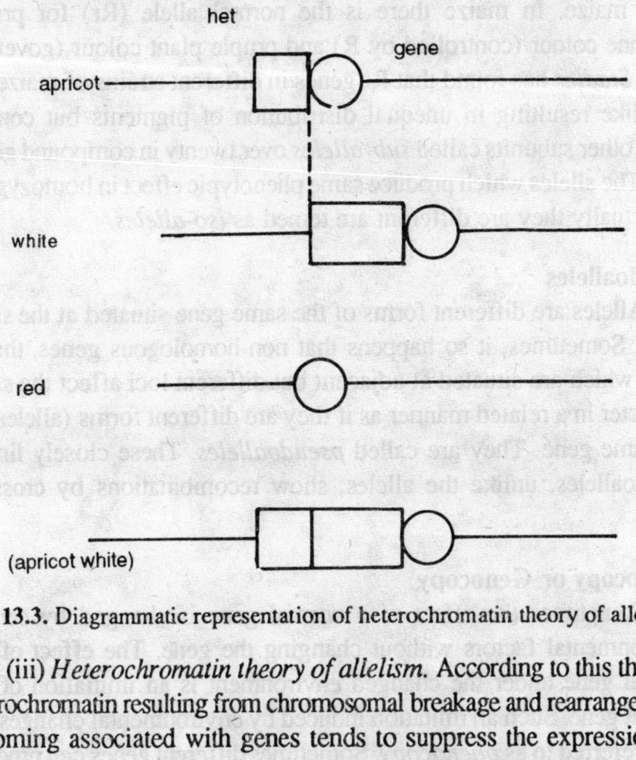

Fig. 13.3. Diagrammatic representation of heterochromatin theory of allelism.

(iii) *Heterochromatin theory of allelism.* According to this theory, heterochromatin resulting from chromosomal breakage and rearrangement becoming associated with genes tends to suppress the expression of the genes due to position effect. Such position effect due to transportation of heterochromatin particles have been observed in maize.

There are also evidences that particles of different kinds of heterochromatin suppress the expression of normal gene to different degrees. In *Drosophila* apricot might be a partially suppressed red (normal) and white completely suppressed red while appricot and white hybrid may give rise to red or intermediate by unequal crossing over.

The above theories in some way or other do not explain clearly the particular case of allelism and it is possible that all the three theories are applicable in different cases.

Sub-alleles and Iso-Alleles

The sub-alleles are the various components of a compound gene as in maize. In maize there is the normal allele (Rr) for pruple aleurone colour (controlled by R) and pruple plant colour (governed by r). *Stadler* has found that Rr genes in different strains of maize are not alike resulting in unequal distribution of pigments but contain many other subunits called *sub-alleles* over twenty in compound genes (Rr). The alleles which produce same phenotypic effect in homozygous but actually they are different are temed as *iso-alleles*.

Pseudoalleles

Alleles are different forms of the same gene situated at the same locus. Sometimes, it so happens that non-homologous genes, that is those which are situated at adjacent but different loci affect the same character in a related manner as if they are different forms (alleles) of the same gene. They are called *pseudoalleles*. These closely linked pseudoalleles, unlike the alleles, show recombinations by crossing over.

Phenocopy or Genocopy

Sometimes the effect of a normal gene is changed by certain environmental factors without changing the gene. The effect of the normal gene under the changed environment is an imitiation of the mutant gene. Such an imitation induced by environmental changes has been referred to as *phenocopy*. Sometimes different genes can produce the same phenotype. This phenomenon is known as *genocopy* a genetic mimic.

Penetrance and Expressivity

Ordinarily, a recessive gene produces its phenotypic effect when in the homozygous condition and a dominant gene produces its effect whether it is homozygous or heterozygous. Some genes fail to produce their phenotypic effects when they should. The ability of a gene to produce its effect is called *penetrance*. A gene that always produces the expected effect is said to have 100 per cent penetrance. If its phenotypic effect is produced only in 80 per cent of the individuals that possess it then it is said to show 80 per cent penetrance. The percentage of penetrance can be altered by changing the environmental conditions such as temperature, moisture, etc. In cotton, a mutant gene produces crinkled leaf. While all the leaves produced in the normal season are crinkled some of the leaves produced late in the season do not show this character and are normal. This means that penetrance is zero or that the gene is non-penetrant. Penetrance is an all-or-none phenomenon. Either the gene produce its effect in an individual or not at all.

Sometimes there is great variation in the manner in which a character is expressed in different plants. In the venturra variety of Lima beans, a dominant gene is responsible for the tips and margins of the leaves of the seedlings to be partially deficient in chlorophyll. Sometimes only the tips are affected and sometimes only the margins. In other words, this single gene can express itself in a variety of ways that many resemble a number of characters. This gene is then said to show variable *expressivity*. Whether a gene is expressed at all is denoted by the term penetrance whereas the term-expressivity denotes the manner or degree of its expression.

Importance of Multiple Allelism

The study of multiple alleles has increased our knowledge of heredity. According to *T.H. Morgan* a great knowledge of the nature of gene has come from multiple alleles. These alleles suggest that a gene can mutate in various ways causing different effects. Multiple allelism also propounds the idea that various amounts of heterochromatin suppress the genes to different extents.

REVIEW QUESTIONS

1. In *Drosophila*, white, cherry, and vermillion are all sex-linked recessive aye colours. A white eyed female crossed with a vermillion eyed male produces white male offspring

and wild type red eyed females offspring. A white eyed female crossed with a cherry eyed male produced white eyed sons and light cherry eyed daughters. Which genes are alleles?

2. In rabbits three of the alleles of the c locus are C, c^{ch}, and c. The various genotypes and their phenotypes are:

C/C, C/c^{ch}, C/c	—full colour
c^{ch}/c^{ch}, c^{ch}/c	—chinchilla
c/c	—white.

A full coloured rabbit is crossed with a white rabbit and produces 3 chinchilla and 4 full-coloured offspring. What are the genotypes of the parents?

3. What offspring would be expected from the mating C/c by c^{ch}/c ?

4. The sex-linked locus, vermillion eye, is now known to be pseudoallelic. One bit of evidence came from offspring of females of the constitution

$$\frac{+ \quad (v^{36}+) \quad + \quad +}{r \quad (+\, v^{1}) \quad m \quad f}$$

where the psuedoalleles are in the parentheses, and their order is unknown. Among the offspring were a few wild type males of the constitution + (++) m f. Is v^{36} to the left or right of v^{1} ? By analogy with the garnet case discussed in this chapter would you expect the parent females to have been mutant or wild type?

5. Do the following data show evidence of position effect? If so, change the numbers so that they don't, or if not, change them so they do.

Genotypes	*Number of eye facets*
+++/a++	120

+++/abc	60
a++/abc	30
+b+/a+c	40
abc/abc	20

6. Suggest a chemical or physiological hypothesis to explain a cis-trans position effect.

7. A pollen-bearing plant with self-sterility alleles s^1 and s^2 is mated with a female plant of genotype s^2/s^3. What genotypes would be found among the progeny, and in what proportions? What if the female plant were s^1/s^2? s^3/s^4?

ANSWERS

1. White and cherry
2. 1/2 C/c^{ch} and c/c
3. 1/2 full colour, 1/4 chinchilla, 1/4 white.
4. Right, Mutant.
5. Yes. The second and fourth should have the same number of facets.
6. 1/2 s^1/s^2, 1/2s^1/s^3. None. 1/4s^1/s^3, 1/4s^1/s^4, 1/4s^2/s^3, 1/4s^2/s^4.
7. The same.

14

Cytoplasmic Inheritance

It is a well-known fact that most of the hereditary traits in an organism are determined by genes residing in the chromosomes and that the bulk of heredity is transmitted through the chromosomes. The nucleus, of which chromosomes and genes are the part, lies in the cytoplam and takes its nourishment from there. The question arises whether the nucleus independently governs the heredity or the cytoplasm also plays some role? It is now clear that some inheritable characters are beyond the control of nucleus and are related with certain self-reproducing cytoplasmic particles. This type of inheritance is called cytoplasmic inheritance.

In several cases of reciprocal crosses it has been observed that the results are not the same. The offspring produced varies in certain characters. It has been, therefore, considered that these characters are governed by the cytoplasmic inheritance. The egg contains a larger amount of cytoplasm whereas a sperm or pollen hardly contains any amount. If all the characters are governed by nuclear material, i.e. chromosomes and genes only, the results of a cross between A male X B female and A female X B male should be similar, as whether one is male or female it will have the same genetic constitution. But the results vary. Why it is so? It is so because certain characters are transmitted through cytoplasm present in larger amount in egg. Cytoplasmic hereditary self-reproductive particles are called cytogenes or *plasmagenes*. All the plasmagenes in an organism taken together represent the totality of cytoplasmic heredity and constitute the plasma, whereas one complete set of nuclear genes contained in a haploid set of chromosomes is known as *genome*. The plasma genes are distributed in the general cytoplasm and are not arranged in a single file, as the chromosomal genes are. That is why the distribution of plasmagenes in the daughter cells is not orderly and regular like that of the chromosomal genes? A cell may contain an indefinite member of them instead of a fixed number and different cells of the same organism may contain different number of these factors.

The role of cytoplam in certain cases of inheritances was first of all suggested by *Sanger* and his colleagues in 1950. They demonstrated non-chromosomal inheritances in *Chlamydomnas*. The same was proved later on in several animal and plant species.

The most recent work on the direct way of investigating the role of cytoplasm and the nucleus towards inheritance has been done in England and U.S,A. The work involves the direct putting in of identical nuclei in different cytoplasms and different nuclei in identical cytoplasms by using a micromanipulator.

J.F. Danielli pushed the nucleus out of an *Amoeba* and pushed the nucleus from another species into it. He found that many characters such as shape, size and longevity, were determined more by cytoplasm than by nucleus and other characters like the amount of certain amino acids were determined wholly by the nucleus.

Some types of *Trypanosomes* show remarkable resistance to drugs. This faculty is inherited and has been shown to be associated with changes in structure of the para-basal body.

Evidences in Favour of Cytoplasmic Inheritances

1. *Reciprocal crosses:* In cases of chromosomal heredity, the results of reciprocal crosses are ordinarily identical but in the cases of cytoplasmic inheritance the results of reciprocal crosses are different.

2. *Maternal influence*: When reciprocal crosses for a certain character are found to exhibit different results, it has been noticed that progency exhibits the characteristics of their female parent. If chromosomal difference can be ruled out, maternal inheritance usually implies transmission through cytoplasm.

3. *Infection-like transmission:* When a heritable phenotype is transmitted without nuclear transmission, it suggests that some particles

from the cytoplasm of the parent/parents have been transmitted to the offsprings.

4. *Indifference to nuclear substitution:* When a particular genotype exhibiting a specific character is replaced by a nucleus having alternative genotype, it does not change the phenotype and suggests the possibility of cytoplasmic inheritance.

5. *Non-segregation and non-mendelian segregation:* Segregation is characteristic of Mendelian and chromosomal inheritance. Failure to show segregation may indicate extrachromosomal heredity. Even if segregation occurs but in a fashion inconsistent with the segregation of chromosomes, the results may suggest the possibility of non-chromosomal inheritance.

Cytoplasmic Inheritance in Animals

Several cases of cytoplasmic inheritance in animals have been studied. Some common examples are given under the following sub-heading:

(i) *Cytoplasmic inheritance in Drosophila: Philippe I' Heritier* found a clear cut case of cytoplasmic inheritance in *Drosophila*. There is one strain which is much more sensitive to carbon dioxide than the others. When the insects are exposed to air with high concentration of carbon dioxide, they soon become anesthetized. This particular strain of *Drosophila* is effected by much smaller concentrations than those required to anesthetize normal strains, and may be killed by stronger concentrations. The character was found to be inherited and a true breeding strain was established.

When females from this strain are mated with normal males, all the offspring show the characteristics of the female parent and are highly sensitive to carbon dioxide. On the other hand, when males from this strain are mated with the normal females, only very rarely are any of the offspring so highly sensitive. The reciprocal crosses by replacing each chromosome had excluded the sex-linked influence beyond any doubt. Thus, it may be safely concluded that the transmitted factor is located outside the chromosomes and in the cytoplasm, since it is transmitted regularly through the eggs and only rarely by the sperms.

(ii) *Shell coiling in Snails:* In some species of snails specially *Limnaea peregra,* two types of coiling having been observed — *dextral* (when the coiling is towards right), and *sinistral* (where the coiling is towards left). The coiling is determined by a pair of alleles showing Mendelian inheritances. The dominant allele D control dextral (right handed) coiling while its recessive alleles d determines sinistral (left handed) coiling. *Boycott* and his colleagues (1930) found that the pattern of coiling is determined not by the genotype of an individual but by the genotype of its mother. Due to maternal predetermination all the offspring of a dextrally coiled homozygous (DD) or heterozygous (Dd) mother show right handed coiling, even if they are homozygous for sinistrally. Similarly, all the progeny of sinistrally coiled (dd) mothers show left handed coiling even if they possess the dextrahty gene (D). The maternal effect, however, lasts for only one generation. This is clear from the fact that dextrally coiled mothers of the genotype (dd) produce progeny with sinistral coiling. Coiling in snail is determined by the orientation of the spindle of the second cell division after the fertilization.

If the spindle is tipped towards left, the coiling becomes sinistral; while, if it is tipped towards right, it becomes dextral. The system is governed by gene, acting on eggs in the wary and thus predetermining the direction of coiling.

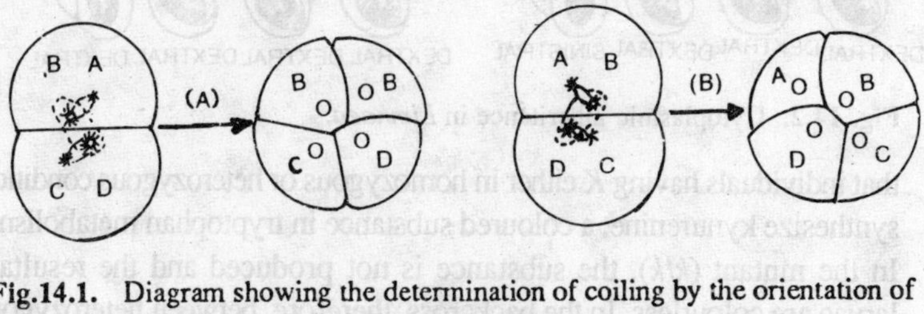

Fig.14.1. Diagram showing the determination of coiling by the orientation of spindle at second cleavage division;

A, cleavage leading to sinistral; B, that leading to dextral coiling.

Eye Colour in Flour Moth, The Ephestia: The expression of colour in the recessive genotype in the larvae of *Ephestia* (flour moth) which is often cited as an example of cytoplasmic inheritance is not also a true instance of this kind. What actually happens in this moth is

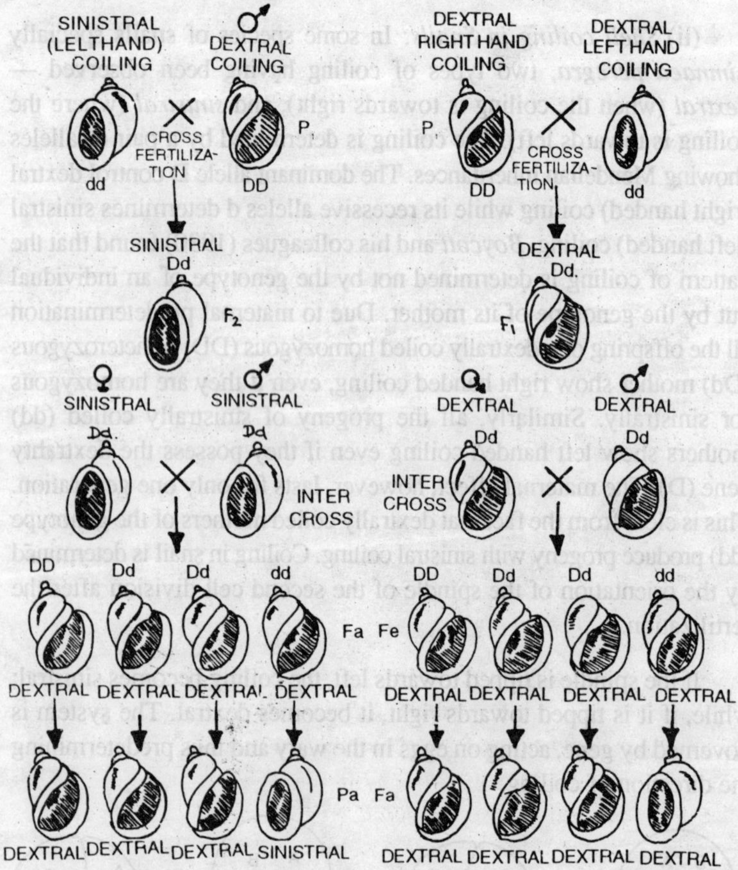

Fig. 14.2. Cytoplasmic inheritance in *Limnaea*.

that individuals having *K* either in homozygous or heterozygous condition synthesize kynurenine, a coloured substance in tryptophan metabolism? In the mutant (*k*/*k*), the substance is not produced and the resultant larvae are colourless. In the backcross, therefore, between heterozygous coloured moth (*K*/*k*) as female and colourless male (*k*/*k*), half of the individuals should be coloured and half without any colour. Initially, however, all the larvae show colour but later, in half of them colour fades out and the pupae and adults mat develop from these larvae do not show any colour. It is not difficult to understand the appearance of kynurenine in *k*/*k* larvae, if we assume that any substance produced by the gene remains diffused in the cytoplasm. Therefore, whatever be the genotype of egg, *K* or *k* resulting from *K/k* heterozygote, the cytoplasm of the egg contains initially kynurenine because the cytoplasm of the heterozygote contains this substance. Later on, this substances used up in *k/k* individuals. Since there is no dominant gene to replenish kynurenine, the larvae which are initially coloured become colourless.

Cytoplasmic Inheritance in Ambystoma meicanum: In axolotl (*Ambystoma meicanum*) a lethal gene O has been discovered which has a maternal effect. The gene O is recessive, so that heterozygote individuals (with genotype ⁺O) are completely normal. Homozygous animals (with genotype (OO) produced by a cross between two heterozyigous parents (⁺O) develop normally in the early stages but in later life show a slight retardation of growth. Their regeneration capacity is severly reduced so that amputated legs are not regenerated as would occur in normal axolotls. The homozygous (OO) mates are sterile—their tests are poorly developed and spermatogenesis does not go beyond the spermatogonial stages. The homozygous (OO) females on the other hand, produce eggs capable of fertilization and normal cleavage, but at the onset of gastrulation the development is retarded and embryos normally die. This abnormal course of development is in no way affected by the genotype of the spermatozoon: the spermatozoon may carry the normal +gene or a mutant lethal O gene (if the male is hetezygous ⁺O). The arrest in development is in both cases exactly the same: this shows that the particular type of abnormal early development does not depend on the genes present in the cells of the developing embryo (⁺O), but exclusively on the OO genes of the mother (maternal effect), *Briggs* and *Justus* (1967) found that eggs of a OO female lack a protein-like substance the *"corrective factor"* which is necessary for normal development and is synthesized by normal genes ++ of homozygous or heterozygous normal females (++ or ⁺O) during oogenesis.

Cytoplasmic inheritance in Paramecium : Sonneborn (1943) carried out extensive mating experiments between various trains of *Paramecum aurelia* and concluded that they can be broadly classified into two types: killer and sensitive strains. Killer strains (e.g., strain No. 51) secete a toxic substance, called *paramecin,* into the culture medium, which proves to be lethal for sensitive strains (e.g., strain No.

47). Fortunately, during mating even the sensitive become resistant and crosses are possible between sensitive and resistant strains.

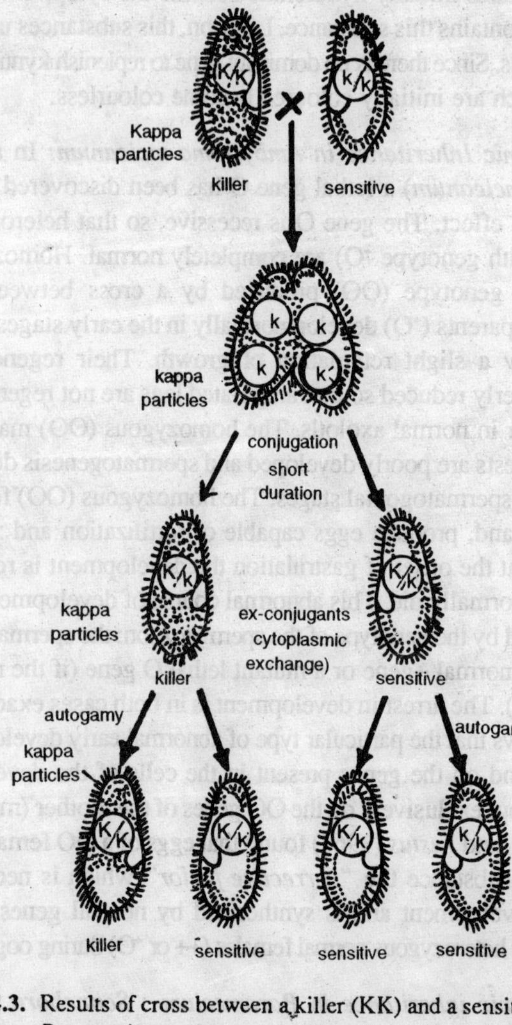

Fig. 14.3. Results of cross between a killer (KK) and a sensitive (kk) strain of *Paramecium*, when no cytoplasmic exchange is allowed.

When *Paramecia* of two different mating types are brought together under suitable conditions, many individuals clump together and ultimately separate in pairs. Conjugation tubes are established at the oral regions and haploid micronuclei are exchanged. Two genetically different micronuclei fuse and individual *Paramecia* separates as exconjugants. As a result of nuclear exchange and fusion both the exconjugants are genetically identical. Under normal condition little or almost no cytoplasm is exchanged during conjugation. Thus, the exconjugants maintain their original cytoplasm but contain similar nuclei. Killer-sensitive matings result into killer and sensitive exconjugants. The killer exconjugants produces a clone of killer *Paramecia* and the sensitive exconjugant produces a clone of sensitive *Paramecia,* although the Mendelian genes segregate during successive generations after self-fertilization or autogamy. Thus, killer and sensitive characters are not controlled by Mendelian genes. When conjugation is prolonged, there is a massive transfer of cytoplasm in addition to the micronucleus and in such cases the killer characters is transferred to the sensitive strain. Therefore, all the progeny of a killer sensitive prolonged mating are killers. This is an additional proof for cytoplasmic inheritance of the killer character. The extrachromosomal particle responsible for killer character has been called *Kappa* by *Sonneborn.*

Although, the kappa particles are cytoplasmic particles and transmitted strictly through the cytoplasm, their maintenance and production of paramecin are, however, controlled by a dominant gene K present in the nucleus of killer strain. The sensitive strain possesses its recessive allele *k.*

By following an appropriate procedure, it is possible to cross the individuals of killer and sensitive races. Observations on the descendants of such crosses have shown that individuals of killer clones carry a dominant gene *K,* whereas those of sensitive clones possess *k.* When killer *KK* conjugates with sensitive *kk,* the exconjugants are *Kk.* These according to their genotype should be killer. In ordinary case of conjugation where only the nuclear material is exchanged and there is no exchange of cytoplasmic material, it has been found that each exconjugant gives rise to the organisms of its type i.e. sensitive produces sensitives and killer produces killers. This suggests that killer character is not govened by gene.

In the second case, when the conjugation is prolonged and cytoplasmic bridge between the conjugants is larger than usual permitting

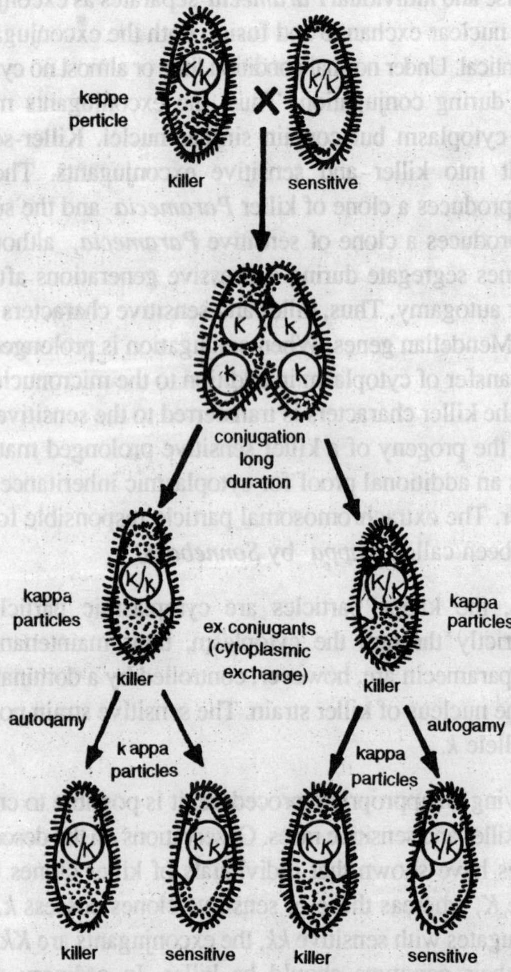

Fig. 14.4. Results of a cross between a killer (kk) and a sensitive (kk) strain of *Paramecium,* when cytoplasmic exchange is allowed.

exchange of cytoplasm, all the offsprings are killer and their cytoplasm contains visible kappa practicles. This means that the kappa practicles from the killer strair are transferred to the sensitive strain through the exchange of cytoplasm so that sensitive individuals have also been converted into the killers. Therefore, a *Kk* individual may be either a killer or sensitive depending upon whether it has received kappa particles or not.

It is possible to convert killer clones into sensitive by making them undergo very rapid fissions. If paramecia of killer strain are overfed they multiply very fast and kappa particles fail to keep pace with the division of the body of the animal, with the result that particles are finally lost of the entirely. The gene *K* present in the genotype is unable to initiate the production of kappa particles, but can simply maintain them if these are already present or are introduced. *Vice versa* gene *k* is unable to maintain the kappa particles. If kappa particles are introduced into a sensitive paramecia having *kk*, these are lost without producing any change.

Milk Factor of Mice

Bittner discovered a case of cytoplasmic inheritance in mice. He had a strain of mice in which almost all developed breast cancer. From these he took some females and bred them to males which did not inherit this tendency. 90 % of the offspring of such crosses formed breast cancer. On the other hand, when he took males from the cancer—susceptable strain and crossed them to normal females, *Bittner* found no cases of breast cancer among the offspring. These results pointed out to cytoplasmic inheritance. Late studies pointed out that it was not the case. When he took the new-born off-spring from a cancer strain female and put them with a foster mother from a normal strain for nursing, there were no breast cancer in the offspring. This pointed out that the milk was responsible for transmission of breast cancer. A virus-like *'milk factor'* was identified and it was discovered that hormones and chromosomal genes play a part by making the mice susceptible to this milk factor.

Cytoplasmic Effect in Habrobracon and Lymantria

In parasitic wasp 'Habrobracon' there are two strains, one is darker than the other. *Kuhn* made reciprocal crosses between these two strains and found that the females were darker when the mother was from darker strain. A similar case has been described in *Lymantria* by Goldschmidt. In reciprocal crosses of a dark and light race, the offsprings in both F_1 and F_2 resemble the female parent more than they do with male.

Cytoplasmic Inheritance in Plants

There are several instances of cytoplasmic inheritance studied so far in plant. However, only few important instances are being described here under the following sub heads:

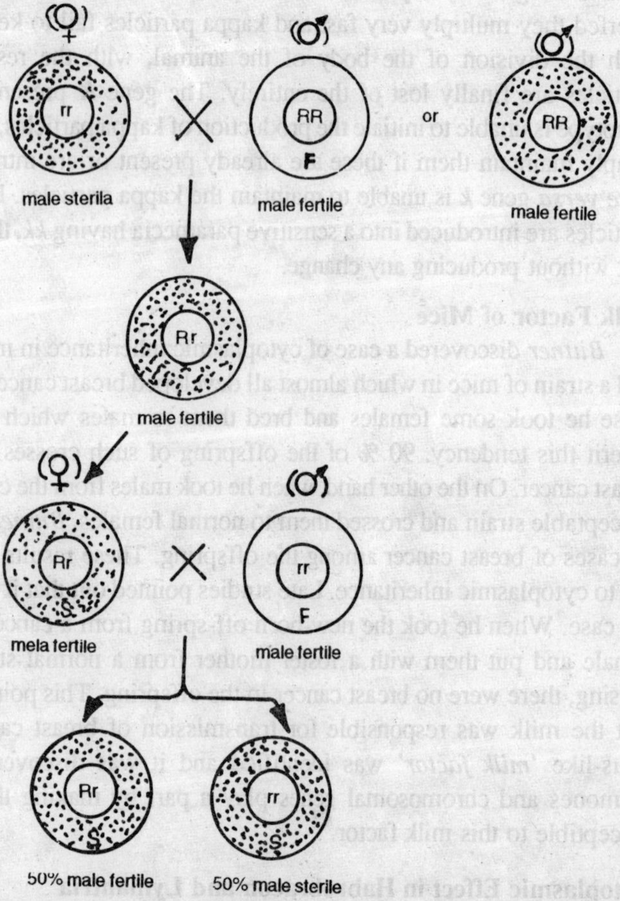

Fig. 14.5. Inheritance of cytoplasmic male sterility and restoration of fertility due to restorer gene.

Males Sterility in Maize

Male Sterility in plants occurs due to pollen failure. It is controlled at least in plant by cytoplasmic factors in the case of maize, wheat, sugarbeet as well as in same other crop plants. The mechanism of transfer of male sterility factor through cytoplasm in maize was discovered and analysed by *Rhoades*. He found out that certain maize plants develop pollen which abort in the anthers, thus the plants are male strile. The female gametic cells (egg) of these plants are fertile and functional.

When the eggs of a male sterile plant are crossed with the pollen from a male fertile strain, the progeny produced are male sterile. The reciprocal crosses produced progeny consisting of all male fertile plants. The reciprocal cross consists of fertilization of egg from a male fertile plant with pollen of a male sterile plant (Some of pollen of male sterile plant are fertile and can be used for pollination). The difference in the character of progeny of the two types of crosses (cross and its reciprocal) clearly indicates the transfer of male sterility character through the maternal cytoplasm. The hybrids are male sterile and male fertile according to the maternal parent irrespective of the nuclear genes. If inheritance of male sterility character is through the nuclear genes then the inheritance pattern should not change in reciprocal crosses. The progeny is similar to the maternal parent as far as this characteristic is concerned, irrespective of which strain (male sterile or male fertile) is used as the female parent.

The male sterile maternal strain, if repeatedly back-crossed to pollen fertile lines until all its chromosomes are exchanged for those of the male fertile line and then crossed with a male fertile male plant still produces hybrids which are male steile, strongly indicating the inheritance of the character through the egg cytoplasm. The inheritance of male sterility in maize is maternal, irrespective of the direction in which the cross is made. The character is inherited through cytoplasmic factors, the plasmagenes. These are transmitted through the female gametes.

Male sterility strains have been helpful in ensuring cross pollination and getting hybrid seeds in many crop plants.

Male Sterility in Wheat

Recently in Japan and in the United States considerable efforts are being made to produce and use male steile lines of wheat in the production of crop showing hybrid vigour. *Kihara* and *Sunesan* are

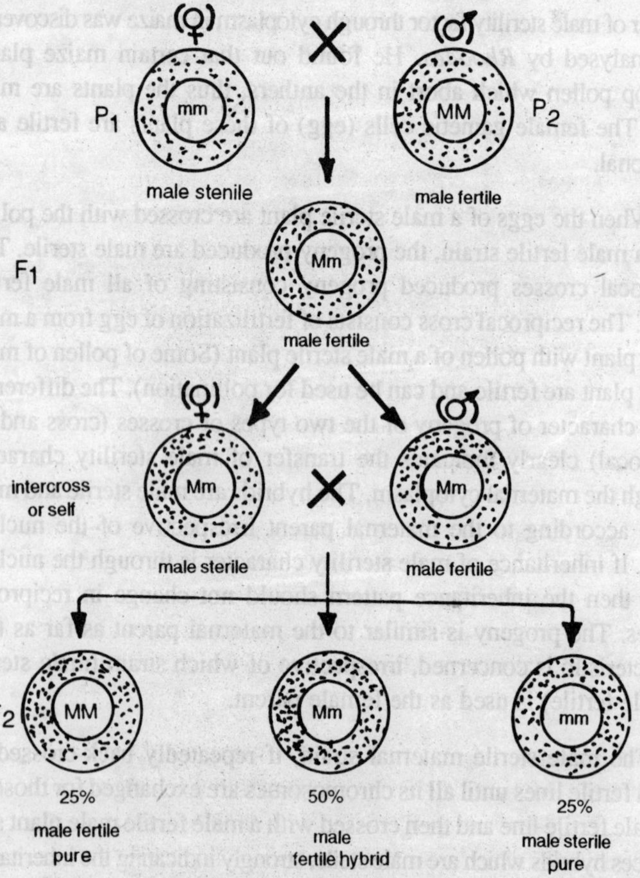

Fig. 14.6. Inheritance of genetic male sterility.

the pioneer in producing male sterile lines in wheat. The techniques of producing male sterile lines used by *Kihara* and his school consist of crossing wheat with another species of the same genus, *Triticum* or a related genus. Quite often in suitable combinations, the F_1 hybrids become male sterile i.e., anthers bearing no pollen. Through repeated backcrossing of this F_1 with the existing variety of wheat, male sterile lines are developed. The next stage of the process consists of finding male restorer lines. By examining a large number of wheat varieties from different parts of the world, male restorer lines are determined through actual crossing. Once some lines are obtained, the task is accomplished because the cross between male sterile and male restorer lines brings about hybrid vigour. The use of male sterile lines increases the yield on one hand and on the other cuts down the expanses necessary to do the emasculation. Cytoplasmic male sterility may induced through interspecific or intergeneric hybridization. By crossing common wheat with *Aegilops triuncialis* cytoplasmic lines have been produced *(Tsunewaki*, 1975).

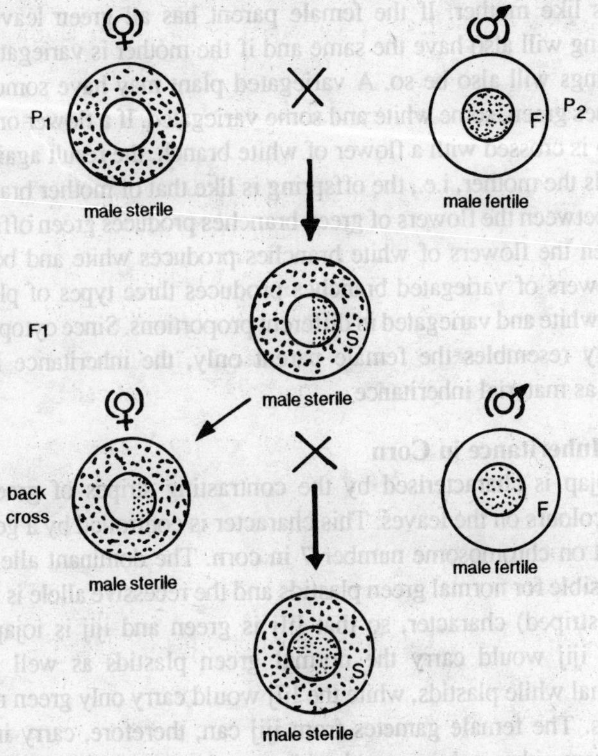

Fig. 14.7. Maternal inheritance of cytoplasmic male sterility.

Variegation in Plants

The cytoplasmic heredity may be explained by an example of plants having variegated leaves. The green colour of the plants is due to presence of chloroplasts, a kind of plastids. The plastids are self-prepetuating bodies and as they are healthy, they maintain the green colour. Sometimes they are deficient in their power to produce chlorophyll; this results into the production of pale green, yellow or white colour. If such deficiency is great and extends to the whole plant, it collapses as there occurs no photosynthesis of which the chlorophyll is a prerequisite.

In *Mirabilis* Jalapa (Four O' Clock Plant)

When a variegated plant of *Mirabilis jalapa* (4 O'clock plant) is crossed with a plant having completely green leaves, the offspring is always like mother. If the female parent has all green leaves, the offspring will also have the same and if the mother is variegated, the offsprings will also be so. A variegated plant may have some of its branches green, some white and some variegated. If a flower on green branch is crossed with a flower of white branch, the result again goes towards the mother, i.e., the offspring is like that of mother branch. A cross between the flowers of green branches produces green offspring; between the flowers of white branches produces white and between the flowers of variegated branches produces three types of plants—green, white and variegated in different proportions. Since cytoplasmic heredity resembles the female parent only, the inheritance is also temed as material inheritance.

Iojap Inheritance in Corn

Iojap is characterised by the contrasting stripes of green and white colours on the leaves. This character is controlled by a gene (ij) located on chromosome number 7 in corn. The dominant allele Ij is responsible for normal green plastids and the recessive allele is for the iojap (striped) character, so that IjIj is green and ijij is iojap. The plants, ijij would carry the normal green plastids as well as the abnormal while plastids, white the IjIj would carry only green normal plastids. The female gametes from ijij can, therefore, carry in their cytoplasm either only green plastids or only abnormal white plastids or both. Consequently, if ijij is used as female parent and IjIj as male parent, the F_1 individuals Ijij) will be green, white or striped. From these F_1 individual Ijij (sriped) is again used as a female parent with IjIj (green) as the male parent, two genotypes, namely IjIj and Ijij would be obtained in equal proportions. Each of these two genotypes would have three phenotypes (green, white and striped) would also be obtained.

It is, therefore, obvious that the iojap character once inherited through the female parent will be able to maintain itself in the cytoplasm irrespective of the genotype of the plant. Therefore, as discussed above, iojap or striped phenotype may have any of the three possible genotypes. IjIj, Ijij and ijij.

In *Primula Sinensis*

In another experiment on a variant of *Primula sinesis* if a yellow leaved female plant is crossed with a green leaved male plant the F_1 offspring will show yellow leaves. But if a green leaved female parent is crossed with a yellow-leaved male parent, the F_1 offspring all show green leaves. It therefore, seems obvious that the leaf colour is determined by the plastids.

In *Oenothera*

Oenothera: The evening primose, provides an interesting variant of chloroplast inheritance. In many species of *Oenothera* chromosomes are transmitted as entire genome so that gametes formed from a plant contain either the chromosome set received from the pollen parent or all those chromosomes derived from the maternal parent. The chromosomes that are inherited in a block are known as *complexes. O. muricata O. hookeri* are two species of *Oenothera* exhibiting the pecularity. Both *O. muricata* and *O. hookeri* are normal green plants. Crosses between *O. muricata* (as pistillate female parent) and *O. hookeri* (as staminate male parent) produce normal green F_1 plants. The reciprocal cross between *O. muricata* male and *O. hookeri* female produce non-green F_1 plants, which do not survice. This is because the muricata plastids are able to develop normally in the presence of muricata-hookeri nucleus, whereas hookeri plastids cannot.

Further, the hybrids of *O. muricata* female and *O. hookeri* male, whose plastids have come from muricata, are found to exhibit variegation in the form of yellow patches. These patches are found to possess hookeri plastids and a hybrid genotype. Further, these hookeri plastids are introduced through the hookeri pollen.

In Funariaeae

Extensive studies make by *Wettstein* on the family Funariaeae also take one towards the same conclusions *Wettstein* observed that in reciprocal corsses between varieties of *Funaria hygrometrica,* the characters segregated in normal Mendelian fashion. But many marked differences appeared when reciprocal crosses were made between the two species. *F hygrometrica* and *F. mediterranea.* In such crosses for most of the traits, the segregating gametophyte offspring tended to resemble the maternal parent, the paternal types often failing to appear. When two different genea, *Funaria* and *Physcomitrium* were crossed, the offspring were all similar to or identical with the female parent and also many of the spores produced were sterile. *Wettstein* interpreted these results as due to the failure of genomes similar to that of the male parent to survive and function in cytoplasm derived from different sources. Further in *Funaria* species the form of the capsule is determined by the cytoplasm of the oosphere.

Streptomycin-resistance in *Chlamydomonas*

The antibiotic streptomycin is inhibitory for wild-type strains of the unicellular green alga *Chlamydomonas*. When the cells are grown at sublethal concentrations of the antibiotic, most of the cells fail to grow, whereas some form colonies. These resistant mutations occur spontaneously but are selected by the antibiotic. About 90 per cent of mutations leading to streptomycin-resistance occur due to hereditary changes in nuclear genes, but about 10 per cent are due to mutations in extra chromosomal factors.

Ruth Sager (1954) made detailed analysis of the inheritance pattern of the streptomycin-resistance in *Chlamydomonas*. Zygotes in this alga are formed as a result of fusion of gametes derived from +(mt$^+$) and — (mt$^-$)mating types. Chromosomal genes determine whether a strain is going to be of + or — mating type. *Sager* found that whenever streptomycin—resistance was determined by extra-chromosomal factors, its inheritance was maternal, i.e., the progeny resembled the + mating type with respect to streptomycin-resistance. All the progeny were streptomycin-sensitive when the + mating type was streptomycin-sensitive and resistant when the + mating type was streptomycin sensitive and resistant when the + mating type was streptomycin-resistant. Segregation of + and — mating types, however, was normal, i.e., + = 50 per cent and — = 50 per cent. Maternal or uniparental inheritance is observed in the progeny of about 99.03 per cent of the zygotes. The rest 0.07 per cent of the zygotes segregate into streptomycin resistant and sensitive individuals. This is because of segregation of cytoplasmic factors during postmeiotic divisions.

Sometimes an extra-chromosomal mutation makes the strain streptomycin-dependent so that such strains grow only in the presence of streptomycin. Like streptomycin-resistance *(sr)* streptomycin-dependence *(sd)* too is maternally inherited. Cells containing both *sr* and *sd* genes are streptomycin-resistant but can grow in the absence of streptomycin, indicating thereby that *sr* is dominant over *sd*. In exceptional zygotes when extrachromosomal genes are transmitted from both the parents to the progeny and show random postmeiotic segregation, there is chance segregation of *sd* and *sr* genes too, as if they are alleles of each other.

Episomes in Bacteria

In *E. coli,* the male cells have the sex factors which are transferred to female bacteria. The sex factor in bacteria is cytoplasmic. Many extra chromosomal rings of DNA are found in bacterial cells. Some of them control antibiotic resistance (RTF or resistance transfer factors) in bacteria. Besides the sex and resistance transfer factors, bacteria may also contain calcinogenic factors. The calcinogenic cells *(Col⁺)* release *Colcin* particles which are capable of killing susceptible strains. *Col⁺* cells themselves are imune to death by colcins. Colcin particles can be transferred from *Col⁺* cells to *Col⁻* cells by conjugation or transduction. Colcin particles have been shown to be supernumeracy DNA particles. All the supernumerary chromosomes of bacteria are known as episomes. They are infectious and multiply independently of the main bacterial genome. Although they are cytoplasmic, they have the capacity to get integrated into the bacterial genome. There is no sharp distinction between the nucleus and cytoplasm in prokaryotes, but episomes can be considered to be extra-chromisomal or extra nuclear because they can exist independently of the main genome.

Cytoplasmic Inheretance in Haploid Organisms

(1) *In Yeast:* Cytoplasmic mode of Inheretance in yeast

Saccharomyces cerevistae by *B. Ephrussi* and others. In the life cycle, diploid and haploid adult alternates, the former reproducting by asexual meiospores called *ascospores,* the later be isogametes Certain strains of yeast are comparatively small-sized and slow growing, producing tiny colonies on agar and so known as littles. The slow growth of littles has been found to be due to the loss of cells aerobic respiratory system *(Ephrussi,* 1953) and the utilization by the cell of less efficient fermentation process. When individuals from tiny colony (littles) are crossed with the individuals from a normal-sized colony, in F_1 little and normal offsprings are in the ratio of 1 : 1. Such tiny strains are evidently due to mutant nuclear genes and are called *segreational petities.* The nuclear genes of normal yeast cell for the formation of normal-sized colony can also be mutated artificially into segregatinal petites or littles by a retain chemical mutagens such as *acridine dye* or *euflavine.* These artificially muated strains are called *vegetative petities.* When vegetative petite cells are crossed with the cells of normal colonies, the petite phenotype does not segregate normally. It shows that nuclear genes segregate in normal Mendelian fashion, but the little phenotype depends upon the kind of cytoplasm and mitochondria present.

In *Neurospara:* In the fungus *Neurospara crassa* the cytoplasmic inheritance is made possible through mitochondria of female. *Mitchell* found a strain of *Neurospara* which was called *poky* on account of slow growth of its filament, exhibiting poorly differentiated mitochondria, due to a mutant gene or genes for such mitochondrial deferences. However, the mitochondrial DNA of poky strain is found to be biochemically similar to mitochondrial DNA of wild species. The poky trait is found to be inherited in a fashion of maternal inheritance. If the protopenthecum derives from a poky hypha and the male conidium from a wild type hypha, the resulting, eight ascospores are all poky, whereas in reciprocal crosses the ascospores are generally all wild type.

Dauermodification : Sometimes the changes induced by the environmental factors are present in the offsprings. These characters may persist for several generations but gradually they diminish and finally disappear. This condition is called dauermodification. These modifications are transmitted maternally through eggs to the offspring. These are not gane controlled characters and do not behave as Mendelian traits. They represent the lingering but transcent effect of some environmental factor on the cytoplasm.

REVIEW QESTIONS

1. Assume that each of the following kinds of paramecium cells undergoes autogamy. Give the genotypes of the two cells which will arise. Indicate whether they are killer or sensitive.

A. KK with kappa.
B. KK without kappa.
C. Kk with kappa.
D. Kk without kappa.

2. In Paramecium, a cell can act as a killer:

A. If it carries kappa particles and the dominant "K" to support them.
B. If it receives through an ordinary conjugation bridge the dominant "K" which can bring about the formation of kappa particles.
C. If it carries kappa in the cytoplasm and is either genotype "KK" or "kk" following autogamy of an F_1.
D. If it arises following the conjugation of any two cells of genotype "KK".
E. As long as both of its parents were killer cells.

3. Which of the following is (are) true about kappa and the killing action?

A. Non-brights are required for the maintenance of brights and the killing action.
B. Nonbrights are eliminated from the cell and cause the killing action.
C. DNA is found in both the brights and the non-brights.
D. The numbers of sensitive cells killed is directly proportional to the number of killer cells.

E. Kappa is essential for the survival of the cell.
F. Kappa cannot mutate independently of the nucleus.

4. In the case of the mate killer trait:

A. A muμ particle is liberated into the surrounding fluid.
B. A mate killer also has protection against the killing action of kappa.
C. A sensitive cell is usually transformed to a mate killer after mating with a mate killer.
D. A cell must carry both dominants, M_1 and M_2, if it is to be a mate killer.
E. A cell lacking muμ particles will manufacture them if the genotype includes either M_1 or M_2.

5. Which of the following is (are) true about the petite story in yeast?

A. Following the cross of a nuclear petite with a normal, the segregation of ascospores from the zygote is 2 normal: 2 petite.
B. Following the cross of a neutral petite with a normal, the diploid cells are petite and give rise to all petites following ascospore formation.
C. Following the cross of a suppressive petite with normal, the diploids will be petite if they are plated on agar soon after mating.
D. Following the cross of normal with suppressive petite, wild-type cells are favoured if subculturing precedes the plating on agar.
E. Petites cannot be induced by any known mutagen.

6. In yeast:

A. Following the cross of an erythromycin-sensitive cell with a resistant one, the sensitive trait disappears, and all cells become resistant.
B. Following the cross of an erythromycin-resistant, Chloramphenicol-sensitive cell (E^RC^S) with an erythromycin-sensitive, chloramphenicol-resistant cell

($E^{S}C^{R}$) asci are produced which contain spores only of the parental types.

C. The cross $E^{R}C^{S}$ x $E^{S}C^{R}$ produces four kinds of asci: $E^{a}C^{S}$ x, $E^{S}C^{R}$, and $E^{R}C^{R}$ $E^{S}C^{S}$.

D. The $E^{R}C^{S}$ x $E^{S}C^{R}$ produces asci, each of which produces four different kinds of spores, the two parentals and the two recombinant types.

E. The determinants for drug resistance may be associated with a physical alteration in mitochondrial DNA.

7. In higher plants:

A. Nuclear genes do not seem to influence the development of the plastid from the proplastid.

B. There is more DNA in a chloroplast than in a mitochondrion.

C. All of the enzymes involved in photosynthesis are apparently formed from transcripts of the plastid DNA.

D. The RNA of the ribosomes of the plastid arises from transcription of the plastid DNA.

E. Replication of the plastid DNA depends on the formation of plastid DNA polymerase on the ribosomes of the plastid.

8. In corn:

A. The iojap mutation is due to an alteration in the DNA of the chloroplast.

B. Altered plastids due to iojap become normal if they are transmitted to a plant carrying the dominant allele "Ij."

C. An individual homozygous for the wild allele (Ij Ij) may have mutant plastids.

D. If a normal, fully green plant (Ij Ij) is used as a female, the offspring will be all green, even though the male parent is striped and is genotype "ij ij".

E. If a striped plant used as female parent (ij ij) is crossed to a fully green one (ij ij) the F_1 may consist of three types of plants: green, striped and non-green.

9. In bacteria:

A. Mutations to drug resistance do not arise on the chromosomes but are restricted to plasmids.
B. The resistance transfer factor (RTF) can bring about conjugation between cells.
C. The RTF cannot be transmitted unless a cell is F^+.
D. Some of the non-chromosomal determinants for drug resistance are not found on the RTF.
E. The RTF must integrate with the chromosome if it is to be transferred.
F. Both the chromosomal and the non-chromosomal determinants for drug resistance alter the structure of the ribosomes and in the this way cause the resistance.

ANSWERS

1. A. KK, killers.
   B. KK, sensitives.
   C. Two KK killers or two kk sensitives.
   D. Two KK sensitives or two kk sensitives.
2. A.
3. A, C and D.
4. None.
5. A, C and D.
6. C and E.
7. B and D.
8. C, D and E.
9. B and D.

15

Linkage

When two or more than two characters are transmitted to the offsprings far for generation without any recombinations, they are called the *linked character* and the phenomena is called *linkage*. This is a deviation from the principle of Mendelian independent assortment. Evidently, random assortment of genes does not occur in such cases. This deviation from the law of independent assortment was first studied by *Bateson* and *Punnett* in 1906 in sweet pea (*Lathyrus odoratus*). Mendel's law of independent assortment is applicable to the genes that are situated in separates chromosomes. When genes for different characters are located in the same chromosome, they are tied to one another and are said to be linked. They are inherited together by the offspring and will not be assorted independently. This phenomena in which there is a tendency of two genes of the same chromosome to remain together in the process of inheritance is called linkage.

In *Drosophila* the brilliant and extensive investigations of *Morgan* and his co-workers have resulted in definitely placing several hundred known genes in four distinct linkage groups, corresponding to the four pairs of chromosomes present in this accommodating fly. It has been shown by reciprocal crosses that the occurrence of linkage is not due to some relation *per se* between the genes, but simply to the fact that linked genes chance to be together in the same chromosome. In other words, if two characters due to two genes in one chromosome enter a cross together from ine parent, they naturally stay together in the offspring, and if they enter from separate parents, they remain separate in the offspring.

History of Linkage

Mendel could not notice the linkage because fortunately the seven pairs of factors or alleles selected by him were found, on different chromosomes. Various hypothetical views have been given under the following headings:

Sutton's Hypothesis

Sutton (1903) predicted that each chromosome bears a number of genes. Since the chromosomes move as units during meiosis to the gametes, all the genes which are situated in same chromosome will be linked together. But he was unable to prove his predictions experimentally.

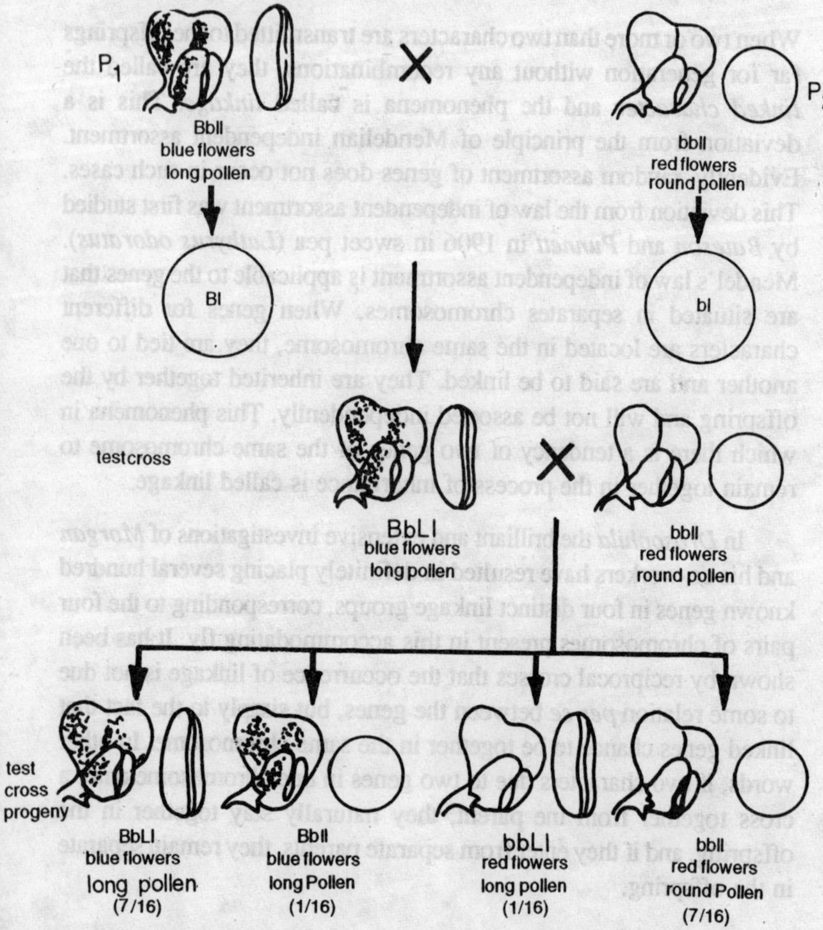

Fig. 15.1. Results obtained by Bateson in Sweet pea when two characters, flowers and long pollens, were present in coupling phase.

Coupling and Repulsias Hypothesis

After the rediscovery of Mendel's laws, *W. Bateson* and *Punnett* in 1906 formulated the hypothesis of coupling and repulsian. They crossed plants having purple and long seeds with the plants having red and round seeds. The first two characters are dominant over round and red. They observed that there was a different in the results of F_2. The following types were obtained in undermentioned ratio:

	Purple Long	*Purple Round*	*Red Long*	*Red Round*	*Total*
Result obtained	4831	390	393	1338	6952
Ratio obtained	11	1	1	3	
Expected Ratio	9	3	3	1	

When these F_1 purple long (heterozygous) hybrids were crossed with the double recessive red and round (homozygous) individuals (test cross), failed to produce expected 1:1:1:1 ratio in F_2 generation. These actually produced following four combination in the ratio of 7:1:1:7.7 purple long :1 purple round : 1 red long : 7 red round.

The 7 : 1 : 1 : 7 test cross ratio of this cross clearly indicated that there was a tendency in the dominant alleles to remain together. Similar was the case with recessive alleles. This deviation was, therefore, explained as *gametic coupling* by *Bateson* and *Punnett*. Similarly, it was observed that when two such dominant or recessive alleles come from different parents, they tend to remain separate. This was called *repulsion.* To understand the phenomenon of repulsion we can consider another test cross performed by *Bateson* and *Punnett*. In another test cross they took a sweet pea plant with blue flowers and round pollens (BBII) and another plant with red flower and long pollens *(bbLL)* and obtained the ratio of 1 : 7 : 7 : 1 in F_2 as has been illustrated in following figure:

P_1:	Blue Round BB ll ↓	X	Red long bb LL ↓
P_1 gametes:	Bl		bL
F_1 :		BbLl Blue Long	
Test cross :	F_1 Blue long Bb Ll	X	P_1 Red Round bb ll

Test cross progeny : 1/16 Blue long (bb Ll) : 7/16 Blue Round (Bb 11) : 7/16 Red Long (bb Ll) : 1/16 Red Round (bb ll) or 1 : 7 : 7: 1.

However *Bateson* and *Punnett* could not explain the exact reasons for conpling and repulsion.,

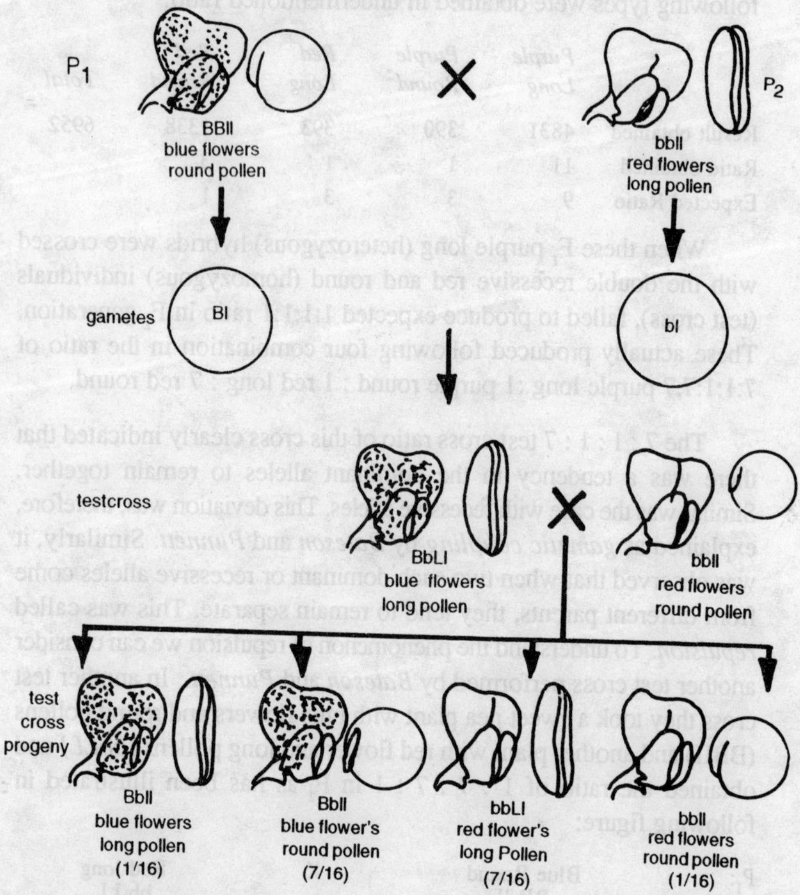

Fig. 15.2. Results obtained by Bateson in Sweet pea when two characters, blue flower and long pollens, were present in repulsion phase.

Morgan's Views on Linkage

While working on *Drosophila, Morgan* stated that the pairs of genes of homozygous parent tried to enter in same gametes and to remain together, whereas same genes from heterozygous parent tend to enter in different gametes and remain apart from each other. According to him both coupling and repulsion are two aspects of linkage.

Differential Multiplication Theory of Linkage

Bateson (1930) assumed that linkage was due to differential multiplication of cell containing different combination of genes which are supposed to be segregated before the formation of gametes. After segregation there is an increase in the production of one set of gametes than the other resulting in deviation from the usual Mendelian dihybrid ratio. This theory has been discarded since the phenomenon of segraggation occurs at meiosis after which only one division occurs.

Chromosome Theory of Linkage

Morgan and *Castle* formulated 'The c' mosome theory of Linkage.' It has following characteristics :

1. Genes that show linkage are situated in the same chromosome.
2. Genes are arranged in a *linear fashion* in the chromosome i.e., linkage of genes is *linear*.
3. The distance between the linked genes is inversely proportional to the strength of linkage. The genes which are closely located show strong linkage, whereas those, which are widely separated, have more chances to get separated by crossing over.
4. Linked genes remain in their original combination during course of inheritance.

The chromosome theory of linkage is widely supported from the cytological studies. It has helped in the construction of linkage maps of chromosomes. (The distance between the genes is determined by the percentage of linkage).

Arrangement of Linked Genes : When two pairs of genes are linked, the linkage may be of either *cis* or *trans* types in an individual heterozygous for both pairs. In the *cis* arrangement two dominant characters (e.g. gray G and long L wings of *Drosophila*) are located on one member of the chromosome pair, with the two recessive characters (such as black *g* and vestigial *l* wings) on the other chromosome. On the other hand, in the trans arrangement one dominant G or L and one recesive *g* or *l* character are located on the same chromosome as shown in figure.

G	L	G	i
lg	i	g	L

cis and *trans* arrangements for two pairs of linked genes in a diploid cells.

Types of Linkage

Linkage has been found to be complete or incomplete depending upon the presence on absence of new combination of the gene.

Complete linkage : Complete linkage is the phenomenon in which two or more characters are inherited together and consistently appear through two or more generations in their original or parental combinations. The complete linkage is due to the fact that there is no break in the gene combinations in chromosomes. In this crossing over does not occur that is why the linked genes do not separate to form new combinations.

Incomplete linkage: Incomplete linkage involves the accidental or occasional breakage of chromosomal segments. It is due to the happening of crossing over and results parental as well as new combinations. It was first of all noticed by *Morgan* in while eyed and miniature winged *Drosophila* flies.

EXAMPLES OF LINKAGE

In Drosophila

Complete cis linkage or coupling : Morgan (1911) crossed an ordinary wild *Drosophila* fly with gray body and long wings with a mutated recessive fly of black body and vestigial wings. All the dihybrids in F_1 generation are with gray bodies and long wings i.e., phenotypically like the wild type parents. If now a male of F_1 generation is back crossed i.e., crossed with a double recessive female (black body and vestigial wings) only parental combination without new combinations are farmed.

Complete trans linkage or repulsian : Morgan has also reported one example of complete linkage at 'trans' position. If black long and gray vestigial flies are crossed in F_1 generation, all flies are or gray-long nature. When such male hybrids are back crossed with double recessive (black vestigial) females, only two types of flies are obtained in place of four (Mendelian independent assortment) as shown in following figure where the genes g and L stay together and the genes *G* and *l* stay together.

Mendelian practice of crossing F_1 hybrids together, (instead by back crossing to recessive) tends to conceal linkage, and perhaps has prevented its earlier recognition. In the reciprocal cross also whatever pair goes in together comes out together and linkage is complete as no independent assortment with new combination appears. Thus, linkage is an exception to the pattern of random segregation of genes.

In Maize

Maize provides a good example for linkage. In this plant recombination between linked genes occurs in both sexes and the characters are easily observed in the seeds. *Hutchinson* crossed a variety of maize having coloured and full seed with a variety having colourless and shrunken seeds. The gene C for colour is dominant over its colourless allele 'C' and the gene 'S' for normal full seeds is dominant over 's' the gene of shrunken endosperm. A cross between CC SS and cc ss, produced seeds with genotype CS/cs (linked genes when enter the zygote, are separated by a vertical line). If C and S assort independently, in F_2 according to Mendel's law, four types of gametes should be formed in equal numbers, and when back crossed with a double recessive plant for the characters concerned, they should produce 4 classes of progeny in equal numbers. But a different result was obtained on back crossing the F_1 generation. The various combinations obtained were as follows:

(i) Coloured full CS/cs—4032 (parental combination) ; (ii) Coloured shruken Cs/cs — 149 (New combination); (iii) Colourless full cS/cs — 152 (New combination); (iv) Colourless shrunken cs/cs — 4035 — (Parental combination). The total number of parental combinants are 4032 + 4035 = 8067 or 96%; while the new or recombinants are 149

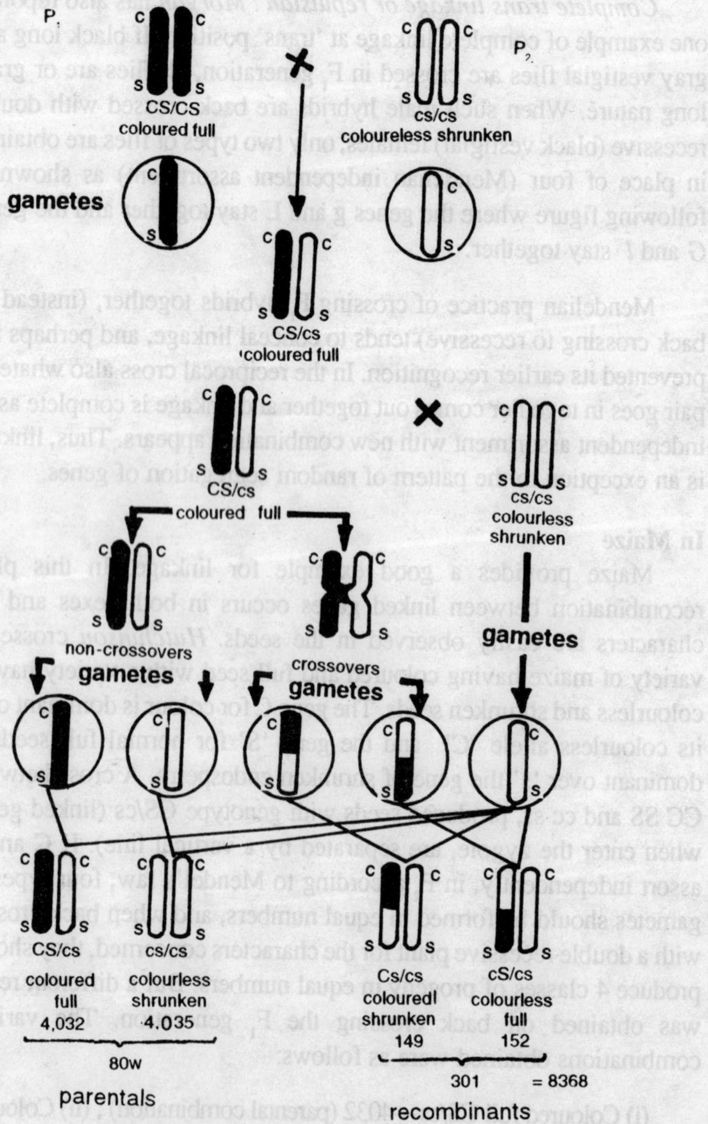

Fig. 15.3. Results obtained in maize by Hutchinson, when characters coloured seeds and full seeds were taken in coupling phase.

+ 152 = 301 or 3.6%. This shows that the percentage or parental combination is higher than the new combination. The genes S and C are linked together on the same chromosomes and do not asort independently and tend to remain together.

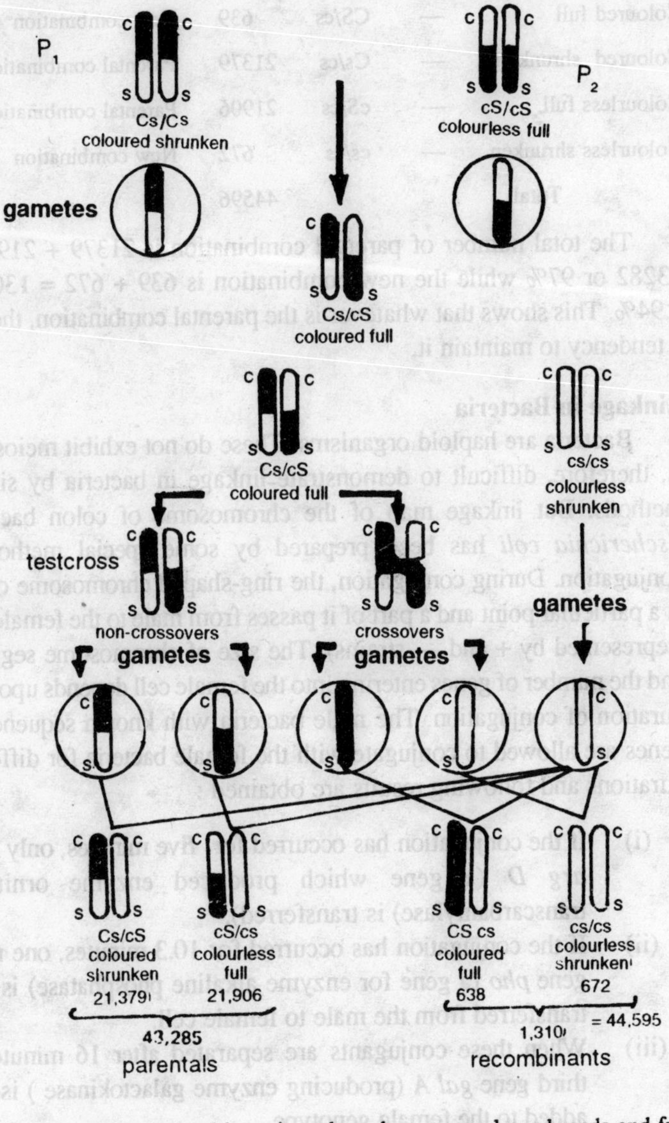

Fig. 15.4. Results obtained in maize when characters coloured seeds and full seeds were taken in repulsion phase.

Another experiment was made with the plants having coloured shrunken and colourless full seeds. Results obtained are as follows:

Coloured full	—	CS/cs	639	New combination
Coloured shrunken	—	Cs/cs	21379	Parental combination
Colourless full	—	cS/cs	21906	Parental combination
Colourless shrunken	—	cs/cs	672	New combination
Total			44596	

The total number of parental combination is 21379 + 21906 = 43282 or 97% while the new combination is 639 + 672 = 1301 or 2.94%. This shows that whatever is the parental combination, there is a tendency to maintain it.

Linkage in Bacteria

Bacteria are haploid organisms. These do not exhibit meiosis. It is, therefore, difficult to demonstrate linkage in bacteria by simple methods. But linkage map of the chromosome of colon bacteria. *Escherichia coli* has been prepared by some special method of conjugation. During conjugation, the ring-shaped chromosome opens at a particular point and a part of it passes from male to the female cell (represented by + and — strains). The size of chromosome segment and the number of genes entering into the female cell depends upon the duration of conjugation. The male bacteria with known sequence of genes are allowed to conjugate with the female bacteria for different durations and following results are obtained :

(i) If the conjugation has occurred for five minutes, only gene *arg D* (a gene which produced enzyme ornithine transcarbamylase) is transferred).

(ii) If the conjugation has occurred for 10.3 minutes, one more gene *pho* (a gene for enzyme alkaline phosphatase) is also transferred from the male to female cell.

(iii) When these conjugants are separated after 16 minutes, a third gene *gal A* (producing enzyme galactokinase) is also added to the female genotype.

(iv) After 24 minutes of conjugation of fourth gene *try C* (gene producing enzyme indole 3-glycerol phosphate synthetase) passes from male to female bacterial cell.

On the basis of these results, and by using time as a unit, the sequence of these genes and the relative distance between them has been determined and a complete chromosome map of bacterial chromosome has been prepared.

Linkage Group

The genes found in one chromosome are usually linked into particular associations. The group of genes showing linkage relationships with one another constitutes a linkage group. The number of linkage groups is proportional to the number of the chromosomes. Thus in *Drosophila* there are four linkage groups corresponding to the four pairs of chromosomes. Each linkage group contains many genes responsible for particular traits. Two linkage groups in this fruitfly are very long; one is too small and fourth is intermediate in size. In human beings there are twenty-three pairs of chromosomes hence there are 23 linkage groups.

Strength of linkage

Strength of linkage depends upon the distance between the genes. If genes are closely placed, the linkage is strong. If they are placed quite apart from one another the linkage is weak. It shows a sort of inversely proportional relationship between the strength of linkage and the distance between the genes.

REVIEW QUESTIONS

1. In the fruit fly, the allele for purple eye colour (pr) is recessive to its allele for red (pr^+). The allele for vestigial wings (vg) is recessive to the wild allele for normal wings (vg^+). The two genes are autosomally linked. Females from a purple stock are crossed to males from a vestigial stock. The F_1 flies are all wild (red eyes and normal wings.) F_1 females are testcrossed with the following results:

purple—210
wild—40

vestigial—215
purple, vestigial—35

A. Diagram the testcross.
B. Calculate the amount of crossing over.

2. Assume that a double crossover takes place in a meiotic cell of the dihybrid. What would be the combination of the alleles resulting from the double crossover between "t" and "u"?
3. In chickens, the dominant allele. I, prevents pigment formation so that the feathers are white. The recessive allele (I^+) permits the feathers to be coloured. This pair of alleles is autosomally linked to a pair which influences the development of the feather. The allele F^+ results in normal feathers, but the allele, F, causes brittle feathers when in the homozygous condition. However, there is a lack of dominance so that the heterozygote has mildly brittle feathers.
A hen with coloured, brittle feathers is mated several times to a rooster with white, normal feathers. Over three dozen eggs are produced which yield chicks with white, mildly brittle feathers. Diagram the cross, giving the genotypes of the parents and the F_1.
4. Using information from Question 4, assume that two dihybrids having the alleles in the cis arrangement are crossed to each other.
   A. Give the kinds of gametes and the frequency of each expected from each parent.
   B. Give the phenotypes and the expected frequencies from a cross of two such hybrids with the cis arrangement.
   C. Assuming 1000 plants are obtained in this cross, how many of them would you expect to have the combination: long fruits and simple flowering shoots?
5. In tomatoes, round fruit (o^+) is dominant to long (o). Simple flowering shoot (s^+) is dominant to branching flowering shoot (s.) Plants from two different pure-breeding varieties are crossed. One variety bears round fruit and has branched flowering shoot. The other variety has long fruits and simple

flowering shoots. F_1 plants were testcrossed, and the following progeny were obtained:

round, simple—23
long, simple—83
round, branched—85
long, branched—19

A. Diagram the testcross.
B. Calculate the amount of crossing over.

6. Give the result to be predicted if an F_1 dihybrid male from the cross described in Question 1 is testcrossed, assuming 500 offspring.

7. F_1 hens from the cross in Question 3 are crossed to roosters having coloured, normal feathers. The following chicks are obtained:

white, mildly brittle—17
coloured, mildly brittle—64
white, normal—61
coloured, normal—15

A. Diagram the cross.
B. Give the genotypes of all the kinds of birds.
C. Calculate the amount of crossing over.

8. In the fruit fly, the allele w^+ for red eye colour is dominant to its allele for white eye, w. This pair of alleles is sex linked, as is another pair which influences body colour: the dominant for gray body (y^+) and the recessive for yellow (y). Females from a pure-breeding, white-eyed, gray-bodied stock are crossed to males from a stock pure breeding for red eye and yellow body. Give the genotypes of the P_1 and the genotypes and phenotypes of F_1.

9. F_1 females from the cross described in Question 8 are crossed to white-eyed, gray-bodied males. From the results presented below, calculate the amount of crossing over between alleles at the white (w) locus and alleles at the yellow (y) locus.

Females : red-eyed, gray-bodied—340
white-eyed, gray-bodied—350

Males : White-eyed, gray-bodied—346
red-eyed, yellow-bodied—324
red-eyed, gray- bodied—9
white-eyed, yellow-bodied—6

10. In the human, two pairs of alleles d^+, d and P^+, P are associated with colour perception. The recessives, d and p, can result in colour blindness. Both loci, d and p, are found on the X chromosome. Linkage is so close that the amount of crossing over is very low.

A. Ignore crossing over, and diagram a cross between a colour-blind woman who is homozygous for the recessive d but who is homozygous for the dominant allele at the p locus and a man who is colour blind due to the recessive he carries at the p locus. Show the expected off-spring.

B. Assume a dihybrid female with the colour perception alleles in the *trans* arrangement marries a man who carries neither recessive. Diagram the cross and show the offspring, ignoring crossing over.

C. In the last case, show the expected results if crossing over occurs.

11. The dominant allele o^+ is required for pigment deposition in the iris of the human eye. Its recessive allele, o, is responsible for ocular albinism. The locus is found on the X chromosome, as is the d locus associated with colour blindness which was mentioned in Question 10.

A. Assuming no crossing-over, what are the possible results of a cross between a woman with ocular albinism who is homozygous for the allele for normal colour vision and a man who has a pigmented iris but who is color blind due to the recessive d.

B. Assuming no crossing over, give the results of a cross

between a woman who is dihybrid at the albinism and colour vision loci (cis arrangement and a man with both normal traits.

12. The amount of crossing over between the linked genes e and s is found to be 11%. The gene r is found to be linked to e and s. Alleles of the r gene engage in crossing over with those of the s gene with a frequency of 7% and with those of the e gene with a frequency of 4% . What is the order of these genes on the chromosome and the distance between them in map units?

13. Given the following portion of a chromosome map, answer the question below:

d 14 m 18 p

A. The amount of double crossing over in this region is found to be 1.5%. What is the coefficient of coincidence?

B. Alleles of the s gene are found to be linked to alleles of the genes shown on the map. A crossover value of 13% is found between the s and the p genes. Where should the s gene be placed on the map?

14. In corn, the genes v, b and I are linked. The data given below summarize the results of a trihybrid testcross. From the data:

A. Give the correct genotype of the trihybrid parent showing the correct sequence of genes.

B. Construct a map segment showing the distances among the genes.

C. Calculate the coefficient of coincidence.

v^+	b^+	I	304
v^+	b	I	119
v	b	I	18
v	b^+	I	70
v^+	b	I^+	64
v^+	b^+	I^+	22
v	b^+	I^+	108
v	b	I^+	295

15. In the Oriental primrose, short style (I^+) is dominant to long style (I). Magenta flower colour (r^+) is dominant to red (r), and green stigma (g^+) is dominant to red (g). A trihybrid testcross is performed. From the data presented below, construct a map giving the positions of the three genes and the distances involved.

I^+	r^+	g	292
I^+	r^+	g^+	153
I	r	g^+	286
I	r	g	139
I^+	r	g^+	36
I^+	r	g	22
I	r^+	g	40
I	r^+	g^+	18

16. In *Drosophila,* assume the three genes, x,y and z, are linked and that the alleles x^+,y^+ and z^+ are dominant to x, y, and z. Crosses of trihybrid females and males showing the three dominant traits, x^+, y^+ and z^+, give the results tabulated below. From the data:

A. Give the genotypes of the P_1 males and females, showing the correct sequence of the genes.

B. Construct a map giving the locations and distances.

C. What can you say about the interference in this map region on the basis of the data?

Females :	$x^+y^+z^+$	1012
Males:	$x^+y^+z^+$	30
	x^+y^+z	32
	x^+yz^+	441
	x^+yz	2
	xy^+z^+	3
	xy^+z	430
	xyz^+	27
	xyz	38

17. The following is a map segment showing the location of loci, r, s and t:

r 12 s 10 t

The coefficient of coincidence in the region is equal to 0.5 the following cross is made in the fruit fly:

$$\frac{r^+s^+t}{rst^+} \times \frac{r^+s^+t^+}{\text{Y chromosome}}$$

Answer the following on the expectation of 4000 offspring:

A. How many will be wild-type females ($r^+s^+t^+$)?

B. How many males will be of the constitution r^+st?

C. How many males will be r^+s^+t and rst^+?

18. There is 21% crossing over between the locus of p and that of c in the rat. Suppose that 150 primary oocytes could be scored for chiasmata within this region of the chromosome. How many of these oocytes would be expected to have a chiasma between these two genes?

19. The longest chromosome in the sweet pea has a minimum uncorrected map length (based on known genetic markers) of 118 units. Cytological observations of the longest chromosome in meiotic cells revealed an average chiasmata frequency of 2.96 per tetrad. Calculate the maximum number of crossover units remaining in this chromosome for the mapping of new genes outside the range already known.

20. The recessive gene *sh* produces shrunken endosperm in corn kerned and its dominant allele sh^+ produces full, plump kernels. The recessive gene c produces colourless endosperm and its dominant allele c^+ produces coloured endosperm. Two homozygous plants are crossed, producing an F_1 all phenotypically plump and coloured. The F_1 plants are testcrossed and produce 149 shrunken, coloured : 4,035 shrunken, colourless : 152 plump, colourless : 4,032 plump, coloured. (a) What were the phenotypes and genotypes of the original parents? (b) How are the genes linked in the F_1? (c) Estimate the map distance between *sh* and e.

21. The presence of one of the Rh antigens on the surface of the red blood cells (Rh-positive) in man is produced by a dominant gene R, Rh-negative cells are produced by the recessive genotype *rr*. Oval shaped erythrocytes (elliptocytosis or ovalocytosis) are caused by a dominant gene E, its recessive allele c producing normal red blood cells. Both of these genes are linked approximately 20 map units a part on one of the autosomes. A man with elliptocytosis, whose mother had normally shaped erythrocytes and a homozygous Rh-positive genotype and whose father was Rh-negative and heterozygous for elliptocytosis, marries a normal Rh-negative woman. (a) What is the probability of their first child being Rh-negative and elliptocytotic? (b) If their first child is Rh-positive, what is the chance that it will also the elliptocytotic?

ANSWERS REVIEW

1. A. $\dfrac{pr \quad vg^{+}}{pr^{+} \quad vg} \times \dfrac{pr \quad vg}{pr \quad vg}$ B. 15%.

2. $t^{+}\ u^{+}$ and $t\ u$, the original combinations from the dihybrid $\dfrac{t^{+}\ u^{+}}{t\ \ u}$.

3. P_1: $\dfrac{I^{+}\ F}{I^{+}\ F} \times \dfrac{I \quad F^{+}}{I^{+} \quad F^{+}}$; F_1: $\dfrac{I^{+}\ \ F}{I \quad F^{+}}$

4. A. $\dfrac{o^{+} \quad s^{+}}{o \quad s}$ $\dfrac{(40\%);}{(40\%)}$ $\dfrac{o^{+} \quad s}{}$ (10%); $\dfrac{o \quad s^{+}}{}$ (10%)

5. A. $\dfrac{o^{+} \quad s^{+}}{o \quad s^{+}} \times \dfrac{o \quad s}{o \quad s}$ B. 20%.

6. Purple flies and vestigial flies in a ratio of 1:1.

7\. A. $\frac{I^+ \; F}{I \; F^+} \times \frac{I^+ \; F^+}{I^+ \; F^+}$

B. $\frac{I \; F^+}{I^+ \; F^+}$ (white, normal);

$\frac{I^+ \; F}{I^+ \; F^+}$ (coloured, mildly brittle)

$\frac{I^+ \; F^+}{I^+ \; F^+}$ (coloured, normal);

$\frac{I \; F}{I^+ \; F^+}$ (white, mildly brittle)

C. 20%.

8\. P_1: $\frac{w \; y^+}{w \; y^+} \times \frac{w^+ \; y}{Y}$:

F_1: $\frac{w \; Y^+}{w^+ \; y}$ (red-eyed, gray-bodied females) and

$\frac{w \; y^+}{Y}$ (white-eyed, gray-bodied males)

9\. 15/685 + approx. 2%. Consider only the male offspring.

10\. A. $\frac{d \; p^+}{d \; p^+} \times \frac{d^+ \; p}{Y}$

offspring: $\frac{d \; p^+}{d^+ \; p}$ (non-colour+ blind daughters) and

$\frac{d \; P^+}{Y}$ (colour-blind sons)

B. $\frac{d^+ \; p}{d \; p^+} \times \frac{d^+ \; p^+}{Y}$;

offspring:

$\frac{d^+\ p}{d^+p^+}$ and $\frac{d\ \ p^+}{d^+\ p^+}$ (non-colour-blind daughter)

$\frac{d^+\ p}{Y}$ and $\frac{d\ \ p^+}{Y}$ (sons colour blind)

C. Daughters all with normal vision:

$\frac{d^+\ p}{d^+\ p^+}$; $\frac{d\ \ p^+}{d^+\ p^+}$;

$\frac{d^+\ p^+}{d^+\ p^+}$; $\frac{d\ \ p}{d^+\ p^+}$

Sons: $\frac{d^+\ p}{Y}$; $\frac{d\,p^+}{Y}$; $\frac{d\ \ p}{Y}$ (colour blind)

$\frac{d^+\ p^+}{Y}$ (normal vision)

11. A. $\frac{o\ \ d^+}{o\ \ d^+}$ X $\frac{o^+\ d}{Y}$

offspring: $\frac{o\ \ d^+}{o^+\ d}$ (daughters normal)

$\frac{o\ \ d^+}{Y}$ (sons with ocular albinism)

B. $\frac{o^+\ d^+}{o\ \ d}$ X $\frac{o^+\ d^+}{Y}$

offspring $\frac{o^+\ d^+}{o^+\ d^+}$ and

$\frac{o^+\ d^+}{o\ \ d}$ (daughters all normal)

$\frac{o^+ \ \ d^+}{Y}$ and $\frac{o \ \ \ d}{Y}$ (sons normal and sons with both ocular albinism and colour blindness)

12. $\underline{e \ \ 4 \ \ r \ \ 7 \ \ s}$ or $\underline{s \ \ 7 \ \ r \ \ 4 \ \ e}$

(both maps are the same)

13. A. 0.18 x 0.14 = 0.025 (expected);

$$\text{coincidence} = \frac{0.015}{0.025} = 0.6$$

B. The locus cannot be placed with certainty on the basis of the information given. It could be located 13 units to the right of "p" or 13 units to the left. Information is needed on crosses involving genes at the "s" locus and those at "m" or "d".

14. A. $\frac{v^+ \ \ 1 \ \ b^+}{v \ \ 1^+ \ \ b}$ B. $\underline{v \ \ 17.4 \ \ 1 \ \ 26.7 \ \ b}$

C. .174 x .267 = 5% (approx.)

$$\text{coincidence} = \frac{40 \text{(obtained)}}{50 \text{ (expected)}} = 0.8$$

15. $\underline{1 \ \ 11.8 \ \ r \ \ 33.7 \ \ g}$

16. A. P_1 females: $\frac{x^+ \ \ z^+ \ \ y}{x \ \ z \ \ y^+}$ males: $\frac{x^+ \ \ z^+ \ \ y^+}{Y}$

B. $\underline{x \ \ 6.4 \ \ z \ \ 7.3 \ \ y}$

C. Little, if any, interference, since five doubles were obtained, and this is about what is to be expected, assuming no interference.

17. A. Would expect one-half of the flies to be females. Therefore, approximately 2000 would be wild females, $r^+ \ s^+ \ t^+$.

B. Would expect 2000 males. Approximately six should be r^+ s

t. This is a double crossover type. The total number of doubles expected on the basis of the map is 2000 x 1.2% or 24. However, the coincidence is 0.5 and so a total of 12 doubles is expected. Of these, 6 would be r^+ st and 6 would be r s^+ t^+.

C. These are parental types. They are determined by subtracting the sum of the doubles (12) and the singles in region I (228) and the singles in region II (188). The singles in region I, for example, are determined in this way: 12% x 2000 = 240 expected from the map information. However, the doubles (12) must be subtracted, since they were included in the construction of the map. This gives 228. The same reasoning gives 188 for region II. The parental types among the male flies, r^+s^+t and r s t^+, would total about 1572 (200 - 428, the sum of the doubles and the singles in I and II).

18. 63.

19. 30 units.

20. (a) sh c/sh c (shrunken, colourless) x sh^+ c^+/sh^+ c^+ (plump, coloured)

(b) sh^+ c^+/sh c (coupling phase) (c) 3.6 map units

21. (a) 2/5 (b) 1/5.

16

Crossing Over

As stated already that the linked genes are inherited together alongwith the chromosome on which they are placed. But this finding of *Morgan* does not always hold true and the linkage is not complete. Sometimes it has been found that linked genes separate out and new combinations are formed. This is due to the interchange of parts between two homologous chromosomes for which a term "crossing over" is used. Thus crossing over may be defined as an interchange of corresponding chromosomal parts between chromatids of a homologous pair of chromosomes resulting in a recombinations of genes. One or two, or more fragments may be interchanged during crossing over. The crossing over takes place during meiosis or maturation division during gametogenesis.

EXAMPLES OF CROSSING OVER

Drosophila

When a wild type fruit fly with gray body and long wings (BBVV) is crossed with double recessive black vestigial fly (Bb Vv) all the individuals of F_1 generation are gray long (Bb Vv) as seen during experiments of linkage. In test cross four kinds of offsprings are grey long, black vestigial, grey vestigial and black long. According to Mendel's law of independent assortiment, these four F_1 types should result in equal numbers or in the ratio of 1:1:1:1. However, in actual experiments, result of extensive and repeated confirmatory crosses of this kind, Morgan obtained 41.5% each of gray long a black vestigial and 8.5% each of the new combinations back long and gray vestigial as shown in the figure.

By arranging data in this way, we see that 87.6% of the back cross progeny are like one parent or the other in both traits (parental types); whereas only 12.4% are new combinations or non-parental types. If there is no linkage between genes B and V, we would expect about 50% each of parental types. In this case, the F_1 hybrid would have produced four kinds of gamates in equal numbers. However, in this case BV and bv gametes are produced with a total frequency of 0.876 and Bv and bV gametes with a frequency of only 0.124. This change in frequency has ultimately modified the common Mendelian raio.

In Sweet Pea

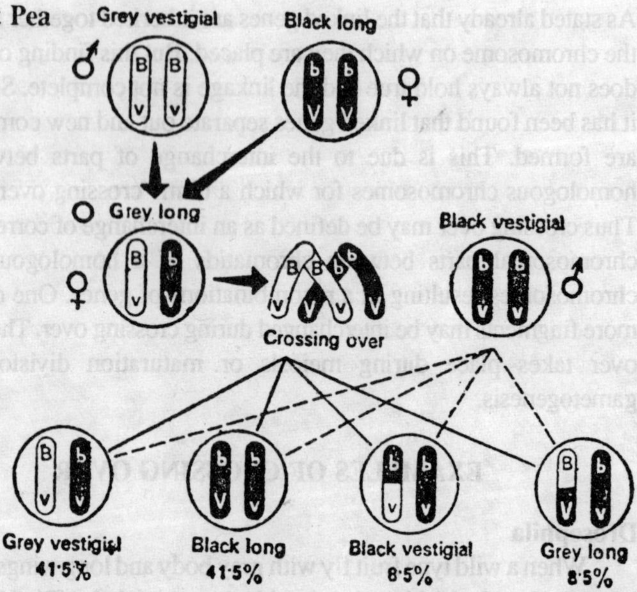

Fig. 16.1. Results in a cross in *Drosophila* showing crossing over.

In the sweet pea, *Lathyrus odoratus,* two pairs of genes purple flowers (RR) and long pollen (LL) are dominant. On the other hand, red flowers (rr) and round pollen grains (ll) are the recessive genes. If a cross of homozygous purple long and red round is made, and F_1 of all purple long (RrLl) is obtained. If these F_1 hybrids are back crossed the results obtained are shown in fig.

However, in actual experiment following results were obtained:

Phenotype	*Genotype*	*Number*	*Frequency*	*Percentage*	*Type*
Purple long	Rs/rl	192	.4496	44.96%	Parental
Red round	rl/rl	182	.4262	42.62%	type 87.6%
Purple round	Rl/rl	22	0.0538	05.38%	New
Red long	rL/rl	30	0.0702	07.02%	Combination 12.4%
Total		427	1.0000	100	

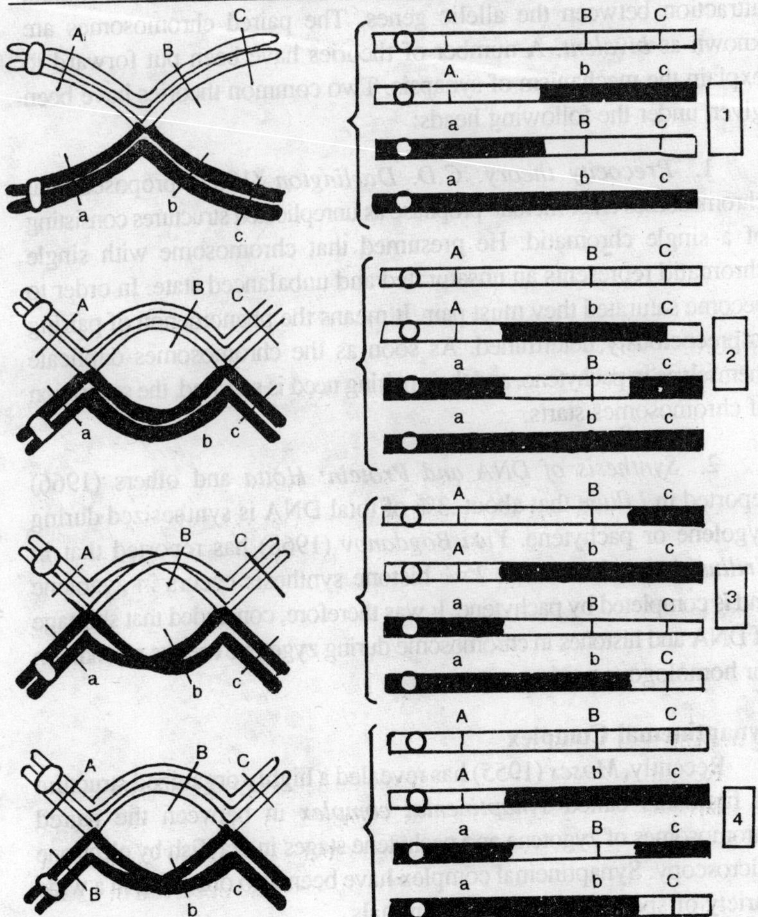

Fig. 16.2. Diagram showing the results of single and double crossing over.

Mechanism of Crossing Over

The mechanism of crossing over can be explained easily under following heads:

(I) Synapsis

During the zygotene stage of the first prophase of meiosis, the homologous maternal and paternal chromosomes pair off and usually come to lie closely side by side. This phenomenon is called *synapsis*. Synapsis provide the mechanical basis of heredity and variations. The pairing of homologous chromosomes is brought about by the mutual attraction between the allelic genes. The paired chromosomes are known as *bivalent*. A number of theories have been put forward to explain the mechanism of synapsis. Two common theories have been given under the following heads:

1. *Precocity theory: C.D. Darlington* (1937) proposed that chromosomes enter meiotic prophase as unreplicated structures consisting of a single chromatid. He presumed that chromosome with single chromatid represents an unsaturated and unbalanced state. In order to become saturated they must pair. It means the phenomenon of pairing is precociously determined. As soon as the chromosomes duplicate themselves in pachytene, and their pairing need is satisfied, the separation of chromosomes starts.

2. *Synthesis of DNA and Protein: Hotta* and others (1966) reported in *Lilium* that about .3% of total DNA is synthesized during zygotene or pachytene. *Yuri Bogdanov* (1968) has reported that in *Crillus domesticus* about 25% histone synthesis occurs in zygotene and is completed by pachytene. It was therefore, concluded that shortage of DNA and histones in chromosome during zygotene creates a condition for homologous pairing.

Synaptinemal Complex

Recently, *Moses* (1955) has revealed a highly organized structure of filaments called *synaptinemal complex* in between the paired chromosomes of zygotene and pachytene stages in crayfish by electrone microscopy. Synaptinemal complex have been also observed in a wide variety of species of plants and animals.

In electron-micrographs the synaptinemal complex appears as three parallel dense lines that lie equally spaced in a plane and are flanked by chromatin. The elements of two lateral lines usually appear densest, while the element of central line is of variable promience. Some fine transverse strands also cross between lateral elements, connecting them with the central element. Though, the morphology of lateral and central elements may vary from species to species, but the basic structure and the spacing of the synaptinemal complex is constant within the species.

Chemically, the synaptinemal complex contains DNA and a specific material known as the synaptinemal complex material of proteinaceous nature.

Functions of Synaptinemal Complex

1. *Moses* considers that it is pre-requisite to chiasma formation and crossing over but, alone it is not sufficient. It is neither responsible for nor necessarily a consequence of pairing, though chromosome pairing precedes its formation.

2. It may serve crossing over by faciliating effective synapsis (i) by maintaining pairing in fixed state, (ii) by providing a structural frame work within which molecular recombination may occur, and (iii) by segregating recombination DNA for bulk of the chromosomal DNA.

3. *Robert King* (1970) suggested that the synaptinemal complex may orient the non-sister chromatids of homologues in a manner that facilitates enzymatically induced exchanges between their DNA molecules.

4. *Comings* and *Okada* (1971) reported that synapsis occurs at two levels, one at the chromosomal level and the other at the molecular level. The synaptical complex pulls homologous chromosomes into approximate association with each other but plays no role in molecular pairing of DNA strands.

(II) Duplication of Chromosomes

Synapsis is followed by the duplication of chromosomes which changes the bivalent nature of chromosome pair into *tetravalent*. During this each of the homologous chromosomes in a bivalent splits longitudinally into two sister chromatids attached to the unsplitted centromeres.

(III) Crossing Over

After the homologous chromosomes have duplicated to their respective sister chromatids, the crossing over takes place. During the crossing over, the non-sister chromatids of homologous pair twist over each other due to the action of an enzyme called *endonuclease*. The chromatid threads are connected with each other at point known as *chiasma*. A segment of one side fuses with segment of opposite side and this is accompanies by the action of another enzyme called *ligase*.

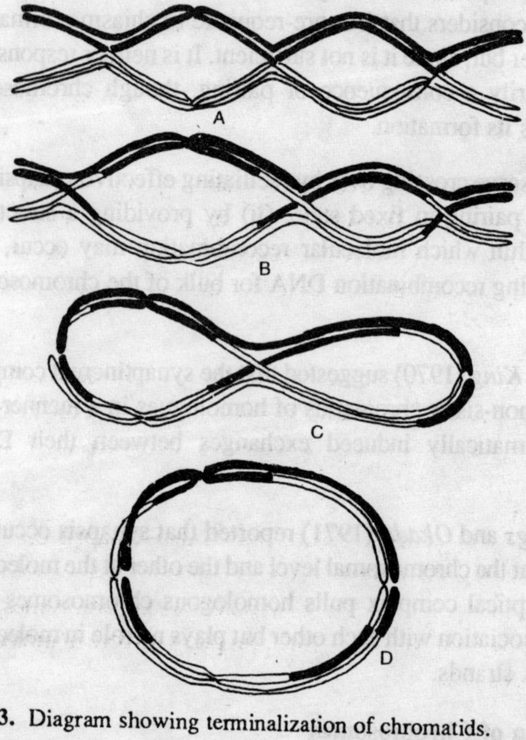

Fig. 16.3. Diagram showing terminalization of chromatids.

The crossing over can take place at several places in one strand and many chiasmata can be formed. The number of chiasmata formed is proportional to the length of chromosome thread, greater the length greater is the number but in one species this number is constant. The genes at distant loci may undergo crossing over but closely placed genes have less chances of crossing over due to this position.

(IV) Terminalization

After the crossing over is completed, the non-sister chromatids repel each other due to lack of attraction between them. The repulsion or separation of chromatids starts from the centromere towards the end just like a zipper and this separation process is named as *terminalization* and thus the twisting chromatids separate so that the homologous chromosomes are separated completely.

THEORIES OF CROSSING OVER

Various hypothesis have been postulated to explain the mechanism of crossing over.

1. Classical Hypothesis

It was put forth by *Sax* in 1932 and is based on the assumption that chiasma does not represent a crossing over, rather crossing over results when a chiasma is rupured and the fusion again of broken ends of chromosomes bring about recombination. According to this hypothesis, there is change of pairing partner of a chiasma.

At one region of the chromosome two paternal chromatids would be associated and like arise two maternal ones, but on the other side of a chiasma each pair of chromatids would consist of one paternal and one maternal thread. This classical hypothesis was preferred by most cytologists because to it did not require breadage and rejoining of chromosomes, with consequent threat to their permanence and individuality.

Matusuura (1940, 1950) and *Haga* (1944) modified the classical hypothesis and proposed a *Neo-classical hypothesis* which postulats that the cytologically observed chiasma bears no relationship to genetic crossing over. Chiasmata are considered to be due to the change opening out of loops between paired chromatids.

2. Janssen's Hypothesis

It is also called *partial chiasma type hypothesis*. Thus hypothesis was put forth in 1909 by *Janssen*. According to this hypothesis of the four chromatids only two cross each other and two do not. Breakage occurs only in two non-sister chromatids. These again rejoin forming the chiasmata so that it is the result of crossing over. *Janssen* believed that prior to the formation of the chiasmata, the paternal and maternal chromosomes in the early diplotene stage were loosely coiled round each other. In order to derive chiasmata from such a loose coil, *Janssen* postulated that the paternal and maternal chromatids made contact at intervals, and that one of each penetrated the other until they broke, whereupon they rejoined in new ways paternal to maternal, and *vice versa*. The other two chromatids (one paternal and one maternal) were considered to remain intact, so giving rise to the typical chiasma with two chromatids crossing each other and two not. *Janssen* also thought that occasionally all the four chromatids may be involved during crossing over.

3. Torsion Hypothesis

Thus theory was proposed by *Darlington* in 1935. According to this hypothesis the two homologous chromosomes are attracted to each other as a pair to satisfy a two by two association. During the formation of chromatids the paired equilibrium and longitudinal cohesion is, however, disturbed because the two by two association will then be between sister chromatids and not the whole homologous chromosomes. The attraction between sister chromatids and not the whole homologous chromosomes. The attraction between the homologues is replaced by a repulsion which forces them apart but they are relationally coiled about each other, a torsion is developed which eventually breaks two non-sister chromatids. The relational coiling is undone and the torsion is relieved by the movement of broken chromatids, and when the broken ends rejoin. If the rejoining is between non-sister strands a cross over occurs. The hypothesis of crossing over is a part of Darlington's *Electrostatic Theory of Chromosome Behaviour* (1937). This theory postulates the existence of electrostatic forces of attraction and repulsion. Such forces have, however, not been demonstrated experimentally.

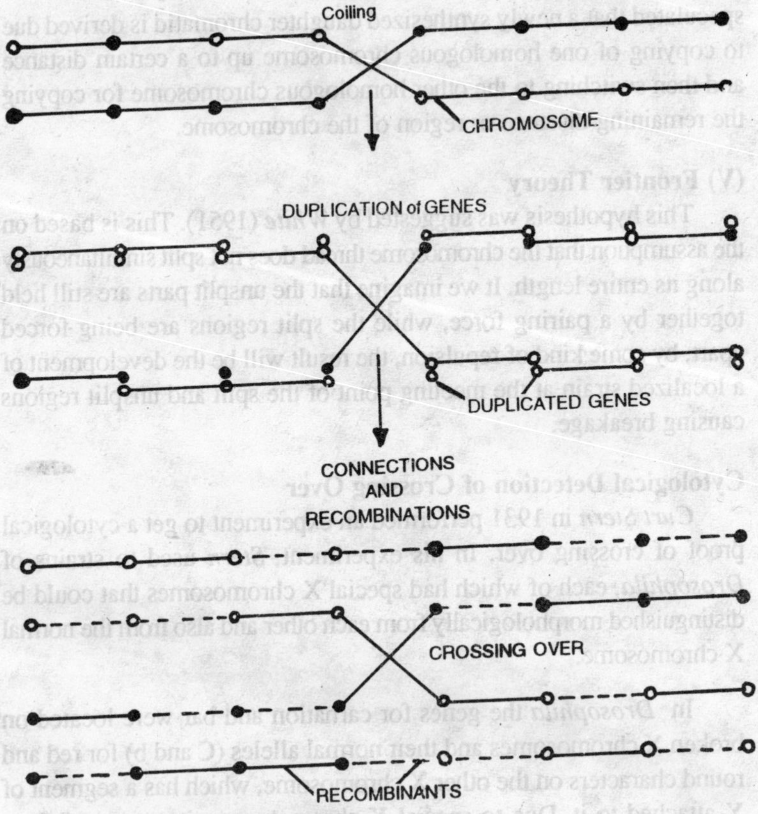

Fig. 16.4. Belling's Hypothesis.

4. Belling's Hypothesis and Copy Choice

It was earlier proposed by *J. Belling*. According to this hypothesis crossing over is really brought about due to novel attachments formed between newly synthesized genes. According to him chromosome duplication consists of two stages—(a) formation of new genes, and (b) the formation of new connections between these newly synthesized genes. This results in the formation of recombinants or cross overs.

To explain recombination in micro-organisms, *J. Lederber* (1955) proposed a modified version of Belling's hypothesis. This is called *copy choice mechanism* of recombination or crossing over, where it is speculated that a newly synthesized daughter chromatid is derived due to copying of one homologous chromosome up to a certain distance and then switching to the other homologous chromosome for copying the remaining distance or region of the chromosome.

(V) Frontier Theory

This hypothesis was suggested by *White* (1951). This is based on the assumption that the chromosome thread does not split simultaneously along its entire length. It we imagine that the unsplit parts are still held together by a pairing force, while the split regions are being forced apart, by some kind of repulsion, the result will be the development of a localized strain at the meeting point of the split and unsplit regions causing breakage.

Cytological Detection of Crossing Over

Curt Stern in 1931 performed an experiment to get a cytological proof of crossing over. In his experiment, *Stern* used to strains of *Drosophila,* each of which had special X chromosomes that could be distinguished morphologically from each other and also from the normal X chromosome.

In *Drosophila* the genes for carnation and bar were located on broken X chromosomes and their normal alleles (C and b) for red and round characters on the other X chromosome, which has a segment of Y attached to it. Due to special X chromosomes, it was possible to distinguish the crossovers from non-crossovers under the microscope. When a cross was made between a female with cB and a male with Cb. The offsprings of such a cross were found to belong to following four groups:

(a) Carnation and bar
(b) Red and round
(c) Red and bar
(d) Carnation and round.

In this the heterozygous female produced four types of eggs due to the occurrence of crossing over. The crossing over results in the broken X getting the fragment of the Y chromosome and the unbroken X not getting it. Before crossing over, the condition was opposite to it.

Thus, the crossover chromosomes can be distinguished microscopically from the non-crossovers.

Kind of Crossing over

Crossing over may be single, double or multiple depending upon the number of chiasmata formed during the process:

(i) *Single crossing over*: If only one chiasma is formed all along the length of a chromosome pair during crossing over, it is referred to as single crossing over. Grametes formed by this type of crossing over are called single cross over gametes.

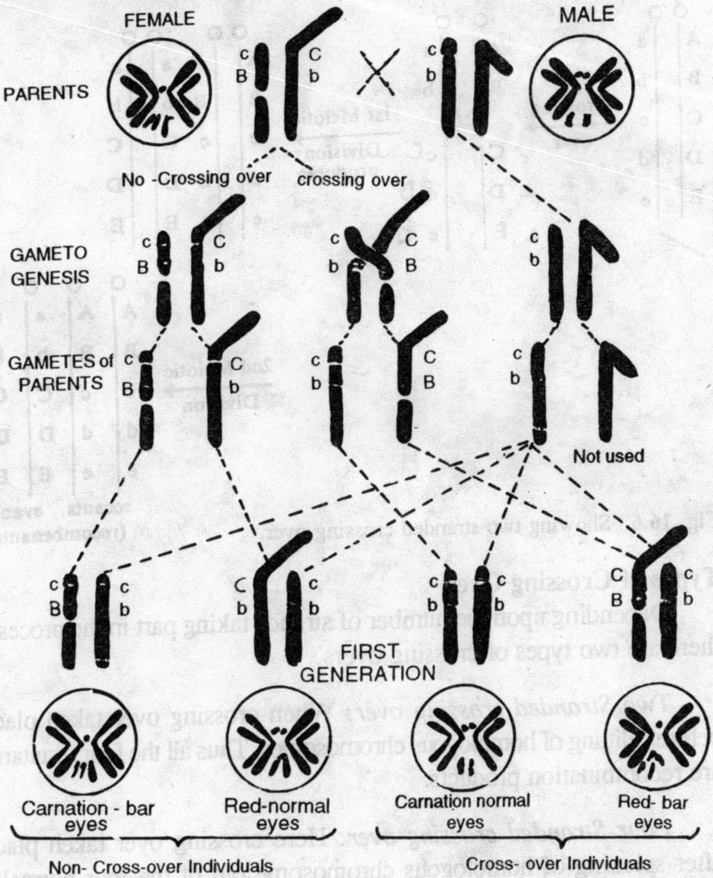

Fig. 16.5. Stern's experiment.

(ii) *Double crossing over*: In double crossing over chromatids break and rejoin at two points. It means two chiasmata are formed along the entire length of the chromosome. The gamates by this crossing over are called double crossover gametes.

(iii) *Multiple crossing over*: When crossing over occurs at more than two places in the same chromosome pair and more than two chiasmata are formed, this type of crossing over is known as *multiple crossing over*. It is very rare because of interference in crossing over.

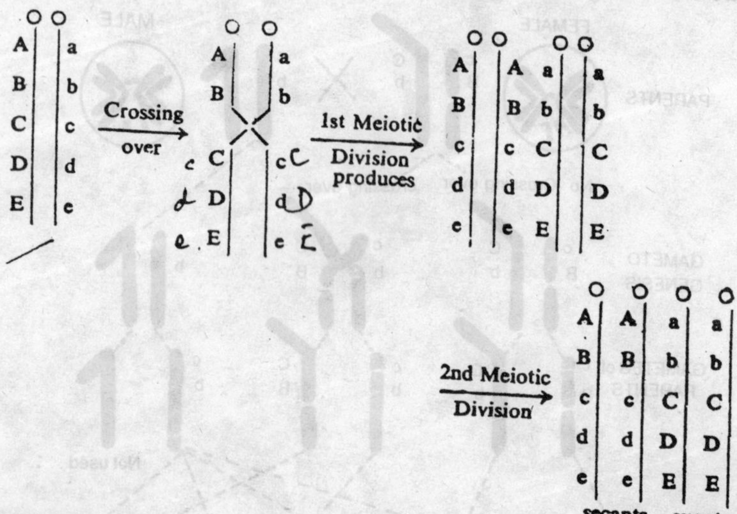

Fig. 16.6. Showing two stranded crossing over.

Types of Crossing Over

Depending upon the number of strands taking part in the process, there are two types of crossing overs.

Two Stranded crossing over: When crossing over taken place before splitting of homologous chromosomes. Thus all the four resultants are recombination products.

Four Stranded crossing over: Here crossing over taken place after spletting of homologous chromosome out of the four normally two non-sister chromatids exchange their segments. Thus, the resultants are two recombination products and two of parential combination.

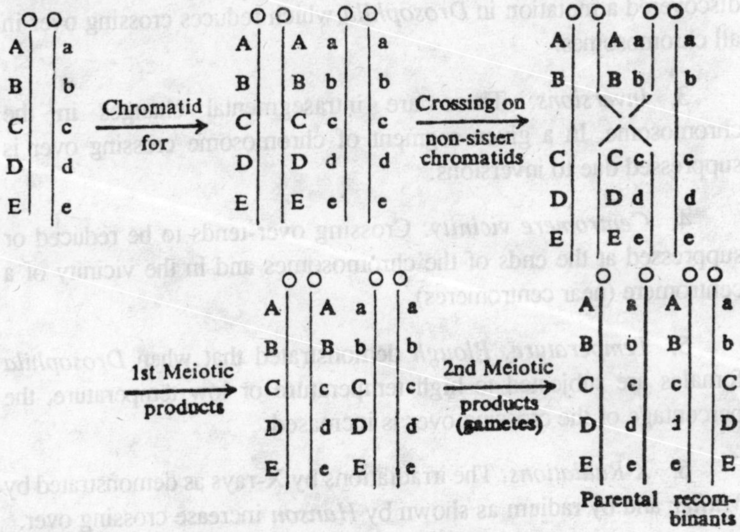

Fig. 16.7. Showing four stranded crossing over.

Frequency of Crossing Over

The frequency of crossing over depends upon the distances of genes (between which crossing over occurs). The frequency of crossing over between any two linked genes increases with the distance between them. Hence, the recombinations are more frequent with the increase in distance between the genes. Such data are used in constructing genetic maps of the chromosomes and the percentage of recombinations between two genes indicates their relative position with respect to other genes in the chromosomes.

Factors Influencing Crossing Over

Following factors show their significant influence over crossing over:

1. *Sex:* In *Drosophila,* crossing over is completely masked or suppressed in male and there is tendency of reduction of crossing over in male of mammals.

2. *Mutations:* These are the gene changes causing visible effects in the organisms. They might reduce the crossing over. *Gowen* has discovered a mutation in *Drosophila* which reduces crossing over in all chromosomes.

3. *Inversions:* These are intrasegmental changes in the chromosome. In a given segment of chromosome crossing over is suppressed due to inversions.

4. *Centromere vicinity:* Crossing over tends to be reduced or suppressed at the ends of the chromosomes and in the vicinity of a centromere (near centromeres).

5. *Temperature: Plough* demonstrated that when *Drosophila* females are subjected to high temperature or low temperature, the percentage of the crossing over is increased.

6. *X-Radiations:* The irradiations by X-rays as demonstrated by *Muller* and by radium as shown by *Hanson* increase crossing over.

7. *Age:* The age of the individual may also affect the frequency of crossing over. *Bridges* demonstrated in *Drosophila* that as the female becomes older the crossing over tends to increase.

8. *Nutritional effect:* Crossing over frequencies are affected by concentration of metallic ions, such as calcium and magnesium. *Levine* fed young *Drosophila* on a high calcium diet and found a decrease in crossing over between X chromosomes. When fed in a medium that removed metallic ions, crossing over was increased. *Neel* found that larval starvations at certain ages increases crossing over.

Significance of Crossing Over

Crossing over is an universal phenomenon, observed in all groups of organisms from viruses to man. Its study is of great significance because of the following reasons:

1. Crossing over affords a proof for the linear arrangement of genes.
2. Crossing over has led to the construction of *linkage maps* or *genetic maps* of chromosomes, which are condensed graphic

representations of the relative distances, expressed in terms of *percentage of recombination* among the genes in a linkage group. Linkage maps of *Drosophila* and Maize chromosomes have been constructed in details.

3. As a result of crossing over, new gene combinations are produced which change the frequency of genes in genetic pool of a population and therefore, play important role in the process of evolution (recombination of genes).

Chromosomal Mapping

Chromosomal mapping is a diagramatic, graphic representation of relative distances between linked genes of a chromosome. It sufficient data about the number of genes, the number of chromosomes present in an individual and their linkage groups is available, it is possible to prepare maps of different chromosomes showing the probable loci of various genes on them. The first chromosome maps were prepared in 1911 by Sturtenant and later by *Bridges* and others on *Drosophila.* Later in 1920 *T.H. Morgan* and his associates prepared extensive maps on *Drosophila* and subsequently chromosome maps were worked out with respect to maize, chick, tomato etc.

The chromosome map can be constructed either on the bars of percentage of linkage groups or frequencies of crossing over. It is known fact that the frequency of crossing over between any two genes depends upon the distance between them. A higher percentage of crossing over occurs between genes that are more apart than those which are close. In other words, the greater the distance between two genes, the higher is the percentage of cross over. Thus, the percentage of crossing over is a measure of the relative distance of genes from one another.

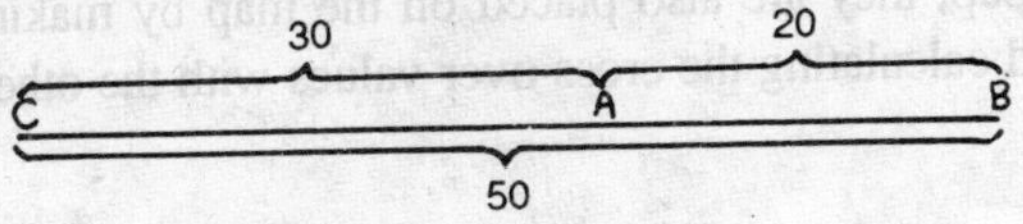

Fig. 16.8. Genes A,B and C on chromosome.

Determination of map distance: The relative distance of genes can be calculated on the basis of percentage of crossing over, because the frequency of crossing over is directly proportional to distance between the genes. If the percentage of crossing over between two linked genes is 10 per cent it means that the distance between two linked genes is 10 units of map distance. This is the *map unit, Morgan* or *Centinorgan.* The percentage of crossing over between two linked genes is calculated by test crosses on which a F_1 dihybrid is crossed with a double recessive parent. Such crosses because involved crossing over at two points, so called *two point test crosses.* Similarly in *three point test cross* a trihybrid test cross (involving three genes) gives us information regarding relative distances between these genes, and also shows us the linear order in which these genes should be present on chromosome.

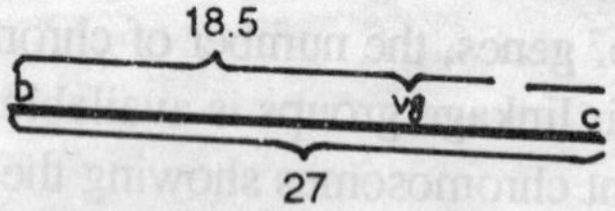

Fig. 16.9. Chromosome map of 11 chromosome in Drosophila for genes, black, curved and vestigial.

Determination of Gene Order: It becomes easy to place genes in the proper linear order after determining the relative distances between the genes. Supposing there are five genes on a chromosome and designated as A,B,C,D and E and the cross over data shows that genes A and E are fathest apart, due to higher percentage of crossing over between them. A can now be taken as over start up point and is placed to 0.0 on the map. Now if 'A' and 'B' show 5% crossing over between them, we can place B at a distance of 5 units from 'A' on the map. If the gene C show 3.5% crossing over with B, and 8.5% with A, we can place it at a distance of 8.5 units from A. In this way all the genes can be placed on the map. As new genes are found belonging to the same linkage group, they are also placed on the map by making suitable crosses and calculating the cross over values with the other genes.

17

Human Blood Groups

The main camponent, of the human blood are (i) cells, which are of three main types, RBC, WBC and platelets., (ii) the plasma, the liquid portion of the blood. *Karl Landsteiner* (1901) collected blood form 22, persons and separated the cells from the plasma. He mixed plasma from one person with the cells from another. He discovered that sometimes a smooth combinations is formed but on other occasions red blood cells clumped together and agglutinated in groups. Thus, these reactions are between the RBC of an individual and plasma of the blood of another.

Four types of the blood groups are discovered terminology designated was suggested by *Moss* and *Janakey* but later on an International system was adopted using letters O, A, B, AB. The types of blood is classified in the presence or absence of *antigens* or *agglutinogens* and *antibodies* or *agglutinins*.

Antigen are proteinaceous substances at least in part. They are capable of stimulate the production of specific antibodies in an animal body is response to foreign antigen. Animal body does not produce antibodies against its own antigens. Clumping takes place if the red blood cells and the plasma belong to the different blood groups having different antigens. Antibodies produced are specific for the antigen which stimulate their production. A reaction between the antigen and its antibody is the basis of the clumping of blood which was observed by *Landsteiner*. Antibodies which are produced in response to an injection of a foreign antigens are termed as *acquired antibodies*. Their production is stimulated only when an antigens makes entry into the body. In certain cases, antibodies are produced in the blood even in the absence of their antigens. These are termed as *natural antibodies* and are present in the human blood. The type of antibody present depends upon the type of blood of an individual.

Identification for Blood Groups

How are these four blood groups recognised? It is done by a test in which it is seen whether the blood of one individual gets clumped or not when mixed with antigen A and B separately. For instinct, if the blood from group A is mixed with the of B, the blood coagulated or is agglutinated.

The agglutination takes place because of antigen and antibody reaction. The blood cells of group A contain antigen A and its serum antibody B; so the moment the blood form group B is transfused into a person of group A, violent reaction starts between the antibody B present in the serum of A and antigen B brought by the blood of B. The patient receiving such transfusion dies of shock. The result of transfusion of blood of A group containing antigen A into patient of group B containing antibody A will be the same.

TABLE 17.1

Blood Groups, their Antigens and Antibodies in Man

Blood Groups	*Antigen present in the RBC*	*Antibody present in plasma*
A	A	anti-B or b or β
B	B	anti-A or a or α
AB	A and B	No
O	No	anti-A and anti-B or a and b or α and β

The blood of AB group contains both antigens but none of the antibodies A or B in its serum. If there were any antibody, say, A or B in the serum of a person having AB blood group, its reaction with antigen A or B respectively would have coagulated the blood. Since the serum of blood group AB does not contain any antibody it can, besides the blood of its own group, tolerate the transfusion of blood from all other blood groups. For this reason, the individuals belonging to this groups are called as *unversal recepients*. The later, however, cannot donate blood to persons other than those of their own group because both antigens A and B are present in their blood. Antigen A will react with antibody B and antigen B with antibody A when blood from AB is transfused into individuals of either A or B group.

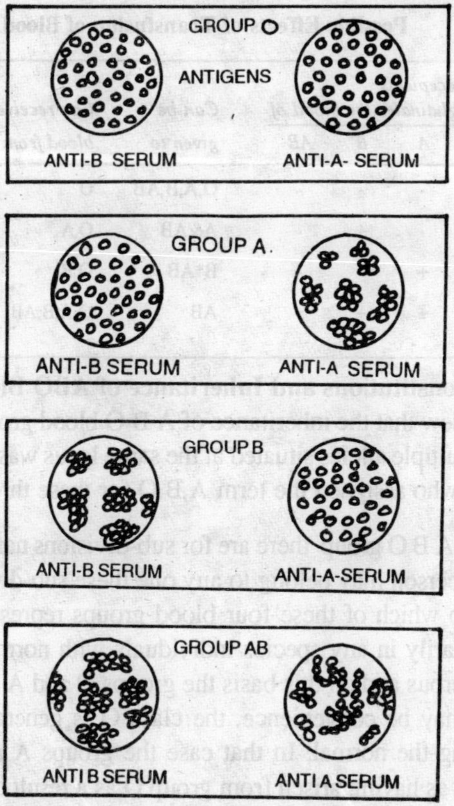

Fig. 17.1. Antigen-antibody reaction between different blood groups. Note that O blood group is acceptable to all (universal donor: last horizontal me). On the other hand, AB blood group can accept all (universal recepient; last but one vertical line).

The group O has both antibodies A and B but no antigen. Because of lack of antigen, it does not coagualte blood when administered in to the blood of other groups, A,B or AB containing respectively antibodies B,A and BA. Since all individuals can tolerate the transfusion O blood group, the later is called *universal donor*. It can however only accept its own blood since all others with one kind of antigen or the other will violently react with the antibodies (A,B) present in the serum of O blood group.

TABLE 17.2

Possible Effects of Transfusion of Blood

Donors Blood group	*Recepients Agglutinates the blood of* O	*A*	*B*	*AB*	*Can be given to*	*Can receive blood from*	*Remark*
O	-	-	-	-	O,A,B,AB	O	donor
A	+	-	+	-	A&AB	O,A	-
B	+	+	-	-	B*AB	O,B	-
AB	+	+	+	-	AB	O,A,B,AB	Universal recipient

Genetic Constitutions and Inheritance of ABO Blood Groups

The view that the inheritance of A B O blood groups is governed by three multiple alleles situated at the same locus was propounded by *Bernstein* who assigned the term A,B,O for these three alleles.

In the A B O group there are for sub-divisions namely A,B,O and AB and a person may belong to any one these sub-divisions. It is not know as to which of these four blood groups represents the normal one. Ordinarily in any species individuals with normal traits are the most numerous and on this basis the groups O and A which are fairly common may be convenience, the class O is generally regarded as representing the normal. In that case the groups A and B might be considered as having arisen from group O as a result of two dominant mutations, one for each group. The mutant genes may be represented by the symbol A and B respectively, both originating in the same locus from one of the normal genes in class O. The normal gene representing class O is indicated by the symbol II by *Dodson,* ii by *colin* and *Snyder,* and *Strandskov* and + + by *Altenburg*. The classes A and B may be conveniently represented by the symbols I^A and I^B. Following *Altenburg* the three +, I^A and I^B which occupy the some locus represent the *multiple alleles*. The genotypes of the four blood groups can be represented as follows:

Genotypes

Class		*As per Bernstein's Theory*		*As per Strandskov's Theory*
O	...	OO	...	ii (or) + + (Altenburg)
A	...	AA; AO	...	$I^A I^A$ (or) I^A i
B	...	BB; BO	...	$I^B I^B$ (or) I^B i
AB	...	AB	...	$I^A I^B$

Since + is a recessive, group O must be homozygous for + (+/+). Since gene A is a dominant allele group A might be either homozygous (A/A) or might be either homozygous (A/A) or might be heterozygous (+/A). Similarly group B might be either B/B or +/B. Group AB is always a hybrid and has the composition A/B. If both the parents in a family belong to class O, they must both have the genotypes +/+. Naturally all their progeny also must be of the genotype + + i.e. class O. On the other hand, if both the parents belong to class A and both are hybrids with the constitution A/+ then the chances are that some of their children might of class O and some of their own class because a cross A/+ X A/+ will produce offspring in the Mendelian ratio of 1A/A:2A/+:1+/A. Therefore, if both the parents belong to group A the children would have to be either of group A or group O.

TABLE 17.3

Showing the Relation of Blood Group Inheritance Based on the Study of Thousands of Families

S.No.	*Blood groups in parents*	*Blood group which cannot occur in children*	*Blood group which can occur in children*
1	OXO	O	A, B, AB
2	OXA	O, A	B, AB
3	AXA	O, A	B, AB

4	OXB	O, B	A, AB
5	BXB	O, B	A, AB
6	AXB	O, A, B, AB	—
7	OXAB	A, B	O, AB
8	AXAB	A, B, AB	O
9	BXAB	A, B, AB	O
10	AB X AB	A, B, AB	O

Sub-divisions of the Blood Group A

On the basis of differences in the degree of behaviour the blood group A has been separated into $A_1A_2A_3$ and A_3 sub-groups, them A_1 is more common and the occurence and the occurence and reactivity of A_2, A_3 and A_4 gradually diminishes. Each of the A subgroups is determined by a separate gene and all these genes are alleles. The genes A_2, A_3 and A_4 usually occur in combination with A_1 as A_2 A_1, A_3, A_1, A_4, A_1, A_1 is dominant to A_2, A_3, A_4, A_2, is dominant over A_3 and A_4 and so on. Homozygous A_1/A_1 are very common, whereas homozygous A_2 A_2 or A_3 A_3 are very rare.

The basis of subdivision of A blood group is its reaction with B-serum which is found to contain two types of antibodies *a* and a_1. The corpuscles of blood group A_1 react and remove both the antibodies a and a_1, while the corpuscles *of* A_1 group combine only with a antibodies, thus leaving a_1 antibodies in the B-Serum.

In a person of blood group AB, there might be A_1, A_2, A_3 or A_4, and hence AB blood group can be divided into sub-groups A_1B, A_2B, A_3B, A_4B. There is some evidence that the B group is also divided into sub-groups but it has not been established so far.

Medical Dispute Parentage

Sometimes dispute arises to the parentage of a child. In such cases the blood group picture may sometimes provide an unmistakable clue as to the parentage of a given child. For instance, if the husband and wife both have *O* group, then the child cannot have A,B or AB. Likewise if both the parents have *A*, then the blood group of their children cannot be B and vice versa. From the marriage between two parents having A and B and blood group, children with O, A, B and AB blood groups may be born because such parents may be heterozygous for both A and B. Therefore, although a very reliable criterion, the blood group data must be used with caution,. The table below gives the genotypes of the parents and the possible genotypes of their offspring:

TABLE 17.4

Genotypes and Phenotypes of Parents and offspring of Different of Blood Groups Showing Impossible Groups That will not Result from the Mating of Such Parents

Parents		*Offspring*		*Impossible Group*
Group (Phenotype)	*Genotype*	*Group*	*Genotype*	
OXO	O/OXO/O	O	O/O	A, B, AB
OXA	O/OXA/A	A	A/O	O,B AB
OXA	O/OXA/O	O,A	O/O/.A/O	B, AB
OXB	O/O/B/B	B	B/O	A, AB, O
OXB	O/OXB/O	O,B	O/O, B/O	O, AB
OXAB	O/OXB/O	A,B	A/O, B/O	O, AB
AXA	A/AXA/A	A	A/A	B, AB,O
AXA	A/AXA/O	A, AO	A/A, A/O, O/O	B, AB
AXA	A/AXB/B	A	A/A, A/O, O	B, AB,
AXB	A/A X B/B	AB	A/B	A,B,O
AXB	A/A X B/O	A, AB	A/O, A/B	B,O
AXB	A/O X B/B	B, AB	B/O, A/B	A,O
AXB	AO X B/O	A,B, AB, O	AO, A/B, B/B	none
AB X AB	A/B X A/B	A,B, AB	A/A, A/B, B/B	O

The Secretors

Some people secrete their antigens in the body secretions like tears, sweat, saliva and milk. Such individuals are called "Secretors." The antigens of secretors can be identified even from their cigarette butts they have smoked. Secretors have the antigens which are water soluble and are thus passed into the body secretians. Non-secretors have antigens insoluble in water, which thus can not pass out the body alongwith water secretians. Secretors have a dominant gene S which enables them to secrete their antigens outside the body. Non-secretors are homozygous for the recessive alleles s.

M-N Blood Groups

Aster the discovery of ABO blood groups in 1927 two additional antigens called M and N were discovered by *Landsteiner* and *Levine* in human blood. Thus human races may posses M, N, or MN blood groups. These three blood groups have no relationship with the ABO groups and the like ABO Groups M-N blood groups do not contain *natural antibodies (agglutinius)* against other M, N groups.

Genetic constructions is like the ABO type. There are two alleles M and N which control the production of antigens *M* and N. Thus given person may be MM, MN or NN. More recently two more alleles have been discovered namely M^3 and N^3 arising at same locus in lesser frequencies.

Rh Blood Groups

Wienen, Levine, Race, Taylar and others discovered another series of multiple alleles producing antigens and this series is called the *rhesus system* or *Rh-system.* If blood from a rhesus monkey is injected into a rabbit's blood against the antigen in the monkey's blood and the monkey's blood agglutinates. The same antibodies produce agglutination of blood cells of some of the humans. Indicating that these persons also have the same antigen, as that or rhesus monkey. Such persons are termed *Rh.* In others these antibodies do not causes agglutinations of the blood cells indicating the absence of antigens and these are termed as *Rh* Agglutinations occur in about 85% of the population. This shows that rhesus monkey and 85% of human population having Rh blood group are also termed *Rh positive* 85% Rh group are *Rh negatives* 15%.

Erythroblastosis Foetalis (Yellow Baby)

The important aspect of Rh blood system is that when Rh^+ blood is given to an Rh person, immune reaction may result and rhesus antibodies may be formed. Subsequent transfusion of Rh^+ blood to the same recipient will result in clumping of the donor blood as the antibodies produced after first transfusion will now react with the Rh^+ blood of the donor and the recipient may die. Thus, such mismatched transfusion must be avoided. A similar situation may arise when there is a rhesus incompatibility between the mother and the child. Since Rh^+ is dominant over Rh^-, the foetus carried by an Rh^- mother by her Rh^+ husband will be Rh^+. Let us consider a case of first pregnancy of such a mother. Since the foetal blood cells may leak through the placenta, they may mix with the maternal blood at birth and stimulate the production of antibodies in her blood. These antibodies are less number during the first pregnancy and do not harm the foetus and the first child will be safe and normal. The antibodies produced during subsequent pregnancies (with Rh^+ foetus) in mother's blood persist and cross over into the circulation of any subsequent foetus, which if Rh^+ will undergo haemolysis or destruction of its cells. At birth this reaction takes the form of anaemia, a condition known as *erythroblastosis foetalis*. It is advisable for an Rh^- woman to marry an Rh^- man. In case a blood transfusion is needed.

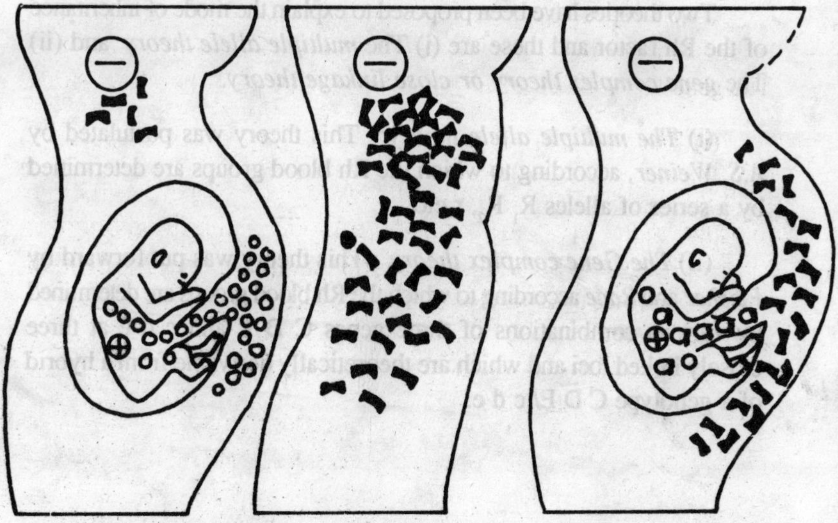

Fig. 17.2. Diagram representing the antigen-antibody reaction between the mother-foetus bloods causing erythroblastosis foetalis.

No Rh⁻ person should ever be given blood of an Rh^+ person as it will lead to erythroblastosis due to incompatibility of the two bloods. Such incompatibility also occurs in some cases of type O woman bearing a type A or B child. She has antibodies that can react with antigens present in her child. The incompatibility occurs very early in pregnancy insted of very late as in Rh system and results in early abortions. This is the reason why such incompatibility cases are not reported as the cause of early abortions is normally not specified.

Genetic Constitution and Inheritance of Rh Antigens

Soon after the discovery of Rh factor, evidences began to accumulate to reveal in this also an instance of multiple allelism, the inheritance of which has been found to be of profound significance in genetical studies. *Weiner* has shown that the gene for the Rh positive antigen is dominant over the Rh negative. Thus Rh positive persons may be Rh Rh or Rh rh but an Rh negative person will be rh rh. Besides, a number of alleles have been discovered which *Weiner* has designated by the symbols Rh_o Rh,' Rh.' Thus, he postulated six alleles as follows:

Rh^o Rh' Rh", Rh_1, Rh_2 and rh.

Two theories have been proposed to explain the mode of inheritance of the Rh factor and these are (i) The *multiple allele theory* and (ii) The *gene complex theory or close linkage theory.*

(i) *The multiple allele theory:* This theory was postulated by *A.S. Weiner*, according to which the Rh blood groups are determined by a series of alleles R_1 R_2, r etc.

(ii) *The Gene complex theory :* This theory was put forward by *Fischer* and *Race* according to which the Rh blood groups are determined by various combinations of three genes C D E and c d e at three closely linked loci and which are theoretically derivable from a hybrid of a genotype C D E/ c d e.

18

Human Genetics

The study of human genetics is complicated by the fact that, unlike other species of animals and plants, humans are not bred experimentally, therefore, we cannot apply the type of genetic analysis that have been discussed in previous chapters. There is difficulty in the study of human genetics due to two reasons (1) Geneticist is not able to arrange the matings of the human individuals he is studying. (2) The long life-cycle of man, production, relatively small number of offsprings; are unfavourable factors in terms of certain standard research techniques. However, the study of human genetis not impossible. In this chapter, we will discuss briefly how family studies can be used to discover the genetic basis of human traits. This involves *pedigree analysis* where the phenotypic records of families extending over several generations are compiled so that gave segreggation patterns can be hypothesized. Naturally the more complex the pedigree, the more accurate the genetic analysis. A modern application of pedegree analysis is called *genetic counseling* in which a human geneticist makes predictions about the probabilities of particular traits occuring among the progeny of a couple in whom or both show evidence of that trait in their family.

Though the branch of human genetics was established as early as 1901, no significant work has been done in the early years as man was found to be an unfavourable object for genetic studies because of the following hundrances:

(1) The life-span of man is so long that it does not permit the geneticist to study more than five or six generations in a family. In *Drosophila,* several generations can be obtained and studied in a few months and in Bacteria, in a week.

(2) The number of offspring produced by man is far less than that produced by most insects and plants. In man, thus, very few samples of the various genetic possibilities can be obtained for study.

(3) Man does not live in a uniform environment throughout his life. His varying natural and emotional environments greatly influence his inherited characters, making the study of the later somewhat complicated.

(4) Marriages cannot be "controlled" by a geneticist who wishes to study the offsprings of a specific mating.

(5) Majority of human beings are genetically heterozygous for many characters. Therefore, it is difficult to get homozygous or pure strain.

Pedigree Analysis

The pedigree analysis is the study of traits as they have appeared in a given family line for several past generations. These days pedigree records are properly maintained and the inheritance of many human traits like polydactyly, syndactyly, idiocy, intelligence, skin colour, haemophilia, colour blindness and so many other diseases have been studied by pedigree analysis. A family pedigree chart conventionally has circles for female individuals and squares for male individuals. The unshaded circles and squares designate the females and males exhibiting normal characters, while the shaded circle and squares are used for those females and males who exhibit the abnormal character or the diseases. A marriage is indicated by a horizontal bar connecting a circle and square end the symbols for offspring are shown suspended from a line drawn prependicular to the marriage bar. Individuals on the same line are from the same generation. Heterozygous customanly are designated by colouring half of the symbol block. Carrier of a sex-linked recessive gene is designated by a black dot in the middle of the symbol.

Identical twins are indicated by a where as fraternal twins by or where as fraternal twins or.

Occasionally, an arrow pointing at a particulars affected individuals indicated that the disease was brought to the notice of genticists by the person indicated by the arrow. Such a person is known as *proband or propositus* (for male) and *proposeta* if female. There are other specialized symbols, but they do not concern as here. The important symbols are put together in the pedigree as shown in the figure given below:

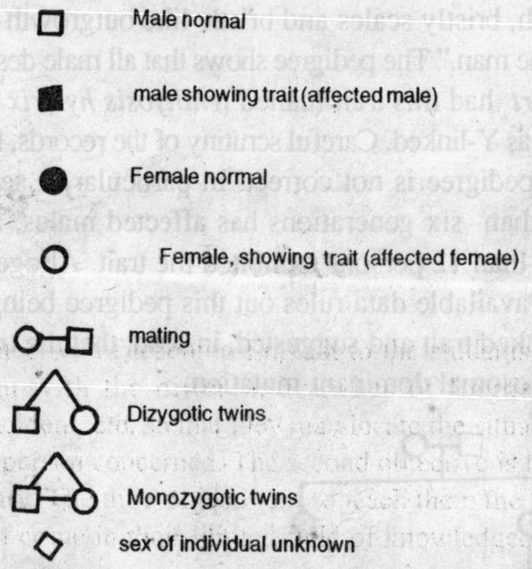

Fig. 18.1. Example of symbols used in human pedigrees. Generation.

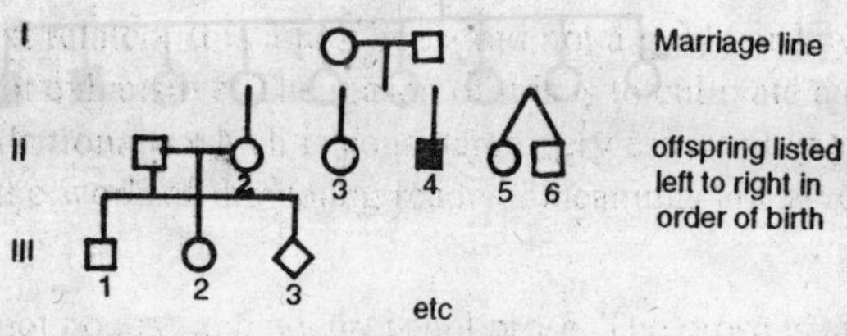

Fig. 18.2. An hypothetical human pedigree showing the typical features of pedigrees.

Examples of Human Predigree

1. The predigree inheritance pattern that is characteristic of a *Y-linked* (holandric) trait, which presumably is the result of mutation on the Y chromosome. Y-lined traits should be readily recognizable since they show father-to-son transmission and no females should ever express them. The predigree is the conventional pedigree given for the *Lambert* family of England that involves the so-called porcupine men. In 1716, *Edward Lambert* was born to two normal parents, *Edward*

was one of many children, and all but *Edward* remained normal throughout their lives. At 7-8 weeks of age, *Edward's* skin begin to turn yellow, and it gradually became black and thickened until his whole body, with the exception of head, face, palms, and soles, was covered with rough, bristly scales and bristle like outgrowth— hence the name "porcupine man." The pedigree shows that all male descendants of *Edward Lambert* had this trait (called *ichthyosis hystrix gravior)* suggesting that it was Y-linked. Careful scrutiny of the records, however, indicates that the pedigree is not correct. In particular, it seems that only three rather than six generations has affected males, and only four or five rather than 12 persons exhibited the trait. Altogether, the revaluation of the available data rules out this pedigree being a valid example of a Y-linked trait and suggested, instead, that the trait is the result of a rare autosomal dominant mutation.

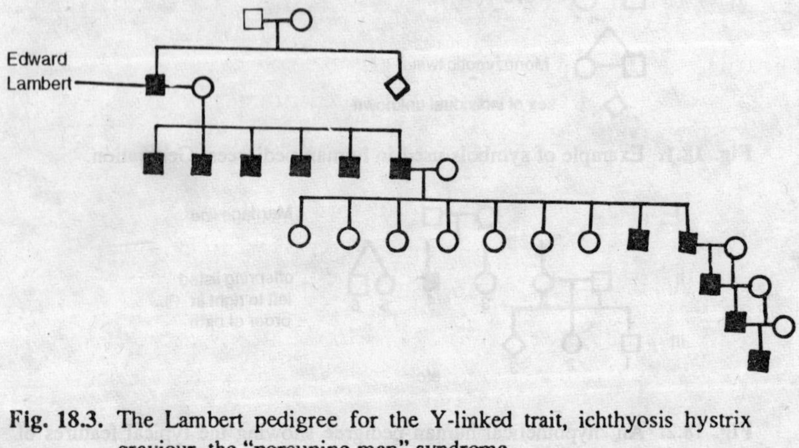

Fig. 18.3. The Lambert pedigree for the Y-linked trait, ichthyosis hystrix graviour, the "porcupine man" syndrome.

One trait from present-day families that is considered to be Y-linked in human is hypertrichosis of the ear. The phenotype is the presence of relatively long hairs on the pinnae of the ears.

One gene that has clearly been localized to the Y chromosome is the H-Y locus, the product of which is the H-Y antigen. (An antigen is any large molecule that stimulates the production of specific antibodies or that binds specifically to an antibody; an antibody is a protein molecule that recognizes and binds to foreign substance — the antigen

— introduced into the organism). The evidence that this locus is on the Y chromosome came from work done in 1955 by *E-J. Eichwald and C.R. Silmser* to determine whether there were detectable differences between males and females that could be related to the sex chromosomes. Their experimental system was strains or inbred mice. The inbreeding (carried out by making brother-sister matings) was done in so many generations that the resulting individuals had essentially the same genotype. These strains are called isogenic. *Eichwald* and *Silmser* reasoned that the only genetic differences between isogenic brothers and sisters would be the result of the sex chromosomal differences between the two sexes. They transplanted skin from females of an inbred mouse strain to male mice of the same strain and *vice versa.* In the former there was no rejection of the transplant, but in the latter experiment the skin transplanted to the female mice was eventually rejected. Since the only difference in males and females of the strain was the presence of the Y chromosome in the males. They concluded that the antigen responsible for the transplant rejection was coded for the Y-linked genes. Since the acceptance of tissue grafts is called *histocompatibility,* the Y-linked gene involved in this phenomenon is said to be a histocompatibility, locus, called the H-T locus. The H-Y antigen product of this locus is a protein found on cell surfaces in many mammals, including humans. It is not known how this protein functions in the sex-determining mechanism in mammals.

2. An example of a *sex linked recessive* trait is haemophilia, a defect in blood coagulation, and the classical pedigree is that of Queen Victoria. Part of her pedigree is presented in Figure. The assumed genotypes, where h is the recessive mutant gene, are given next to the people.

It seems that the h mutation arose in the germline of Queen Victoria. Note that only males show the trait since they are hemizygous h Y. Females would have to be hh to have haemophilia and, since it is a rare gene, this is unlikely as it requires the pairing of a haemophilic male with a carrier (heterozygous) female. Such a pairing does occur from time to time, the characteristics of X-linked recessive inheritance are : (a) males are hemizygous and, therefore, if they express the trait, they transmit the mutant gene involved to all their daughters but to none of their sons; (b) for rare X-linked recessive traits, many more males than females exhibit the trait owing to the difference in the number of X chromosomes between the sexes and (c) there should be approximately 1 : 1 ratio of normal individuals with the trait among the male progeny of heterozygous (carrier) mothers.

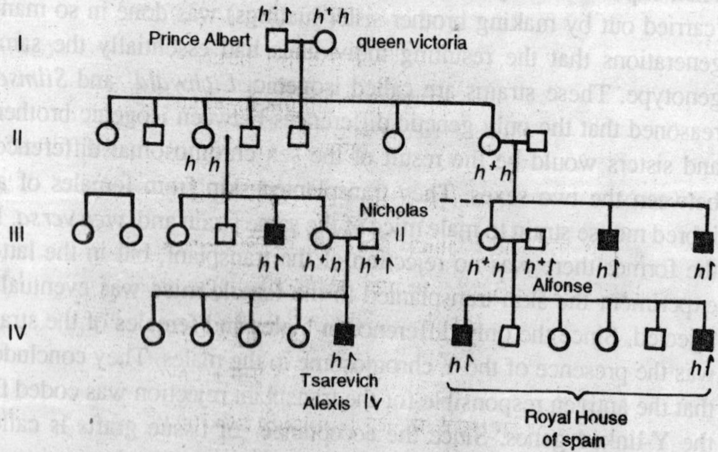

Fig. 18.4. A pedigree for a sex-linked recessive trait: part of Queen Victoria's pedigree showing the inheritance of haemophilia in the Royal families.

3. Not many *sex-linked dominant* traits have been identified. An example is brown enamel. Such traits show the same type of inheritance as sex-linked recessive traits except that heterozygous females express the trait. Since females have two X chromosomes and males have only one X, sex-linked dominant traits are more frequent in females than in males. For rare traits, most affected females in pedigree would be heterozygous.

4. An example of an autosomal dominant trai is *Huntington's chorea,* a disease that results in involuntary movements, progressive central nervous system degeneration, and eventually death. The mean age of onset of this genetic trait is between 40 and 45 years, but the disease may appear early in life and sometimes after 60. The American folk singer *Woody Guthrie* was afflicted by this disease.

The characteristics of an autosomal dominant trait are : (a) every affected person in a pedigree must have at least one affected parent;

(b) each generation in a pedigree should have individuals who express the trait; (c) since the X-chromosome is not involved, father-to-son and mother-to-daughter transmission should occur as frequently as father-to-daughter and mother-to-son transmission; and (d) approximately equal numbers of males and females in a pedigree should express the trait.

5. An example of an autosomal recessive trait is *galactosemia.* Galactosemia is a disease where the biochemical defects known. The disease is manifested in gg babies when they are fed mild: they do not grow well and develop permanent brain damage as well as other problems. If they are removed from a milk diet, the children can develop normally. The defect in individuals with galactosemia is a non-functional enzyme galactose 1- phosphate uridy tranferase (GUT). This enzyme is essential for the conversion of galactose to UDP-glucose. Galactose is produced when the disaccharide lactose in milk is enzymatically cleaved to glucose and galactose. In the absence of the GUT reaction. Galactose 1-phosphate accumulate and this damages cells.

Galactose———>Galactose 1-phosphate

↓ <——GUT (galactose 1-phosphate uridyl transferase)

Uridine diphospho-galactose (UDP-gal)

↓

UDP-glucose

Kinds of Hereditary Traits

The hereditary traits can be grouped into six categories on the bases of their visibility or expression. These are:

1. *Dominant traits* : Dominant traits always appear in almost every generation possessing them. Various skin and exoskeletal traits like peibald (white spotted skin); tylosis (thickened skin): ichthyosis (scaly skin); epidermolysis (blistered skin); red, beaded hairs, hypotrichosis (hairlessness) are some of the dominant traits. In the chart these are classified well.

2. *Autosomic recessive traits :* These are invisible characters characterised by the irregularity of their appearance in pedigree. These recessive characters when become double (homozygous), only then appear otherwise they may skip one or several generations and appearing when recessive gene becomes homozygous. If booth parents are heterozygous for the recessive gene then upon mating they will show usual Mendilian ratio 3 dominant to 1 recessive. These traits include albinism, cretinism, alkaptonuria, etc.

3. *Sex-linked recessives:* These characters pass from generation to generation only through gametes. The colour-blindness, haemophilia and various diseases come under this category which have been described in the chapter of sex-lined inheritance.

4. *Physiological traits :* This category includes various excretoy traits (lime alkaptonuria pheny keotonuria) blood groups traits (like erythroblastosis foetalis) and other biochemical disorders.

5. *Mental traits:* These traits affect the development and work of numours system *feeble mindedness, epilepsy, idiocy etc.* belong to this category.

6. *Abnormal, or pathological traits:* The pathological traits (concerning itself with the acute disease) belongs to two categories namely *teratological* in which defects or deformities are due to developmental upset and *nosological* which occurs as diseases of one kind other or the other.

Autosomal Abnormalities

Autosomes are the chromosomes bearing genes for the somatic characters. There are forty-four autosomes in human beings arranged in twenty-two pairs. A duplication or deficiency of a part or whole of an autosomal chromosome results in a change phenotype is termed as *autosomal abnormality*. Some important autosomal abnormalities related to man are being discussed here:

Down's sydrome: The individuals with Down's syndrome symptoms have 47 chromosomes instead of 46 in their somatic cells. This condition was first described by *Down* in 1968, and was referred it as *Mongolian idiocy* by him, in this the mental growth is retarded and the person have characteristic facial features resembling mongols. This condition may occur in all races. In this almost all organ are defective in growth and development. Mongol females have underdeveloped sexual characteristics. They are not sterile. Mongol males have undescended testes and their semen contains a reduced number of sperm count. The life expectancy of mongols is estimated as 18 years by *collmann* and *Stoller* in 1963.

This syndrome is a congental originating from the non-disjuction of chromosomes of pair 21 during meiosis. The frequency of production to abnormal eggs per 1500 among mothers under 30, are per 750 at maternal age 30-34, are per 600 at the maternal age 35-39 and one per 300 among mothers of 40-45.

Patau's Syndrome: Patau et al. (1960) described a trisomy of the chromosome number 13(D-group) that is why this abnormality is called *Patau's syndrome.* It is characterised by multiple and sense body malformation as well as profound mental deficiency. Hare lip and cleft palate is very common among the suffers. Polydactyly is most conspicuous. The internal organs are severly malformed and the patients often suffer from disorders of kidney and heart. Death usually occurs soon after birth. About one out of 1000 newly born babies exhibit trisomy.

Edward's Syndrome: This syndrome was first described by *Edward et al* in 1960. It is associated with the trisomy of chromosome number 18 (E. group). This produces receding clin, malformed ears and defective nervous system. The hands are short and show little development of second phalanx. The life expectancy is very poor not more than a year or so.

The CRI-D4-CHAT or Cat cry Syndrome : This syndrome was first described by *Lejeune et. al.* in 1963. This syndrome is due to deletian in chromosome number 5. The most prominent symptom is a curious, faint, mewing cry. The babies have small average birth weight and suffer from considerable mental defect. Cardiovascular anomalies are common. The condition is quite compatible with survival.

Sex Chromosomal Abnormalities in Man

An increase or decrease in the number of sex chromosome in the normal complement of female (XX) and male (XY) results in syndrome of sex chromosome in human beings. The common sex linked chromosomal abnormalities are :

(i) *Kleinfelter's syndrome*
(ii) *Turner's syndrome*
(iii) XXX Syndrome or Super female.

These syndromes have already been discussed under separate head— the non-disjunction given elsewhere in the book.

S. No.	*Character Dominant*	*Recessive*
A. Autosomal Dominant Trains		
1.	Brachydactyl	Normal
2.	Syndactyly	Normal
3.	Polydactyly	Normal (5)
4.	Ectordactyly (abortive fingers)	Normal
5.	Cartilaginous growth of bones	Normal
6.	Curly hair	Straight hair
7.	Hairlessness	Normal hair
8.	White forelock	Normal black hair
9.	Ichthyosis (scaly skin)	Normal skin
10.	Dark skin colour	White skin colour
11.	Absence of enamel of teeth	Normal teeth
12.	Progressive muscular atrophy	Normal
13.	Hare lip and clefted plate	Normal
14.	Congenital catract	Normal Eye
15.	Glucoma	Normal eyes
16.	Astigamatism (cornea not spherical)	Normal eyes
17.	Mongoloid fold of skin around eye	Normal, no fold

18.	Long eye lashes	Short eye lashes
19.	Large eyes	Small eyes
20.	Near sightedness and far sightness.	Normal vision
21.	Hypertension (high blood pressure)	Normal
22.	Huntington's chorea	Normal
23.	Polycystic kidney	Normal
24.	Hypospadias	Normal
B. Autosomal Recessive Characters		
25.	Red body hair	Non-red hair
26.	Blond hair	Dark hair
27.	Absence of sweat glands	Normal
28.	Blue or grey eyes	Normal brown eyes
29.	Microphthalmus	Normal
30.	Attached ear lobes	Normal free ear lobes
31.	Sickle cell anemia	Normal
32.	Phenylketonuria	Normal
33.	Schizophrenia	Normal
34.	Alkeptonuria	Normal
35.	Xeroderma pigmentosum	Normal
36.	Retinal glaucoma	Normal
37.	Microcephali	Normal
C. Sex-linked Characters (Recesive)		
38.	Diabetes mellitus	Normal
39.	Haemophillia	Normal
40.	Colour blindness	Normal Colour vision
D. Sex-linked Dominant Character		
41.	Hypophosphatemic rickets	Normal

	vit-D resistance	
42.	Xg^2 blood group (+)	xg^2 (—)
E. Characters Controlled by Multiple Alleles and Multiple Factors		
43.	A,B, AB and O blood groups (Multiple alleles)	
44.	Skin colour (Multiple factor)	
45.	Intelligentia	
46.	Hepatoglobin (a blood scrum protein)	

Twinning

Lower animal usually produce large number of young ones while higher animals show a tendency towards reduction in multiple births. Cats, cattle and dogs etc., give birth to more than two young ones. This phenomenon of production of many young ones is called *superfetation* while the two young ones produces at a time as called *twins.*

In human being the phenomenon of twins is as it were, a kind of natural experience which permits us to distinguish between the influence of heredity and environment, both of which play essential parts in the lives of all organism.

Kinds of Twins

Identical Twins: Such twins are extremely rare. They develop from a single zygote and are also known as *monozygotic* or *one-egg twins.* Each of the two cells resulting from the first division of the zygote, behaves like a zygote and develop into an individual. Identical twins are always of the same sex. They have the same genotype and phenotype and, at times, even the mother cannot distinguish between the two. However, patterns of finger—prints and palm — prints (or sole-prints) that appear on the right hand (or foot) off one are repeated on the left hand (or foot) of the other and *vice versa.*

Origin of Identical twins : The identical twins are considered to be monozygotic as they have developed from single ovum or egg fertilized by a single sperm. This fertilized egg than divides two blastomeres each of which develops into identical twin. Thus these identical twins originating form the same egg would be of the same genotype and same traits in the beginning. In the later stages they may show variations due to the differences of environment on them. That is why these are also called monoval twins (one egg) or monozygotic twins. *Newman* and *Patternson* stated that in *Armadio* a single fertilized egg divides in to four of five segments each of which develops into new individual. Thus, this organism develops quadruplet identical twins.

Fraternal Twins: Such twins are more common than the identical twins. Fraternal twins may be both males, both females or one male and one female. They are just like brother and sister produced by the same parents except with the different twins. They show many dissimilarities and are of different genotypes.

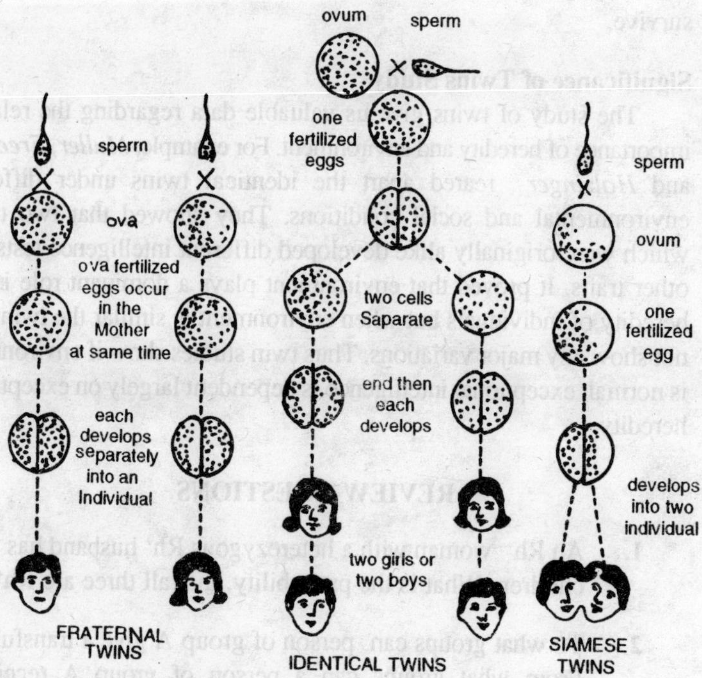

Fig. 18.5. Origin and kinds of human twins.

Origin of Fraternal twins : Fraternal twins develop by the chance fertilization of two or by two sperms and are thus also known as *dizygotic* or two egg twins. The two eggs develop simultaneously in the uterus and form fatemal twins having similar appearance but differing in various anatomical characteristics.

Simese Twins: These twins orginate like identical twins but the separation of the two individuals is incomplete i.e., they remain attached one or more part of the body. They usually do not survive of the birth, though some instances of their survival are well-known. In recent years several cases of successful separation of siamese twins by surgery, have been recorded.

If the degree of union of twins is symmetrical that is if there is one head and two bodies or two heads and single body resulting in abnormal traits thus such twins are called *double monster*. They usually do not survive.

Significance of Twins Study

The study of twins give us valuable data regarding the relative importance of heredity and environment. For example, *Muller, Freeman* and *Holzinger* reared apart the identical twins under different environmental and social conditions. They showed that two twins which were originally alike developed different intelligence tests and other traits. It proved that environment plays a dominant role in the heredity of individuals but when environment is similar the twins did not show any major variations. Thus twin studies show if environment is normal, exceptional intelligence is dependent largely on exceptional heredity.

REVIEW QUESTIONS

1. An Rh^- woman with a heterozygous Rh^+ husband has three children. What is the probability, that all three are Rh^+?

2. To what groups can person of group A give a transfusion? From what groups can a person of group A receive a transfusion?

3. An Rh^- woman has given birth to an erythroblastotic infant and has anti-D antibodies. If the husband is heterozygous there is a probability of 1/2 that the next child will be

normal. What kind of serum would tell whether the husband is homozygous or heterozygous for the factor D?

4. What proportion of the population are O, Rh^-?

5. What genotypes could react with all six Rh antisera?

6. The following data are from a recent court case.

Woman	A, M, cde/cde
Child	A, MN, cde/cde
Man 1	O, M, CDe/CDe
Man 2	A, N, cdE/cde.

Which man could be the father?

7. In another case, the Rh data are as follows (the AB, MN and P tests were in conclusive).

Mother	cde/cde
Child	cde/cde
Man 1	CDE/cde, Cde/cDE, CdE/cDe, or CDe/cdE
Man 2	cDe/cDe or cde/cDe.

Can a decision be made on this data? Man number 1 also had two children by a cde/cde woman, both cDE/cde. Does this additional data answer the question as to which is the father?

8. In cattle, as elsewhere, the presence of antigen is dominant to its absence. Recently a case came up in which a serological laboratory was asked to determine which calf went with which after they had gotten mixed up. The blood groups were as follows:

Sire	A B D G J K O S
Cow L3	A B D G H I J K M S
Cow L4	A B C D E F G H K M N O S

Calf L1 A B C D E F G H K O S
Calf L2 A B D G J K O S

Which calf is the offspring of Cow L3?

9. If an anti-d serum becomes available, predict how the following cells will react with it: cde/cde, cdE/cde, CDe/CDe.

10. Assume that you are in charge of an outrageously careless lying-in hospital. Four children are born with the following blood groups, but the nurses don't know which parents they go with. Assign them to the proper families.

			Father	*Mother*
Baby A.	O, MN, Rh$^-$	Parents 1.	B, N, Rh$^-$	A, MN, Rh$^-$
Baby B.	AB, N, Rh$^+$	Parents 2.	B, N, Rh$^+$	AB, N, Rh$^-$
Baby C.	B, MN, Rh$^+$	Parents 3.	AB, MN, Rh$^-$	O, MN, Rh$^+$
Baby D.	O, N, Rh$^-$	Parents 4.	A, M, Rh$^+$	A, N, Rh$^+$

11. Suppose you cross two inbred strains of mice which differ from each other with respect to five pairs of histocompatability genes. What proportion of transplants would be accepted in each of the following:

	Donor	*Recipient*
a.	inbred strain 1	inbred strain 2
b.	inbred strain 2	F_1 hybrid
c.	F_1 hybrid	inbred strain 1
d.	F_2	F_1
e.	F_1	F_2
f.	F_2	inbred strain 1
g.	F_2	F_2
h.	inbred strain 2	F_2

ANSWERS

1. 1/8.

2. A or AB. A or O.

3. Anti - d.

4. .45 x .15 = .07.

5. CDE/cde, CDe/cdE, Cde/cDE, or CdE/cDe.

6. Number 2 could be , number 1 could not.

7. No decision, though the fact that CDE/cde is very rare suggests that number 1 is not the father. The additional data confirm this by ruling out number 1. Number 2 could be.

8. L2.

9. Positive, positive, positive, negative.

10. Parents 1 - baby D; Parents 2 - baby B; Parents 3 - baby C; Parents 4 - baby A.

11. a. none.
b. all.
c. none.
d. all.
e. $1/2^5$, or 1/32.
f. $1/4^5$, or 1/1024.
g. $(10/16)^5$.
h. $(3/4)^5$.

19

Eugenics

The science of eugenics was founded in 1883 by *Sir Francis Galton,* a cousin of *Charles Darwin.* The idea itself is very ancient and can be found in the works of *Plato* and the sages of qutiquity.

It as much as many diseases, defects, and abnormalities are hereditary and some are regularly transmitted by parents to children, it was not unreasonable to propose that children unfit for life should not be brought into the world. This, of course, amounts to preventing persons who are seriously defective from reproducing.

Legislation to sterilize such persons exists in many countries. In the United States the first law of this kind was adopted in 1907 by the State of Indiana. By 1940 similar laws existed in thirty states, being more strict in some than in others. In all thirty states mental deficiency is subject to sterilization; in twenty-nine mental diseases are liable to the same treatment. In twenty-three states there are laws controlling epilepsy. In other states sterilization can be applied in certain cases of physical malformation and nervous diseases. From 1907 to 1950 more than 50,000 such operations were performed, two-fifths of the sterilized being men. It is still difficult to weigh the results of these eugenic experiments which have not in all cases been based on sufficiently solid genetic and medical grounds. Similar legislation has been proposed in certain provinces of Canada, in Tasmania, and in New Zealand.

In Europe sterilization laws were first introduced in 1929 in the Swiss canton of Vaud, and in the same year in Denmark, Norway, Sweden, Finland, and Iceland. In Germany and in Estonia the law, after being blindly applied, has been repealed. Laws have been proposed, without being passed, in England, Holland, Hungary, Czechoslovakia and Poland. At the moment such laws are retained only in the Scandinavian countries and in some Swiss cantons.

The Scandinavian laws provide that the patient himself can ask to be sterilized and that the operation cannot be performed without his freely given consent. During the period 1929—1950 more than 6,000 operations took place in Denmark, two-thirds of those sterilized suffering from mental deficiency, and two-thirds of these being women. In Sweden, for the period 1935—1948, some 15,650 people were sterilized, women again outnumbering men. The number of those sterilized continues to rise and *Professor Kemp* estimates that in Denmark today there are 600 operations a year, distributed equally between the mentally deficient and those suffering from other afflictions. At this rate an increasingly rapid elimination of mental deficiency may be anticipated. Denmark not only has laws providing for sterilization but also others covering legal abortion, the prevention of certain marriages, and eugenic measures such as isolation, the obligatory declaration of disease, etc.

Of course, these various measures encounter moral, social, and religious opposition, and the question arises whether their efficacy is genuine. The answer will vary according to the way in which the gene responsible for the affliction is transmitted.

Eugenics can be studied under two heads:

(i) Negative engenics.
(ii) Positive engenics.

Negative Engenics

In this the defective germ plasm is not allowed to mix. This restriction can be brought about in the following ways:

1. *By controlling immigrations:* Most of the advanced nations are very selective in permitting foreigners to enter their lands and mingle with their populations. They believe in encouraging only the *cream* of other nations, i.e. people with high intellect. This is an attempt to shut out germplasm which may be undesirable.

It is compulsory to get vaccinated and injected as an immunity measure, against certain contagious and epidemic diseases before travelling to a foreign country.

But these measures are not sufficient to check undesirable germplasm because a person may have high intellect and may possess sound health but in his genome he may be carrying an undetected recessive gene which when homozygous may be fatal.

2. *Marriage Restrictions:* Every country or society has the customs or laws which tend to restrict marriages. According to biological aspects marriage is an experiment in breeding, not human courtship as described by various novels. In several countries the marriages of mental defectives, habitual drunkards, idiots, feeble-minded, insane persons, epileptics, alcoholics and persons having venereal diseases are prohibited. As a result of these regulations the genes necessary for these defective traits do not have chance to mix in the generation. Only marriages between undefective persons should occur. Even the cousin marriages are extremely useful if they are without set-linked diseases.

3. *Sexual Separation of the Defective:* The defective persons may have various sex-linked diseases such as night blindness, haemophilia, colour blindness, etc., and various other defective traits which may be regulated by dominant or recessive genes. The increase of germplasm of the persons having such defective traits in the population can be checked by keeping them away and separated from the society. Different states have wisely adopted the restricted measures in segregation the mental defectives from the society and to place them in mental hospitals.

4. *Sterilization :* It is a drastic step of restricting defective germplasm from meeting with the other germplasm. By this method, ducts carrying sperms or eggs from gonads are removed surgically thus blocking the way of undesirable germplasm to external world. This is of two types:

(a) *Vasectomy.* It is a minor operation in males in which vasa deferentia (sperm ducts) are surgically removed or rendered ineffective by some means. In the *castration* complete *testes* are removed thus putting the sexual life to an end.

(b) *Salpingectomy.* In this type oviducts are gutted or removed surgically in females and thus avoid the chance of fertilization due to non-availability of ovum.

Thus, these operations fall into two categories namely temporary and permanent. In the temporary sterilization, vasa deferentia or fallopian tubes are tied-off with the help of surgical thread in males or females. In the latter method, these ducts are cutted permanently.

Criteria for Negative Eugenics (Genetic Variables): In practice the effectiveness of marriage restrictions, segregation, control on immigrations and sterilization as eugenic measures depends on several genetic variables like:

(a) The mode in which undesirable trait is inherited;
(b) The rate of frequency of the gene or genes concerned in the populations;
(c) The age at which the defective trait makes its appearance; and
(d) The extent to which environment may check or inhibit the expression of the gene. Regarding the mode of inheritance of these trails the following possibilities may exist;
 (i) The trait may be due to dominant gene.
 (ii) It may be due to a gene without dominance (i.e. heterozygote intermediate).
 (iii) Sex-linked recessive gene may bring about it.
 (iv) It may be due to autosomal recessive gene.
 (v) Finally, a trait may be produced as a result of cumulative genes.

Positive Eugenics

1. *Subsidizing the fit and sperm banks:* Because the highly endowed persons lead a well-planned life and to avoid unnecessary difficulties in nursing the children they often prefer to have small number of children. Therefore, the selected young men and women of best eugenic value should be encouraged to increase their birth rate. Moreover, *H.J. Muller* has suggested that such persons not only should increase their family size but through artificial insemination the outstanding man can serve as fathers to many more children than

would be otherwise possible. The artificial insemination is already widely practiced to permit those women whose husbands are sterile or have some serious hereditary afflictions to bear children. The sperms and eggs of outstanding persons can be stored for future use by quick freezing and storing them in deep freeze. These germ cells thus can be stored for 100 or more years.

Recently in U.S.A., a "sperm bank" was floated, primarily for collection and preservation of sperms of Nobel Prize laureates and probably other exceptionally gifted intellectuals. Its aim is to produce children of "high mental quality" through artificial insemination. The initial response to the sperm bank was poor but recently *Robert Graham* has claimed that the response is now fairly good. It is also claimed that about 15 children have already been produced with the help of sperms thus acquired. However, it is yet to be seen how far this revolutionary idea succeeds in its aim.

2. *Promotion of genetic research:* The mechanism of inheritance of many specific traits can be clearly understood provided basic principles of heredity and distribution of genes are known. The physiological action of genes is another problem and with its complete knowledge, many diseases can be cured. The nature of mutation, their effects and possible control may bring about the better offsprings having high learning capacity.

Besides, genetic research on plants and animals is receiving much support from different state governments. By this encouragement, good varieties of plants and animals highly beneficial to man have been produced.

3. *Early marriage of those having desirable traits:* It is most commonly observed fact that the highly placed persons of the society often have great ambitions for the future life. In achieve their ambitions goals, they often devote the best part of their youth and they are able to marry in their mature age (i.e., 30 to 35 years). The biological and psychological investigations have revealed that the aged persons often lack in necessary amount of emotional warmth for the sexual activities and moreover, their germplasm also lost its vigour. Therefore, some

laws should be formulated to prevent the late marriages of highly endowed persons by applying high taxation on them and at the same time the young persons having best hereditary traits should encouraged for early marriages.

4. *Genetic counselling:* To produce healthy progeny should be the endevour of man as this alone can ensure a better future for humanity. Genetic counselling can make a significant contribution in this direction. For families with history of genetic diseases, genetic counselling can provide the much needed releif.

By careful examination the genetic counseller can detect genetic abnormalities like *Down's syndrome* and even recessive genes for disease like *sickl-cell anaemia.* Knowledge of the possibilities of transmitting these diseases to offspring can help a person in choosing a marriage partner.

Genetic counselling is equally useful after marriage. An Rh^- woman with an Rh^+ partner, when aware of the implications in advance, can go in for suitable medical aid well in time. The study of the amniotic fluid (obtained by a technique called *amniocentesis* involving withdrawal of the fluid with the help of a hollow needle) can reveal the *sex* of the foetus as well as *congenital diseases* (like Down's syndrome) and metabolic disorders that it may possess. This information can help the couple to decide whether or not to retain the pregnancy, or at least to be mentally prepared for the inevitable.

5. *Protection against mutagens:* We now know that mutagens can induce mutations, some of which may be deleterious and even lethal. Harmful mutations can result in deformed offspring and increase human misery. It is, therefore, essential to check exposure of human germ cells to mutagens like high-energy irradiations, chemicals, etc. Medical X-ray should be advised and carried-out with utmost care. Chemical mutagens should be experimented with, and handled with all possible precautions. The December 3, 1984 gas tragedy in Bhopal, (M.P.), India, (leakage of MIC—methyl isocyanate) and the subsequent discovery of "dangerous" experiments carried out in Bhopal should be sufficient to warn world governments of the possible hazards of chemical

warfare in which chemical mutagens may be used. These hazards will in no way be less than those of exposure to high-energy irradiations frigerred by atomic bombs.

6. *Improvement of Environmental Conditions:* Environment plays a dominant role in the heredity of man. Better facilities of training, schooling, living and studying should be given. These all are euthenic programmes for the benevolence of human race.

Generally sexual fertility is correlated with intellectuality reaching maximum very rapidly in idiots and stupids while more slowly in intelligent and genius persons. As shown earlier idiots as well genius are homozygous genes responsible for particular traits while average are heterozygous having greater fertility. Their selection may yield good eugenic results.

Euphenics

Euphenics deals with the treatment of genetic diseases of man. The credit for the goes to A.C. Pai (1974). More specifically it deals with the control of several inherted human diseases especially in born errors of metabolism in which the missing of defective enzyme has been identified. The excellent example of two category is the *phenylketonuria* or *PKU*. In the abnormally the patients are unable is metabolize an amino acid—the *phenylalanine* properly, the resulting product causes severe mental retardation. Such patients are advised to take phenyalanine free diet.

Although a number of inherited diseases can be treated in a similar euphenic manner, but these constitute only a small fraction of known inherited diseases. For the most part, biochemical geneticists could not identified the biochemical errors of many genetic diseases. In other, cases, such as *albinism*, even though the metabolic block leading to an abnormality is known, but, it is not possible to correct it.

20

Inbreeding, out Breeding and Heterosis

Inbreeding and outbreeding are the two systems of matings in animals and plants. Inbreeding is the phenomenon in which offspring are produced by the mating of the close relatives. Self-fertilization in plants is one of the extreme type of inbreeding while the marriage between sister and brother is closely form of inbreeding in human beings. In general practice, biologists have been always curious to know whether an individual arose from inbreeding or outbreeding. In many earlier days, in Egypt the pharaohs used to marry their sisters and no harmful effects were recorded. It is now established by the plant and animal breeders that inbreeding often leads to the degeneration of the race. *Kalereuer* (1766), *Knight* (1799), *Naudin*(1865) and other pioneer experimentalists found that hybrids between races and species of plants are luxuriant, large than either parent. *Mendel* (1965) too studied hybrid vigour in his pea crosses.

Inbreeding

The process of mating of individuals which are more closely related than the average of the population to which they belong, is called *inbreeding*. The self-fertilization in plants as in peas and beans is an example of inbreeding. The marriage between brothers and sisters or first cousins is the another familier example of inbreeding among higher animals.

Artificial selection always accompany close inbreeding for the betterment of plants and animals both. In sexually reproducing plant and animals inbreeding is done so as to retain the desirable varieties of the individuals. But in plants where there is a sexual reproduction, if any suitable variety is obtained it can be retained indefinitely. If artificial selection in the breeding individuals is done then individuals with the desirable characters are produced in small number and they will be closely related. Hence close inbreeding becomes a necessity and in order to retain the desirable quality of individuals close inbreeding is a must in order to keep together the genes that have made the individual outstanding.

Effects of Inbreeding

Inbreeding shows following effects:

(1) *Homozygosity under different Degrees of Inbreeding:* Self-fertilization occurs only among a limited group of plants and is not found in the higher animals. In them the inbreeding includes mating between brothers and sisters or mating between double first cousins (i.e. progeny of parents the are brothers and sisters of each other). *Sewall Wright* has contributed significantly to the theoretical aspects of the problem with his studies on the effect of inbreeding in guinea pigs. Wright's coefficient of inbreeding (F) represents the proportion by which inbreeding reduces heterozygosity.

Percentages of homozygosis compared with the original degree of heterozygosis have been calculated by *Wright* and others for successive generations under different systems of inbreeding.

The self-fertilization produces homozygosity most rapidly. After eight generations, nearly all the genes that were heterozygous before inbreeding, bean would theoretically be homozygous. Brother and sister matings are somewhat less efficient introducing homozygosis. Under this system 95% of the genes which were heterozygous originally would become homozygous in eleven generations of inbreeding. In case of double first cousins the inbreeding would field about 65% homozygousity in fourteenth generation. The continued mating of second cousins would produce homozygosis form 50 to 51%.

(2) *Pure Lines:* If a population of pea is hybrid for a pair of alleles, red and white flower colour, continued self-fertilization will change it into two pure races for red white colour. If the plants are hybrid for another pair of alleles, tall and dwarf also, inbreeding

would result in pure tall and pure dwarfs. The population would consists of four homozygous races: (i) red tall (ii) red drarf (iii) white tall, and (iv) white dwarf. Similarly, the original plants may be hybrids for any number of alleles and self-fertilization would result in the production of classes pure for each allele. The various pure races produced after continued self-fertilization are 'pure lines' each pure line continues to produce its own kind if self-fertilized.

In 1900, *Johannsen,* a Danish botanist, studied to effect of selection in pure lines. He carried out his experiments on the seed weight in bean, which is a naturally self-pollinated plant. *Johannsen* selected progenies of 19 distinct bean plants and weighed the seeds. Each of the progeny was called a *line* and was made homozygous by self-pollination. He then selected the lightest and the heaviest seeds from each line and planted them separately to see if the heaviest would produce heavier seeds than the progeny of the lighter ones. The experiment was continued for several years and he found that within a single line, the weight of the mother seed had no influence on the weight of the progeny. The offspring of the heavier as well as the lighter seeds of a pure line, consisted to seeds of average weight.

Johannsen defined a pure line as consisting of individuals descended from a single self-fertilized individual and having the similar genotype. He showed that selection in ineffective in a pure line because genotype is not altered by environmental factors.

Inbreeding Depression : Vigour loss following the mating of closely related individuals is known as inbreeding depression. This happens in cross-fertilized crops when genetically close plants are mated. Inbreeding depression is the result of uncovering deleterious recessive that are masked in the heterozygous state. *East* and *Shull* first documented and described inbreeding depression in plants when they observed loss of vigour in corn progeny produced from selfing of a naturally selfed plants since in these plants a high level of Homozygosity already exists.

The inbreeding depression in calculated by comparing the frequencies of homozygous recessive inbred and outbreed population.

The frequencies of genotypes, in an inbred population as already calcuated are:

$$XX = p^2 = pq^F$$

$$Xx = 2pq - 2pq^F$$

$$xx = q^2 + pq^f$$

where F is the coefficient of inbreeding. The frequencies of genotypes in an outbred population are:

$$XX = p^2$$

$$Xx = 2pq$$

$$xx = q^2$$

The ratio of homozygous recessives in inbred to outbred frequencies is:

$$\frac{q^2 + pq'}{q^2} = \frac{q(q + p^F)}{q^2} = \frac{q + p^F}{q}$$

The inbreeding increment is p^F and its value depends upon the values of q and F.

Outbreeding and Hybrid Vigour

The crosses between two unrelated pure breeding population are called *out crosses or outbreedings*. The out breeding usually result in stronger, healthier, larger and vigorous individuals in F_1. The superiority of the hybrid over the best parent is known as *heterosis*. Such a hybrid may have unproved general fitness, resistant to diseases and may show remarkable growth and vigour. Such improvement observed in F_1 are also called *hybrid vigour*. *Shull*(1914) coined to term heterosis for hybrid vigour to avoid the implication that hybrid vigour was entirely Mendelian in character.

The phenomenon of outbreeding and heterosis has been utilized from time immemorial, even since man has been using mules. Heterosis in plants was noticed by *Kolbeutr* in 1768, who noticed that hybrids of tobacco were more vigoures than either of the parents. *W.J. Beal* (1880) demonstrated the same phenomenon in the hybrids of corn. *Shull*(1908) wording a Cold spring Harbour in New York, *E.M. East* (1908) at the Connecticut Agricultural Experiment Station and later *D.F. Johes, H.K Hayes* and others carried out extensive studies on corn and compared the progeny obtained from self and cross pollination. Outbreeding has been employed extensively in plants for producing improved varieties of maize, wheat, sugarcane, potato and a number of flowering plants. However, the application of outbreeding and crossbreeding is limited in animals.

Hybrid Vigour in Maize: Maize is a cross-pollinated plant in the field but can also be self-polinated. Inbred progenies (pure lines) for various traits of maize produce low yields and are of short stature. When two inbred line are crossed, the hybrid shows more vigour with high yields. This was confirmed by *Shull* and *East* independently in 1905. They found that continued selfing deteriorates the yields due to recessive homozygosity of various genes affecting yield. A point is reached when no further deterioration is possible. This establishes an inbred line. The inter-crossing of two inbred lines (for different genes) restores the vigour in the very next generation following the cross. Exploitation of heterosis as experimented by *Shull*(1910) poses one problem. Most inbred lines are so infertile that they do not produce enough seeds for commercial plantings. Shull's method was modified by *Jones*(1917), a method called the *double-cross* method and is diagrammed below.

Inbred *A*Inbred *B* Inbred *C* Inbred *D*

Single cross A x B — Single cross C x D

Double Cross

(A x B) x (C x D)

A double cross consisted of producing a hybrid between two different unrelated single crosses which resulted from different unrelated single crosses which resulted from different parental inbred lines. Inbred a x B produced a single cross hygrid AB. Similarly, inbred C x

D produces a single cross hybrid CD. The double cross hybrid has the gene combinations from four separate inbred lines.

The hybrids between some inbred line give yield as high as that of the cross-bred parental varieties but some inbred lines produce hybrids which are exceptionally high yielding. These inbred lines possess good *combining ability*.

Certain breeding lines which when crossed among limited lines produce a high proportion of productive progeny. This is known as *specific combining ability*. This includes both aditive and non-aditive (epistasis, dominance, etc.) types of gene action. Certain crop plants when crossed with several other genotypes produce a high proportion of very vigorous, high productive offsprings, a result of aditive gene action. These plants are said to have good general combining ability. Combining ability is, therefore defined as the ability to produce desirable progeny.

Hybrid corn is the most spectacular contribution of science of agriculture to the world. For the first time in 1959 the production of field corn in the United States exceeded four billion bushels. Hybrid maize is superior to the open pollinated varieties because in this there is much better control over the heredity. The hybrid maize is much more uniform and produces practically all of the plants of the same superior type. The same results can be repeated year after year. In the case of open pollinated varieties some of the plants will be as robust and productive as the hybrid plants but the results cannot be duplicated in the following years as the open pollination will lead to progeny plants with many genetic constitutions due to segregation of genes and uniformity is not possible to maintain.

Hybrid Vigour in Drosophila: Heterozygotes of *Drosophila*, like the *Curly-Lobe-Plum* stock, are true breeding because of balanced lethals, *curly, lobe and plum* genes are located on the second chromosome. All the three mutations are dominant. *Curly* mutation results in an upward bending of wings. *Lobe* mutation causes lobed eyes with distorted instead of round shape. *Plum* flies have grey coloured eyes. An inversion between genes curly and Lobe suppresses crossing-over between the two. Thus, heterozygous flies of genotype Cy+L+Pm+give rise to only two types of gametes; Cy+L and +Pm+. Homologous chromosomes contain lethal genes at different loci. Consequently, homozygous zygotes (Cy+L/Cy+l or +Pm+/+Pm+) are non-viable and only heterozygous files, like the parents, are produced. Many strains of the fruitfly are self-perpetuating due to the presence of balanced lethals and chromosoal aberrations. Such strains are also used for assigning and locating autosomal genes.

Hybrid Vigour in Other Organisms

Hybrid vigour has been utilised in the improvement of tomato, tobacco, squash, cucumber, cotton, sugar beet, onion, forage crops and cereals like maize, millet, sorghum and other useful plants. It has also been demonstrated in many ornamental flowering plants. It has also been demonstrated in many ornamental flowering plants and in the fungus, *Penicilium,* in the production of *Penicillin.* It has also been exploited in animals. Hybrid hen lays many more eggs in a year than the average standard strains; and crosses between the Yebu and English type cattle have been produced to suit the hard climate of North Australia. Crosses between the Red Sindhi breed of Indian cattle and the Jersey breed of America give 30% more butter fat per cow and are heat resistant. Another high yielding cow is a Jersey Sahiwal cross. Red Dane and Holstein cows and bulls are other examples of imported exotic breeds used in producing hybrid cattle. Red Dane cows give a yield of more than 30 liters of milk a day. Pork yield in pigs and silk production in silk worms have similarly been greatly increased. A careful exploitation of hybrid vigour in cultivated plants and domestic animals can bring about a great increase in the world's production of food and clothing, and can contribute towards the solotuion of many a problem of man.

Crossing of dissimilar plants does not always lead to Heterosis. To this category belong plants like wheat and barley which are mostly self-pollinated in nature Breeding inthese plants also does not result in degeneration of vigour. The explanation for this is as follows. Self-fertilization sorts out the recessive genes and in a few generations produces forms which are pure breeding recessives. The deleterious recessive forms are weak and are quickly eliminated by natural selection because they must stand or fall on their own merit. Only favourable genes survive and they are also in homozygous forms (AA BB CC. etc.) that have adapted themselves successfully to the environment in the course of evolution. Self-pollination of these successful strains only perpetuates their homozygosity and leads to no loss of vigour and outbreeding causes no heterosis.

The majority of our valuable fruit trees, vegetables and garden plants are the result of so many cross pollinations that they are heterozygous for most of their desirable genes and possess hybrid vigour to a marked degree. On account of the very fact that they are hybrids they seldom come true by seed because of segregation and recombination and resulting homozygosity for many genes which causes loss of hybrid and the appearance of undesirable traits. That is why many of them are propagated vegetatively. Thus, apples and pears are multiplied by grafting; peaches and plums by budding; grapes by cuttings and strawberries by runners. Each new individual in such cases is really a part of the old plant and has all its qualities, good and bad. All 'Golden Delicious' apples are alike, as if borne on the same tree and ail the plants retain the vigour and productiveness of the original hybrid form which they are derived. Fruit trees liked apples, pears, plums (green-gage) and mangoes, ornamental plants like dahlias, chrysanthemums, roses, and the farm plants like white potatoes, sweet potatoes, and sugar-cane rarely come true by seed because they are hybrids for many traits. They are, therefore, propagated vegetatively. Hybridization produces valuable plants and vegetative propagation preserves their valuable characters and vigour.

Male Sterility and Hybrid Vigour

Male sterility occurs in bisexual plants as well as hermaphrodite animals and ensures the production of hybrids which more often than not show hybrid vigour. Male sterility has been studies in detail in plants and it has been found that the disturbing event occurs anywhere along the sequence of steps from premeiotic stags through to the production of normal-appearing, but inviable pollen grains. Male sterile lines of maize, wheat, tomato, etc., are used to produce hybrid plants on a large scale because in such strains there is no danger of self-fertilization and emasculation is not necessary. In most of the cases male sterility is determined by a single recessive gene. In some cases, it is determined by cytoplasmic factors.

There are many other effective mechanisms for maintaining hybrid vigour in successive generation, notable amongst which are vegetative propagation and allopolyploidy. As a result of allopolyploidy different genomes are brought together and are passed on from one generation to another.

Genetic Basis of Heterosis

Several hypothesis have been put forward to explain the basis of heterosis. These are:

1. *Overdominance hypothesis:* This hypothesis was proposed by *Shull* and *East*(1908). The superioity of heterozygote is attributed to the metabolic advantage of the gene products produced by two different kinds of alleles over the products of only one kind in a homozygous form. Sometimes dominant and recessive monozygotes in such cases produce both the substances simultaneously and reap the advantages of the presence of both. Sickle cell heterozygosis (Hb^a/Hb^s) exemplify this situation. Such persons possess both normal and sickle haemoglobin, the former preventing hemolytic anemia and letter malarial infection.

2. *Dominance hypothesis:* It was proposed by *D.F. Jones* (1917) and later elaborated by *Fisher*, *Mather* and many others. According to this theory the increased vigour and size in a hybrid in due to the combination of favourable dominant genes by crossing two inbreed races. The useful characters are determined by dominant alleles. In their absence or in the presence of recessive alleles, the desired characters are lost and vigour decreases. In a crossbreed if the parents contained different dominant genes, all or most of them are inherited by the hybrid, so that the number of dominant genes in a hybrid is more that either of the parents. Most of those characters are quantitative and are controlled by a number of genes. Therefore, the phenotypic expression of these characters is the cumulative effect of dominant genes.

For example, if genes A, B, C, D, E, F and G determine the yield of a plant, an individual with at least one dominant allele of each gene will have maximum vigour and the yield will decrease with increasing number of homozygous recessive genes.

Parents	A A B B cc D D ee ff	X	aa bb C C dd E E F F
		↓	
F_1 Hybrid		Aa Bb Cc Dd Ee Ff	

In this cross either of the parents have dominant genes at only three loci but the hybrid has dominant genes on all the six loci. If all the genes have full dominance, the more dominant genes lead to hybrid superiority over either of the parents.

Biochemical Basis of Heterosis

Robbins in 1941 selected two varieties of tomato and studied the effects of vitamins on the growth rates of excised roots of these two varieties and those of their hybrids. He found that nicotinamide enhances the rate of growth of exicised roots of the Red Currant variety, but this vitamin has no effect on the growth of excised roots of the Johannesfeuer variety. On the other hand, pyrodoxin enhances the growth rate of excised roots of the latter, but not of the former. A great surprise was in store for *Robbins* who found the neither nicotinamide nor pyrodoxin had any effect on the rate of growth of excised roots of a hybrid between the Rd Currant and Johannesfeuer variety. These studies indicated that Red Currant can synthesise its own pyrodoxin but no nicotinamide, Johannesfeuer can synthesise its own nicotinamide but no pyrodoxine and nicotinamide, respectively and that their recessive alleles (p and n) lose this function, results of *robbins* can be explained in the following manner:

parents	Red Currant	X	Johannesfeuer
Genotypes	PPnn	↓	ppNN
Genotype of the hybrid		PpNn	

The hybrid is able to synthesis both the vitamins due the presence of both of the dominant alleles. The hybrid vigour, therefore, is expressed due to accumulation of suitable genes in the same variety.

Heterokaryotic Vigour in Fungi

The cells of *Neurospore mycelium* are typically multinucleate. If the mycelium has arisen from a single ascospore, all the nuclei are genotypically similar. But sometimes mycelia from two different nuclei may fuse and a mycelium containing nuclei of two different genotypes may be included in the common cytoplasm. This is called a heterocaryon. *Beadle and Coonradt*(1944), studied two strains of Neurospora called *nicotinicless* and *pantothenicless*, each of which depends for its growth on the addition of either the nicotinic acid or pantothenic acid. They found that a heterocaryon of these two strains could grow in a medium without nicotinic acid as well as pantothenic acid. A heterocaryon shows better growth and vigour than either or its parents and this is comparable to heterosis.

Heteropoloidy

Each species cf a plant or animal is characterized by a particular number of chromosomes forming a haploid set. A diploid organism has two haploid sets of chromosomes in its body cells and are in genetic cells. A species is recognised by its fixed number of chromosomes and a specific morphology of each of its chromosome. At each nuclear division these regularities in chromosome number and form are duplicated, which is an exact phenomenon. Sometimes, irregularities occur during cell division or some external factor like radiation produce aberrant chromosomes. Changes in the number of chromosome may be reflected in high inviability and phenotypic anomalies in those that survive. Chromosome changes are classified in terms of addition or elimination of parts of chromosomes, whole chromosomes, a whole sets of chromosomes (genomes).

Kinds of Heteroploidy

On the basis of variation in the number of chromosomes we can divide heteroploidy into:

1. Euploidy

The term euploidy consists all the changes involving entire set of chromosome. The euploidy may be of following types:

(1) *Monoploidy:* The organism with only one set of chromosomes is called *hemizygous* or one *genome* i.e. n in the nuclei of their body cells. This stage may be normal in a particular species, or may arise as an abnormality. Bacteria and fungi pass most of their life cycle in the haploid stage. The gametoplytes of lower plants are haploid. The monoploid plants are often weak and sterile as seen in *Sorghum*,

Tritucum, Hordeum, and *Datura.* Among animals, haploid forms ocean in insects specially in honey bee and wasps. Rotifers also exhibit this stage. These forms reproduce parthenogenatically.

Morphology of haploids: Haploids are characterized morphologically by a reduction in size of all vegetative and floral parts. They are less vigour than animals. Only rarely a haploid may equal or exceed the diploid in size. When haploids were studies in *Nicotiana* by *D. Kostoff,* it was found that the leaves, flowers and overall plant size were smaller. The seed size, stomata size and pollen diameter were also generally smaller in haploids than in the diploids. In many cases in haploids the size of nucleus measured as nuclear volume was half as much as in diploids.

Cytology of monoploids. In a monoploid or haploid set, the chromosomes are non-homologous and lack any homologue to pair with, therefore, they are found as univalents at metaphase I of meiosis. During anaphase I, these univalents distribute at random. For instance, a haploid in maize (2n=20) will have 10 chromosomes and the number of chromosomes in a gamete can range from 0-10. Consequently, considerable sterility will be found in a monoploid maize. Further, some gametes will contain complete haploid set of chromosomes, and thus, will act in a normal fashion.

Origin and production of haploidy: In insects specially males of group Hymenoptera are produced due to parthenogenasis. Their all members of the colony except males are diploid females. The origin of haploidy is still uncertain, in most of the cases. Nevertheless certain causes of their origin may be; an unfertilized egg develops into a new plant, the stimulus for such development being provided by the pollen (female parthenogenesis); the male nucleus without involving itself may stimulate the egg to develop (male parthnogenesis). The presence of pollen has been found to be influential in the parthenogenetic development of embryo in *Zea mays (Chase,* 1949). *Kostoff* (1929) produced a haploid *Nicotina* with the characteristics of the diploid parent. It developed from the male nucleus after it had entered the embryo sac and the female nucleus of the egg, failed in its function. Sometimes alongwith the normal development of diploid zygote, an haploid embryo may develop in the same seed. The haploid embryo may arise from any other cell in the embryo sac than the egg. Simultaneous development of diploid and haploid seedlings from the same seed has been reported in rice. Certain environmental factors, low temperature etc., or some sort of shock, may also induce haploidy.

Haploids can be artificially produced by any one of the following methods: (i) X-rays treatment, (ii) delayed pollination, (iii) temperature shocks, (iv) colchicine treatment, (v) distant (interspecific or intergeneric hybridization, (vi) anther or pollen culture.

(2) *Diploidy:* This is the normal stage with two genomes or two sets of chromosomes. Most of living organisms including animals and plants belong to this category.

(3) *Tiploidy* : Organism with three sets of chromosomes or genomes (3n) in the nuclei of their cells. Triploids were first reported in *Oenothrea* by *Gates* 1908. Then after *Ohno* (1963) and *Pennock* (1965) have reported in reptiles, birds and other vertebrates. Triploidy has also been observed in many crop plants like rice, millets and mustard.

Origin of Triploidy: A triploid may originate by the union of a monoploid gamete (n) with a diploid gamete (2n). The extra set of chromosomes of the triploid is distributed in various combinations to the germ cells, resulting in the genetically unbalanced gametes. In general the triploidy formed as a result of fusion between the diploid egg and a haploid male gamete are more functional and strong than in the reverse case. The cross between tetraploid and diploid would also give rise to a triploid.

Effects of triploidy: The triploids can, however, reproduce by asexual methods. Many triploid varieties of apples and pears are perpetuated by vegetative methods, grafting and budding. They maintain their desired characters. Many of the triploid grasses, vegetables and garden flower varieties of the plants are not as good stable as their diploid relatives. The triploids resemble their ancestors morphologically but they are comparatively more robust and have a high degree of infertility. Because of the presence of more vigour in the triploids, they have a better chance of survival than their diploid counterparts, provided they could reproduce themselves by an asexual mode. The triploids may even replace (displace) the diploids, for example, the tuberous species of *Solanum,* the diploids and triploids survive side by side, but in *Tulipa,* the triploids have displaced the diploids.

(4) Teraploidy

The organisms with four genomes (4n) in the nuclei of their somatic cells are called tetraploids. The tetraploidy is arised by somatic doubling of the chromosome number. Doubling is accomplished either spontaneously or it can be induced in high frequency by exposure to chemicals such as *colchicine, acenaphthene* and *veratrine* and also by exposure to heat or cold, e.g., wheat, *Chrysanthemum Solamm,* man (*Carr,* 1967), salamander, etc.

Effects of tetraploidy: The physiological characters of tetraploids are also somewhat different than the diploids. Their development is slow though they are more harder. The vitamin 'C' content of tetraploid potatoes was found to be almost double than the diploids. Similarly the nicotine content of tetraploid tobacco was double than in the diploids.

The tetraploids usually possess low fetility. Tetraploidy has been exploited in the production of cereals like rye (*Muntzing,* 1951), ornamental flowers and fruits like grapes, apples etc.

The tetraploids appear to be more adjusted to grow in variety of habitats than the diploids. Though they may grow alongwith their diploid relatives, but still, they have a wider range of a geographical distribution and distinct ecological preferences. Even with all these variations, there is tendency to consider autopolyploids as cotypes or sub species of the diploid form rather than distinct specific entities.

POLYPLOIDY

History of Polyploidy

The scientific study of polyploidy started in the first decade of this century. The older reference are obscure. But in nature, polyploidy has been operative since Pre-cambrian or rather, since the process of evolution started. The study of polyploidy was on its way when *Lutz* (1907), and *Gates* (1909) recognized that one of the *de Vries* original mutation of *Onenothera lamarckiana* mut. *gigas* was a tetraploid. *Tahara* (1915) observed the phenomenon in *Nicotiana tabacum* and *Chrysanthemum. Winkler* (1916) worked on the artificial production of a teraploid form of *Solanum nigrum*. An important breakthrough came in 1917 when *Winge* formulated a hypothesis regarding the origin of species. According to him hybridization followed by polyploidy is a method for the origin of species. *Longley* (1924) carried on studies of polyploidy on *Euchlaena* and *Crane* and *Darlington* (1927) on loganberry. The work of Russian geneticist, *Karpechenko* (1926, 27, 28) can be considered an important milestone in the study of polyploidy. He crossed radish with cabbage and bred a new polyploid species, *Rhaphanus brassica. Karpechenko's* polyploid was perhaps the first artificial species, created by man, in all history.

Much of our knowledge about polyploidy is indebted to the classical work of Swedish scientist *Muntzing*. His contribution towards the different aspects polyploidy can neve be easily forgotten. In 'Hereditas' of 1936, he contributed a remarkable paper on "The Evolutionary Significance of Autopolyploidy." *Muntzing* also carried elaborate studies on the distribution and frequency of polyploidy in different groups of plants.

A great impetus, to the study of polyploidy was provided by *Blakesle* and *Avery* (1937) when they found out a method including polyploidy (doubling of chromosomes) in plants by using a chemical colchicine. This work opened new vistas in the field of breeding and improvement of crop plants. Immediately after, *Nebel* and *Ruttle* (1938) applied colchicine treatment on *Tradescantia, Petunia* and many other plants and they tried to evolve new varieties. The work of *Lewis* (1941) can be considered important in that he found out the relationship between polyploidy and the fruiting habit of the plants. The relation, that many of our crop plants are polyploid, may be quite old, but the decisive, evidence towards this aspect was proveded by many workers in the forties. *McFadden* and *Sears* (1946) confirmed this for *Triticum aestivum*, while *Stephens* (1946) confirmed this for *triticum aestivum*, while *Stephens* (1947) studied the polyploidy in *Gossypium hirsutum*. It shall not out of way to mention here that *Janaki Ammal*, one of the foremost Indian cytologists, did a commendable work on the subject in the fifties. *While Riley* (1961) was studying the diploidization of polyploid wheat, *Muntzing* did the work on genetics in relation to polyploidy. Credit for induced polyploidy must also be given to Indian Scientist H. Swarup for his valuable contribution in this field. He induced polyploidy in vertebrate species.

An organism with more than two genomes is called a polyploid. There are mainly three different kinds of polyploidy:

(i) Autopolyploidy
(ii) Allopolyploidy
(iii) Autoallopolyploidy.

(i) Autopolyploidy: The term antopolyploidy refers to those organisms which have orginated within a single species (intra-specific). Here chromosome sets of same species become duplicated to the original basic number.

They occur in nature to some frequency. According to *Clausen, Keek* and *Hiesey* (1945) and *Stebbins* (1950), autopolyploids are quite rare in nature. Their view is in contrast to once prevalent thought that they are quite common (*Muntzing,* 1936). The artificial induction, by treatment with colchicine, has resulted in the production of many autopolyploids in large number of species. Many of the common plants like maize, rice, cotton, Chrysanthemum are autopolyploids.

(a) Origin and production of autopolyploids: The autopolyploids may occur in nature or may be artificially produced. When they are found in nature, their autopolyploidy relvant to Northern India, pertains to common '*doob*' grass (*Cynodon decytylon*). In U.P. and Bihar, common '*doop*' grass was found to be an autotriploid as inferred from its meiotic behaviour (*Gupta* and *Srivastava,* 1970). It is perhaps successful due to efficient vegetative reproduction because, as will be seen, autotriploids are normally sterile and set no seeds. Autotriploids are also known in water-melons, sugarbeet, tomato, grapes and banana, although in several of these cases these polyploids have been artificially produced. Similarly autotetraploids are known in rye (*Secale cereale*), corn (*Zeamays*) red clover (*Trifolium pratense*), berseem (*Trifolium alexandrium*), marigolds, snap dragons, lilly, phlex, grapes apples,

etc. In *Oenothera lamarckiana*, an American plant, a giant mutant described by *de Vries* was later discovered to be an autotetraploid.

Effects of autoploidy: Autotetraploids usually differ from their diploid relation in phenotypic characters. The tetraploids are more robust, tall, have more thick stems and larger leaves of more dark green colour. The reproductive parts like flowers, fruits and seeds are also larger. The cells and the nuclei of tetraploids are larger. The variations also occur in the size of the stomata and the epidermal cells.

(ii) Allopolyploidy: These are produced as a result of hyloridization i.e., by the introduction of new genomes. For example if two diploid species or varieties having AA and BB chromosome complements respectively are crossed, in normal cases a diploid variety having AB chromosome complement will be produced. But due to irregularity in cell division or anyhow one of the parent varieties form a diploid gamete is fertilized by a haploid gamete (AA X B) or BB X AA), the resultant (AAB or ABB) will be a allotriploid. In similar manner (AABB) will be allotetraploid, (AAABB) allopentaploid and so on.

The best known example of this type is *Raphanobrassica* belonging to family cruciferae. This allotetarploid was obtained by *Karpechenko* (1926) by making an intergeneric cross between *Raphanus sativus* (Radish) and *Brassica oleracea* (Cabbage). Both the species hybridized are diploid and each has eighteen (18) chromosomes number. The gametes formed by each of these have 9 chromosomes i.e. n = 9. When the two species were crossed, the F_1 had 18 chromosomes-9 from *Raphanus* and 9 from Brassica. The hybrid was sterile because there was no or very little homology between the chromosomes coming from the two different genera. This resulted into the failure of the formation of normal gametes. However, some spore mother cells without undergoing normal reduction division, formed diploid gametes having the same chromosome number i.e. 18. The result between the fertilization of two such gametes was the production of F_2 individuals with 36 (18+18) chromosomes i.e. 9 pairs of homologous chromosomes. This tetraploid had 18 chromosomes of *Raphanus* and other 18 of *Brassica* type.

This teraploid had 18 chromosomes of *Raphanus* type i.e. 9 pairs of homologous chromosomes and other 18 of *Brassica* type froming

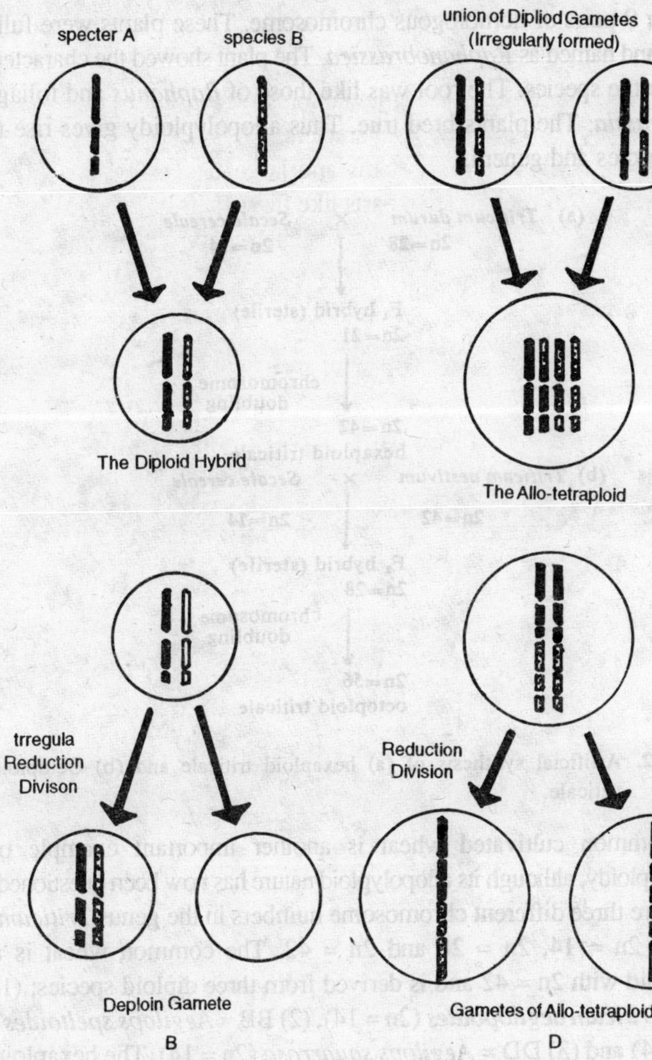

Fig. 21.1. Diagram to show mechanism of allotetraploidy.

A—Diploid hybrid of species A + B
B—A diploid gamete is formed from the hybrid species
C—Union of two diploid gametes of the hybrid (formation of allotetraploid)
D—Gametes formation in allotetraploid.

another 9 pairs of homologous chromosome. These plants were fully fertile and named as *Raphanobrassica.* The plant showed the characters of both the species. The root was like those of *Raphanus* and foliage like *Brassia.* The plants bred true. Thus allopolyploidy gives rise to new species and genera.

(a) ***Triticum durum*** × ***Secale cereale***
2n=28 **2n=14**
↓
F_1 hybrid (sterile)
2n=21
↓ **chromosome doubling**
2n=42
hexaploid triticale

(b) ***Triticum aestivum*** × ***Secale cereale***
2n=42 **2n=14**
↓
F_1 hybrid (sterile)
2n=28
↓ **chromosome doubling**
2n=56
octoploid triticale

Fig. 21.2. Artificial synthesis of (a) hexaploid triticale and (b) Octoploid triticale.

Common cultivated wheat is another important example of allopolyploidy, although its allopolyploid nature has now been questioned. There are three different chromosome numbers in the genus *Triticum,* namely 2n = 14, 2n = 28 and 2ń = 42. The common wheat is a hexaploid with 2n = 42 and is derived from three diploid species; (1) AA = *Triticum aegilopoides* (2n = 14), (2) BB = *Aegilops speltoides*? (2n = 14) and (3) DD = *Aegilops squarrosa* (2n = 14). The hexaploid wheat, therefore, is designated as AABBDD, the tetraploid (2n = 28) as AA BB and diploid 92n = 14) as AA. There is, however, evidence available now which suggests that A, B and D genomes from three diploid species mentioned above are not much different from one another, so that it is believed that the three diploid progenitors of common hexaploid wheat were derived from a common ancestor. For this reason, the common hexaploid wheat is now considered an autopolyploid rather than an allopolyploid. At best, it may be segmental allopolyploid.

Synthsized allopolyploidy: To find out the origin of naturally occuring allopolyploids some cytogeneticists produced certain allopolyploids in laboratory by employing artificial means. Common hexaploid wheat and teraploid cotton furnish two such exampies.

(a) *Triticum spelta,* is a hexaploid wheat which was artifically synthesized in 1946 by *E.S. McFadden* and *E.R. Sears* and also by *H. Kihara.* They crossed an emmer wheat, *Triticum dicoccoides,* (tetraploid : 2n = 28) with goat grass, *Aegilops squarrosa* (diploid: 2n = 14) and doubled the chromosome number in the F_1 hybrids. This artificially synthesized hexaploid wheat was found to be similar to the primitive wheat *T. spelta.* When the synthesized hexaploid wheat was crossed with naturally occurring *T. spelta,* the F_1 hybrid was completely fetile. This suggested that hexaploid wheat must have originated in the past due to natural hybridization between tetraploid wheat and goat grass followed by subsequent chromosome doubling.

P_1 : *Triticum dicoccoides* X *Aegilops squarrosa*
(Tetraploid emmer wheat) (Diploid goat grass)
AA BB DD
(2n = 28 : 14 bivalents) (2n = 14 ; 7 bivalents)
F_1
↓
ABD

Triploid hybrid (wn = 21; 21 univalents)
Colchicine
↓
AA BB DD
Synthesized hexaploid wheat
(*Triticum spelta*)
(2n = 42; 27 bivalents)

(b) In the primrose, *Primula kewensis,* the diploid number of chromosomes in 36. The species is an allotetraploid formed by a cross between *Pimula floribunda* and *Primula veticillata,* both with the haploid chromosome number of 9. The hybrid between *P. floribunda* and *P. verticellata* were sterile. However, a hybrid plant had a branch containing 36 chromosomes, which must have arisen by doubling. Propagation of this branch resulted in a fertile primrose plant, the allotetraploid *P. kewensis*.

(iii) *Autoallopolypoidy*. These are the types which are found only at the level of hexaploidy or above. In such polyploids one of the genomes has at least 4 sets of chromosome complement. The doubling of allotetraploid would give an octaploid having the characteristics of both auto and allopolyploid with karyotype AAAA $B_1B_1B_1B_1$. This could easily arise by hybridization between two auto-tetraploids *AAAA* and $B_1B_1B_1B_1$ followed by doubling. Mostly these polyploids are hexaploid, having a general formula of AAAA BB.

The term *autoallopolyploidy* was first used by *Kostaff* (1939). He has cited the example of *Helianthus tuberosus*. It is an hexaploid with 2n = 102. After a cross with the diploid *H. annuus* (n=17), this hexaploid forms a tetraploid hybrid having 34 bivalents at meiosis. The genomic formula of H *tuberosus* in $A_1A_1A_1A_1A_1A_1$ and that of H. *annuus*, B_1B_1 *(Stebbins*, 1950). The genomes with the same type of letters, have much in common, so they can pair normally.

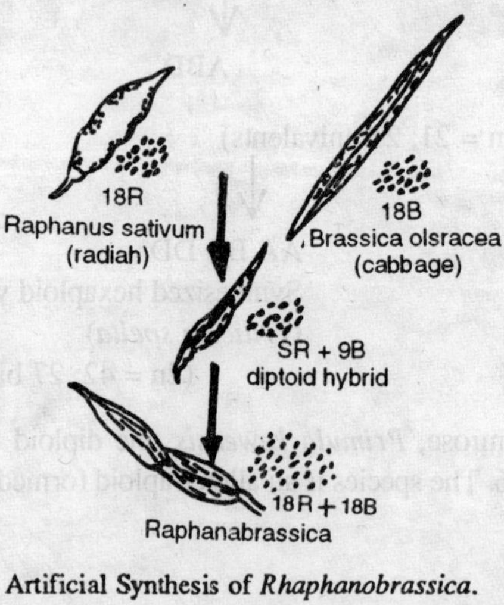

Fig. 21.3. Artificial Synthesis of *Rhaphanobrassica.*

The other two well-known examples of such polyploids are *Phleum pratense* and *S. nigrum. P. pratense* was considered to be an allopolyploid (*Gregor* and *Susome*). But *Nordenskiold* adjudged it to be an autopolyploid. But the chromosomal studies, by *Levan* in one species and by *Nordenskiold* in the other, indicated the formation of 7 bivalents and 7 univalents (7_{11} 7_{1}). So the genomic formula should be AAB and the diploid complement as AAAABB. In case of *S. nigrum, Jorgenson* (1928) considered it to be allopolyploid, but *Nakamura* (1937) adjudged it to be an autoploid. The haploids, of this species (n=36) we found to form 12 bivalents and 12 univalents (21_{11} 12_{1}). So this species in an autoallopolyploid, deriving four genomes from one species and two from the other. Such autoallohexaploids are though to have originated through the union of a diploid gamete AA and B gamete, followed by the doubling. Such an hypothesis for the origin of autoallopolyploid could be held that most of the polyploids, at the octaploid level or higher, are not ture autopolyploids and various combinations of the auto and allopolyploid conditions exist.

Segmental Allopolypoidy

This term was coined by *Stebbins* and may be defined as a polypoidy containing two pairs of genomes having in common many homologous chromosomal segments but differ from each other in respect to number of genes. Examples are *Primula kewensis, Crepis foetida-rubra* etc. These genomes may be designated $A_1A_1A_2A_2$ and individual chromosomes as $1A_1$ $1A_1$, $2A_2$ $2A_2$... $nA_1nA_1aA_2nA_2$ where n is haploid number. The pairing between two $1A_1$ and $1A_2$ chromosomes is called *heterogenetic association* and between $1A_1$ chromosomes is *homogenetic association.*

Induced Polypoidy

Polyploidy has been induced experimentally is a number of plants and animals by following agencies:

1. *By decapitation*: In this processs the tips of branches are cut off near the ground surface and stems and roots are left to grow. The surface of the cut stem becomes covered with callus tissue which contains polypoid cells.

2. *By temperature: Blakeslee* subjected the *Datura* plant to cool temperatures is summer and he obtained some polyploid plants. Similarly *Randolph* used electric current on young corn plants and obtained polyploids.

3. *Chemicals: Blakeslee, A.G. Avery* and *B.R. Nebel* in 1937 used an alkaloid *colchicine* in various concentrations to produce polyploids. When root tips or plant tips were placed in its solution they showed polyploidy. Similarly other chemicals like *indole-acetic acid, acenapthene, podophyllin* and *sulfanilamide* are effective in polyploid production.

4. *Raditions:* Both radium and X-rays are effective in the production of polyploidy in plants by interfering in their cell divisions.

5. *Hybridization:* By crossing various euploid races hybrids may be produced of varying degree of polyploidy. By a cross between haploid and diploid race a triploid race can be obtained.

II. Aneupolyploidy

The polyploids contain more than two sets of chromosomes i.e., all the chromosomes are present more than two times in each cell. Sometimes in addition to normal number of chromosomes one or more extra chromosomes are present or one or more chromosomes are missing. This condition is termed aneuploidy. This condition is produced due to irregularity in the cell division. If in spore mother cell (2n), two chromosomes of a pair, some how do not separate form each other during reduction division; two types of gametes will be formed, one having an extra chromosome (n+1) and other having a less chromosome or lacking one chromosome (n-1). If n+1 gamete fuses with anormal gamete n, the zygote will be 2n+1 that means one of the chromosomes will be in triplicate. Such individual is called trisomic for the particular chromosome which is present in triplicate. The union of n-1 with n will form 2n-1 zygote. In this one chromosome is present only once and not in duplicate. Such organism is called monosomic. Similarly individuals having more extra chromosomes or lacking more chromosomes are produced.

Kinds of Aneupolyploidy

On the basis of loss on gain of one or two chromosomes the aneupolyploidy can further be divide as follow:

(i) *Monosomic*. The monosomics lack one chromosome of any pair and more devoted by a general formula 2n-1. This condition is developed by *non-disjunction* during gametogenesis and results the formation of two types of ova one ovum having both the sex chromosomes and one without sex chromosomes. If the ovum which is devoid of any sex chromosome is fertilized, the monosomic condition occurs. In plants gametes having n-1 genome do not survive while in animals, that may cause genetic imbalance and results in either high mortality of reduced fertility, *Bridge* has reported two kinds of monosomic aneuploidy, e.g., aneuploidy of sex chromosome and, aneuploidy of the autosomes. Examples are *Drosophila*, wheat and tobacco. Monosomic condition may be one single (2n-1), double (2n-1-1), or triple (2n-1-1-1).

(ii) *Nullosomic*. Here the organisms lose a pair of chromosome. They have 2n-2 genomic formula. Nullosomic type can arise from the union of two monosomic gametes or from non-disjunction of chromosomes. Nullosomic individuals usually do not survive but their polyploid forms (nullosomic polyploid) may survive but they remain weak and lose their fertility. Wheat (6n-2) is the excellent example.

(iii) *Trisomic*. Trisomics are those organisms, which have an extra chromosome (2n+1). Since the extra chromosomes may belong to any one of the different chromosomes of a haploid complement, the number of possible trisomics will be equal to the haploid chromosome number. For instance, we know that haploid chromosome number in barley in n=7, consequently seven trisomics are possible. Trisomics, where the extra chromosome is identical to its homologues, are called *primary trisomics*. Besides these, there are secondary and tertiary trisomics. While the secondary trisomic means that the extra chromosome should be an isochromosome (both arms genetically similar), a tertiary trisomic would mean that the extra chromosome should be the product of a translocation. Trisomics were obtained for the first time in *Datura stramonium* (jimson weed) by *A.F. Blakeslee* and his co-workers. Since the haploid chromosome number in this species is n=12, 12 primary trisomics, 24 secondary trisomics and a large number of tertiary trisomics are possible. Most of the trisomics were identified by the size, shape and other morphological featues of the fruit.

(iv) *Tetrasomic.* In tetrasomy condition two complete sets of chromosomes plus two extra chromosomes of one particular chromosomes are present. The genomic formula is 2n+2. Here a particular chromosome is represented in form doses. The double tetrasomic is obtained when two homologous to one pair and homologous to a second pair of chromosomes occur. The genomic formula is 2n+2+2.

Aneuploidy in Human Beings

Tjio and *Levan* (1956) first of all demonstrated that somatic cells of man have 46 chromosomes each. Out of 23 pairs 22 are *autosomes* and one sex *chromosome.* Normal females are XX and males are XY. Different autosomal pairs have distinct morphology and karyotypic analysis is simple. Aneuploidy is man can be seen cytologically and results into many physical, mental and reproductive abnormalities. *Barr* and *Bertram* (1949) discovered that besides the chromosomal differences, male and female nuclear differ in their morphology too. Chromosomal differences can be seen only in dividing cells. Resting nuclie of females cell have a dark staining spot whereas nuclei of male cells lack this chromatin positive body.

Named after its discoverer, the dark staining spot is called Barr body. many cases of sexual abnormalities are associated with the absence or presence of Barr bodies. Males with Klinefelter's syndrome have one or more Barr bodies in their cells whereas females with Turner's syndrome jack the Barr body. There is a direct relationship between the number of Barr bodies and the number of X chromosomes. Thus by just looking at the number of Bar bodies in a cell one can find out the number of X chromosomes. According to the "Lyon hypothesis", which was first presented by *Lyon* and *Russel* (1961), all but one of the X-chromosomes become inactivated or under go heterochromatisation and the inactivated chromosomes are present as *Barr bodies*. In normal females one of the two X chromosomes is inactivated and, therefore, the ratio between the active X chromosome and autosomes is the same in normal males and females. In contrast to *Drosophila,* therefore, the human Y chromosome is essential for maleness. Females with more than two X chromosomes are viable because all of them except one are inactivated. Females without an X chromosome are not viable. Inactivation of the additional X chromosome or chromosomes is random. If different X chromosomes contained different alleles, random inactivation will result into tissues which will be mosaic for different characters (genes). This is evident in some mammals like cats in which a heterozygous individual (Yy, where Y controls black fur and y controls yellow fur) produces random patches of yellow and black fur. In leucocytes of mules too it is a matter of chance whether the X chromosome of horse or donkey will be active.

Individuals with Turner's syndrome have only 45 chromosomes in each of their somatic cells. They have only one X chromosome. Such females are generally mentally retarded, sterile and have hearing defects from the very birth. The adult symptoms are faulty or incomplete development of ovaries, short stature, abnormal jaw formation, webbd neck, low set ears, shieldlike chest and puffy hands and feet. Some of these symptoms can be recognised even in the infants. Females with more than two X chromosomes have normal growth and development but suffer from mental abnormality and sterility.

REVIEW QUESTIONS

1. Let the following represent a pair of homologous chromosomes in another inversion heterozygote. (The 0 represents the centromere.)

HIJ KLMN hlk jimn

——0—————— ——0————

Assume that a crossover takes place between the loci "k" and "l". Give the two products resulting from the crossover event.

2. In the inversion heterozygote in Question l, what will be the two products, assuming a double crossover occurs between the same two strands, the first one between c and d, and the second crossover between e and f?

3. Let the following represent a pair of homologous chromosomes in an inversion heterozygote. (The 0 represents the centromere.)

```
   ABCDEFG              abfedcg
—0————————————     ——0——————————
```

Assume that a crossover takes place between the c and the d loci at meiosis. Give the two products which will result from the crossover event.

4. In the inversion heterozygote in Question 1, what will be the two products, assuming a double crossover between the same two strands, one crossover between "i" and "j", the other one between "k" and "l"?

5. Assume that an individual has one chromosome with the gene order as follows:

```
    R S T U V W X Y Z
—0————————————————————
```

In the homologue, a shift resulted in the insertion of the segment TUV between "X" and "Y" Show how this situation can lead to the origin of a chromosome with a duplication and one with a deficiency.

6. A certain race of *Oenothera* carries two complexes of seven chromosomes each. One complex of seven is called alpha; the other set of seven is beta. Following self-fertilization, what will be the composition of the zygotes? Indicate which ones will survive.

7. Assume that a deletion arises in the X chromosome of the mouse and removes a gene which is required in at least single dose for the survival of the embryo.

   A. What would be the result of crossing a female carrying the deletion and an ordinary male?

   B. Assume that the deletion of an autosomal gene required in single dose accumulates in a population. What will be the outcome when carriers of the deletion mate with each other?

8. Species A has a diploid chromosome number of 24; so does another plant, Species B. After many attempts a highly infertile F_1, is produced from this F_1 a fertile allotetraploid is eventually derived.
   - A. Give the chromosome number of the F_1 plant.
   - B. Give the chromosome number of the fertile derivative.

9. Assume plant species No. 1 has a diploid chromosome number of 16. Give the following chromosome numbers:
   - A. An autotriploid derivative of species No. 1.
   - B. A trisomic member of species No. 1.
   - C. An allotetraploid derived from a cross with species No. 2, which has a haploid chromosome number of 8.
   - D. A monosomic member of species No. 1.

10. Assume that three separate populations of plants A, B and C have many similar phenotypic features. However, many combinations of traits make each population distinct from the other. When plants from Population A are crossed with those of Population B, the F_1's are intermediate in phenotype, but they are completely sterile. When plants from Population B are crossed with members of population C, very fertile intermediate F_1's arise. Plants from Population A crossed with those from C give F_1's which are intermediate between A and C, but they are sterile. Give an explanation of these observations.

11. Give at least one way to account for the origin of humans with the following chromosome constitutions:
   - A. XXY.
   - B. XO.
   - C. XYY.
   - D. A person with only one normal metacentric chromosome No. 1 but with an unusual chromosome the size of No. 1 but which is acrocentric.
   - E. Three complete sets of chromosomes.
   - F. A person with only one chromosome No. 21 and only one No. 13 plus one unusually large chromosome.

G. XXX.
H. XX/XO mosaic.
I. XXY/XX mosaic.
J. XY/XYY/XO mosaic.

12. Match the disorder in Column A with the letter of an appropriate chromosome anomaly in Column B (there may be more than one possible choice in a given case and an item in column B may be used more than once):

A	B
— Turner syndrome	1. polyploidy
— Klinefelter	2. aneuploidy
— syndrome	3. monosomy
— Down's syndrome	4. triploidy
— *cri-du-chat*	5. inversion
— syndrome	6. translocation
— Philadelphia	7. deletion
— chromosome	8. non-disjunction.
— XYZ male	

13. Occasionally males with Klinefelter syndrome are found to have an XX sex chromosome constitution. No Y chromosome is present! Offer an explanation using information on the sex chromosomes.

Select the correct answer or answers, If any, which best complete each of the following (Question 18-24):

14. The dominant "P" in Drosophila results in red eye colour. Its recessive allele "p" is associated with purple eyes. The locus of this pair of alleles is in the euchromatin. A shift places the "P" allele next to heterochromatin. A fly arises with the following genotype:

$$\frac{\text{P (heterochromatin)}}{\text{P(euchromatin)}}$$

A very possible result would be that:

A. Purple eye colour must be expressed, since red eye colour cannot be expressed at all.
B. The fly must have red eyes, since purple cannot be expressed due to the chromosome change.
C. The fly could have eyes with patches of red and patches of purple.
D. The fly would die as a result of genic unbalance.

15. Allow "A" to represent a haploid chromosome complement in a plant species. Suppose that an individual is AAAA. The following can be said:

A. The individual is certain to be larger than the diploid.
B. The individual may possibly grow more slowly than the diploid.
C. This is an example of euploidy.
D. There my be some reduction in fertility due to irregular chromosome segregation.
E. This is an example of tetrasomy.

16. A diploid gamete (AA) fertilizes a normal haploid gamete (A) of the same species. The following can be said about the resulting offspring:

A. The offspring is an allotriploid.
B. The offspring will be completely fertile.
C. The offspring will be quite infertile due to the random segregation of trivalents.
D. The offspring will be quite infertile due to the lack of chromosome homology.
E. The offspring is an autotetraploid.

17. Tetraploid races are rare among animals because:

A. Diploid gametes cannot arise in animals.
B. The sexes are usually separate in animals.
C. Polyploidy is lethal in all animals.
D. Tetraploidy produces individuals which are not normal in the balance of sex chromosomes to autosomes.

18. An individual is heterozygous for a pericentric inversion. Which of the following would be true:

A. The individual would probably be abnormal in some way, since certain genes are deleted.

B. The inversion can be recognized cytologically by the formation of cross figures or circles during meiotic prophase.

C. Chromosome with two centromeres will result if crossing over occurs within the inverted segment.

D. A complex loop will be seen cytologically, since one inversion overlaps the other.

E. The position of the centromere will be shifted if breaks on either side of it are not equidistant.

19. Reciprocal translocations:

A. Will be widespread in nature if there is no crossing over in the male.

B. Result in unbalanced gametes if alternate chromosomes in the multivalent move to the same pole.

C. Result in the formation of acentric fragments at first meiotic prophase.

D. Do not occur in animal cells.

E. Can result in shifts of chromosome arms and bring about new linkage arrangements.

20. In *Oenothera*:

A. Typically no bivalents form at meiotic prophase.

B. The pure-breeding individuals are homozygotes with very favourable gene combinations.

C. At least two lethals are present, and yet the plants are vigorous.

D. All the chromosomes from the male parent go to the same pole at meiosis, and all from the female parent go to the opposite pole.

E. No chromosome in the complement is completely homologous ot any other one.

ANSWERS

1. H I J ——0—— K l h and

   n m i j ——0—— k L M N

2. A B C de F G ——0—— and

   ——0—— a b f E D e g

3. The dicentric ——0—— A B C de f b a ——0——

   and acentric g c D E F G

4. H I j ——0—— k L M N

   h I K ——0—— J i m n

5. At meiosis, each chromosome will form a loop which includes the genes TU and V.

6. alpha-alpha (dies); alpha-beta (survives); beta-beta (dies).

7. A. One half of the male embryos would die so that the sex ratio would be 2 females: 1 male. One-half of the females carry the deletion.

   B. One fourth of the offspring from two carrier parents will fail to survive, just as if a recessive lethal gene were present.

8. A. 24.

   B. 48.

9. A. 24.

   B. 17.

   C. 32.

   D. 15.

10. Two species can be recognized. Populations B and C, while

phenotypically distinct, have not accumulated genetic differences which limit free gene exchange. The two populations may be considered varieties of the same species (let us say, species B). Population A is genetically isolated from members of the other two populations (members of Species B), as seen by the lack of fertile hybrids when A members are crossed to B and C members. Population A is thus composed of individuals of a different species (let us say, Species A).

11. A. Non-disjunction of X chromosome at first anaphase of oogenesis giving an XX egg which becomes fertilized by a Y-bearing sperm. Also non-disjunction of the X and Y chromosomes at first anaphase of spermatogenesis, giving an XY sperm which fertilizes an egg.

B. Non-disjunction of sex chromosomes at spermatogenesis or oogenesis, producing gametes with no sex chromosomes. Also loss of one X chromosome at first mitotic anaphase of an XX zygote.

C. Nondisjunction of Y chromosome at second anaphase of spermatogenesis.

D. A pericentric inversion.

E. Complete failure of anaphase movement at meiosis producing an unreduced sperm or egg which unites with a haploid gamete.

F. A translocation.

G. Non-disjunction of X chromosomes at oogenesis producing an XX egg which becomes fertilized by an X-bearing sperm.

H. Loss of an X chromosome during mitotic divisions of an XX embryo.

I. Loss of a Y chromosome during mitotic divisions of an XXY embryo.

J. Non-disjunction of Y chromosome during mitotic divisions of an XY émbryo.

12. Turner syndrome : 2, 3, 8
Klinefelter syndrome : 2, 8
Down's syndrome : 2, 6, 8

cri-du-chat syndrome : 7

Philadelphia chromosome: 6

(previously considered a deletion)

XYY male: 2,8.

13. Testis determination seems to depend on very few (possibly just one) Y-linked loci. This small region could have become inserted into an X chromosome at the time of chromosome pairing at meiotic prophase. The rest of the Y chromosome could have been lost following this at meiotic divisions. The outcome could be a gamete lacking a visible Y but containing the tiny, testis-determining region.

14. C.

15. B,C,D.

16. C.

17. B,D.

18. E.

19. E.

20. A,C,D,E.

22

Mutation

Inheritance is the transmission of characters from one generation to the other. It is based on genes, which are located in chromosomes. These are duplicated and passed on to the progeny through the gametes during sexual reproduction. Thus the genes of children are the copies of the genes of their parents. The copies are exact and are made by an exact and correct method of gene duplication. Occasionally this exact and accurate method of gene duplication goes wrong, a copy of the gene being not an exact copy but slightly different from the original is produced. The gene now is a modified gene. It has undergone a structural change. This is called gene mutation. The original gene has mutated to a modified form. Thus the term mutation refers both to the change in the genetic material and to the process by which the change occurs. Mutations are the drastic change in the genotype of organisms arising at any stage of the life cycle, resulting into the production of true breeding varieties.

Darwin's and Batesons Variations

Darwin noticed several examples of suddenly arising characters of new hereditary type in plants and animals. To such characters he called *sports*. He did not give much importance to these variations. He was in favour of the view that the continuous variations, caused by the effect of polygenes, have main role in the evolution. *Bateson* (1894) called such changes discontinuous variations and regarded these to be the chief source of evolution. He said that in the absence of such large variations, evolution would not have progressed so far.

De Vries' Variations

The term 'mutation' was applied by *Huge De Vries* (1901-1903) for large discrete changes in the genotype of the organisms. He strongly supported the view of *Bateson* and proposed his mutation theory to account for the organic evolution. He was highly impressed by the suddenly arising changes in the genotype of a plant of *Oenothera lamarckiana* (evening primrose). This plant is a native of America which *De Vries* grew in Holland, and marked unusual variations in this plant.

Modern Concept

According to modern concept, the mutation is considered to be sudden change in the genotype of an animal or a plant. *Morgan*(1909) studied mutations in *Drosophila* and showed that mutations may be of all magnitudes. Small mutations are called the *macromutations*; here the mutant differs very slightly from its parents and the difference can only be studied by special techniques. The large mutations are called the macromutations; in these cases the difference between the mutant and parents is large. Both the types of mutations have an important role in the evolution. They result into the production of new varieties which breed true in the next generations. Now-a-days by mutations we usually mean the *point mutations* which change the internal arrangement of a DNA molecule including small deletions, etc.

The mutations, when they occur, may be responsible for the origin of new species, which differ both in phenotype and genotype from their ancesotors. It was the view of *de Vries* that mutations are often indifferent, neither harmful nor helpful. Gradually both of these types of mutations become fixed in the populations. The presence of non-adaptive characters lends support to the theory of mutations; because natural selection brings about the change in a definite way.

Range of Mutations

Although the changes due to single gene mutations may involve sudden and large changes (like that of *Ancon* breed of sheep), they can also be so small that one way not be able to notice them. Therefore, the earlier concept that mutations could involve sudden and big changes only, was not correct. Mutations could also be observed which do not involve any change in the morphology but only modify the nutritional requirements as in case of *Neurospora*. Similarly, mutations could be observed in pathogens, as a result of which the pathogen could lose or acquire virulence against a particular host.

Mutation Rate

The frequency of mutation is different in different genes. Genes with rapid mutation rates are called *mutable* or *unstable*. Genes which do not mutate so easily are called *stable genes*. In man, genes associated with muscular dystrophy are estimated to mutate once in 10^5 people. Bacterial cells mutate in the order of 1 in 10^8 and a particular locus in *Drosophila* mutates in the order of 1 in 100,000 to 2000,000 flies. *Emerson* found that the R gene in maize mutates much more frequency than other, an the rate of 50 per 100,000 gametes. In some organisms some genes mutate so frequently that the individuals are simply mosaics of mutated and unmutated genes. Such mutable genes are more common in plants than in animals and are more frequent in somatic tissues than in germ tissues.

Stage at which Mutations Occur

Mutation can occur at any stage during the development. Depending upon this stage the following situations will be met;

i. If mutation occurs before the differentiation of gametes, it would influence several gametes and will thus influence the individuals derived from these affected gametes.
ii. If mutation occurs in a gamete or a zygote, a single individual will carry the mutation.
iii. If mutation occurs in a cell after the zygote has undergone division, part of the body will show the mutant character. Such mutations will be called somatic mutations.

Mutator Genes

In certain cases some genes influence the mutation rates of other genes. Such genes are called *mutator genes*. In 1938 *Rhoades* reported a gene which produces its effect by influencing another gene. Mutator genes may frequently result into mosaic plants and animals because in different cells, tissues and organs. They bring about different mutation. variations in flower colour in roses, peas, larkspur, etc., and variegated petals of many ornamentals are due to the presence of mutator genes.

Specific as well as non-specific mutator genes are known in micro-organisms too. Mutator genes are known as *controlling elements, modulators etc.*

Mutation and Gene Symbols

Most of genes are *pleiotropic*. A gene is named after the character that is most affected by its mutation. Genes are named by alphabetical letters derived from their descriptive names. Thus wx is used for the mutation producing waxy endosperm in maize. Small letters of alphabet denote that it is a recessive mutation. The wild type gene for starchy endosperm is called Wx. Similarly, vermilion (v) eye colour in *Drosophila* is recessive to the brick red (v) wild type. Sometimes the wild type gene is symbolically designated as + and mutant as—. Thus W/+ means a fly with hetrozygons far the white eye colour gene or w/W: Sometimes + and—signs are used as super scripts to designate the wild type and the mutant respectively; e.g: wx^+/wx^-.

Types of Mutations

Mutations, besides being dominant or recessive, can be sex linked or autosomal, germinal or somatic, lethal or non-lethal, useful or deleterious, etc. Most of mutations are recessive and deleterious. Sometimes, the mutation are neither useful nor deleterious, but neutral and very frequently go unnoticed.

Somatic Mutations

Mutations may occur at any stage in cell cycle and any cell may act as the seat of mutations. The effects of mutations are mostly determined by the type of the cell in which they occur. A somatic cell, after the mutation has occurred in it, will give rise to the mutant tissue. That is, if it possesses the power of reproduction. The other parts of the organism will not be affected.

The 'Delicious' apple and the naval orange have been produced as a result of somatic mutations. The changes, that ultimately produced the desirable characters in these points, originated in a single cell. That cell reproduced and formed a shoot. The mutant shoot could give rise to the new plants by vegetative propagations. Mutations may arise in buds of the fruit trees. These buds give rise to new varieties. These are generally referred to bud sports. Such somatic mutations have been observed and detected in many plants and animals. The formation of white sectors, which sometimes appear in the otherwise red eyes of *Drosophila*, can be cited as the example of somatic mutation.

Germinal Mutation

If the mutation originate in the reproductive cells of gonads then these are termed germinal mutation. Germinal mutation further may be *gametic* occuring in the gametes of individuals or *zygotic* originating in the fused diploid gamte. Various sex linked mutations are of these types and pass from generation to generation. If the mutation occurs in the dominant gene of the germ cell, the effect would appear in the next progeny. The effects on the recessive genes would not express themselves because they may be masked by other genes. The germinal mutations are not broad based like the somatic mutations. They occur only in the specific germinal cells and at the particular stage of the life cycle. Generally they occur just before or after the gametogenisis . If they occur much earlier in the germinal tissue all the gametes formed will be affected and so all members of next progency would receive the quota of mutations. If, on the other hand, the mutations occur in only one gamete, only one individual of the next generation will show the change whatever the stage of mutation, the expression in the next generation will always be affected by the dominance of the gene. A classical example of germinal mutation is that of Ancon breed of sheep. These sheep have short legs and are more advantageous than the long legged sheep. They cannot cross over the fence. The first mutant with short legs evolved in a farm in U.S.A. in 1791. Since then, further breeding has led to the development of this race in various countries of the world.

It has been advocated that the germinal cell are more susceptible to mutation than the somatic cells. The germinal cells undergoing reduction division are more vulnerable. The abnormalities during crossing over many also lead to mutation.

Spontaneous Mutations

These gene mutations occur in nature and originate in the individuals by natural agencies like sunlight etc. Various Mendelian traits are spontaneous gene mutations. In *Drosophila,* there are many normal or wild type genes and their mutants, like white eyes, pink eyes, black body colour, yellow body colour and vestigial wings, etc. In mice and rodents spontaneous gene mutations determine coat colour which may be black, white, brown and spotted etc. Similarly, in man, many characters, as hair colour, eye colour, skin pigmentation and several body deformities are due to mutant genes. A classical example of gene mutation is for haemophila appeared in the germin cells of one of the parents of Queen Victoria.

It has been found that some of the natural occuring mutations are caused by the natural radiations emitted by the radioactive substances in the rocks, soil, water or atmosphere of the earth. The organisms are also affected by the ultra-violet rays and the cosmic rays coming from the outer space. They may undergo mutations because of such effects but the relations between the dosage of natural radiations and frequency of spontaneous variations is not established.

Paramutation

In heterozygotes, according to *Mendel,* the two allelic genes (recessive and dominant) are unaffected by the presence of each other. In maize the locus R has a number of alleles associated with anthocyanin pigment production. R is of self-colour, R has a stripped effect, and R has a marbled effect. One dose of R produces dark mottling and two or more doses produce full colour. Brink found that R extracted from heterozygous condition with either R or R alleles produces lighter colour than the one extracted from homozygous condition. Brink named this as the paramutation effect and the alleles of the gene R as the paramutagenic.

Spurious Mutations

These are hidden mutations appearing in the generation as a result of crossing over or other means. In *Drosophila,* the gene for pink eyes remains usually hidden but it comes to surface through crossing over. The appearance of recessive genes produced by crossing over constitutes *superious mutations*.

Biochemical Mutations

Such mutations affect the chemicals produced by an organism or prevent their production. These gene changes causing defects (as a result of prevention of some enzyme formation) constitute *biochemical* mutation. They have been studied in great detail in *Neurospora* by *Beadle* and *Tatum. Alcaptonuria* and *phyenylketonuria* described under gene nature are also biochemical mutations.

Autogenous and exogenous mutations

The autogenous mutations are caused by internal factors like changes in metabolism, nutrition, certain radiations of internal organ and so on. The studies by *Darlington* and *Mather* on mutations in *Drosophila* have revealed that mutations are affected (influenced) by temperature and nutrition conditions. They found, in the larva of *Drosophila,* and expression of variations changes according to the nature of food. Vitamin B_2 has been found to help the flies to get over some of the difficulties. The nutrition or rather the lack of nutrition has been found to cause mutation in Bacterial and Fungi, less of food would make Bacteria and Fungi more resistant to bacteriophages.

Reserve Mutations

Most mutant events consist of a change from normal or wild type to a new genotype (recessive or dominant). Such mutation events are called *forward mutations,* while if the same mutant genotype changes back to the wild type, it is *back or reverse mutation.*

Reverse mutations can in some cases be confused with suppressors. The reverse mutation occurs at the same locus as the forward mutation, whereas a suppressor mutation occurs at some other locus but it suppresses the effect of the original mutant genotype. The suppressor mutation can be distinguished from reverse mutation by back-crossing the reversed stock to wild type. If mutant types appear in the offspring, it means that it is not reverse mutation. The reappearance of mutant type means that the suppressor gene has separated from the original mutant. If no mutant forms are seen, it shows that the suppressor genes are absent and the mutant gene has reverted back.

Reverse mutations are easy to be identified in microorganisms, especially the nutritional reverse mutations. the occurrence of reverse mutations usually helps in distinguishing the points mutations from large mutational effects. Point mutations caused by base substitution without a significant loss or gain of genetic material are more easily reversible than the mutations caused by deletion or duplication.

Point Mutations

When heritable alterations occur in a very small segment of DNA molecule, i.e., a single nucleotide or nucleotide pair, then this type of mutations are called *"point mutations."* The point mutations may occur due to following types of subnucleotide changes in DNA and RNA.

1. Deletion Mutations

The point mutation which is caused due to loss or deletion of some portion (single nucleotide pair) in a triplet codon of a cistron or gene is called deletion mutation. Deletion mutations have been frequently reported in some bacteriophages (Phage T^4)

2. Insertion or Addition Mutation

The point mutations which occur due to addition of one or more extra nucleotides to a gene or cistron are called *insertion mutations.* The insertion mutations can be artificially induced by certain chemical substances called *mutagens* such as acridin dye and proflavin. A proflavin molecule, it is believed, insert between two successive bases of DNA strand, there by stretching the strand length-wise. At replication, this situation would allow the insertion of on extra nucleotide in the complementary chaing at the position occupied by the proflavin molecule.

Both types of these point mutations (deletion and insertion) alter the code words and cause radical changes in the rest of the message since every code word after the affected (deleted or inserted) triplet will also be modified. This can result in the production of an inactive protein and can lead to the death of the cell.

3. Substitution Mutation

A point mutation in which a nucleotide of a triplet is replaced by another nucleotide, is called *substitution mutation.* The substitution mutation affect only a particular triplet codon. Such an altered code word (triplet codon) may designate a different amino acid and many result in the production of a protein with a single amino acid substitution. The substitution mutations alter the phenotype of an organism variously and of great genetical significant. They may be of following types:

(a) *Transitian:* In these one purine is replaced by another purine or one pyrimidine is replaced by another pyrimidine.

(b) *Tansversions:* In these a purine is replaced by a pyrimidine or *vice versa.*

Induced Mutations

Before the end of the Seccard World War, only radiations X rays etc; were considered to be mutagenic. The discovery mode during the War by *Auerbach* that chemical agents e.g. mustard gas can also cause mutation became known to the world soon after the war. The agents used to induce mutation are as follows:

(a) *Radiation Induced Mutation.* Although X-rays were discovered in 1895, it took a good many years for the geneticists to learn that X-rays constitute a very potent mutagenic agent.*Gager* between the years 1905-1910 showed the effect of X-rays on plants; between the years 1920-1929, *Mavor* showed their effect on *Drosophila;* their methods were not, however, refined enough to detect a mutation or to measure their rate. It was *Mull* in 1927, who, by means of the powerful method developed unmistakably demonstrated that mutation caused by X-rays is several times higher over the rate of spontaneous mutation. His outstanding contribution towards the enhancement of human knowledge earned for him the most coverted Nobel Prize in the field of Medicine. It was soon discovered that both non-ionizing, such as, ultraviolet rays and ionizing i.e., those those emitting electromagnetic waves are mutagenic. The rate of mutation was, however, found different with different types of radiations. For instance, the mutagenic effect of ionizing radiation, X rays (0.05A°) or gamma rays from 6 Co or radium (0.10A°) is 1000 times more powerful than the non-ionizing radiation e.g., ultraviolet rays (2000 to 3000 A°).

X-rays, being of shorter wavelength (0.1 to 10A), are very penetrating and produce ions by colliding with atoms. In this process electrons are released, which in turn impinge on other atoms, releasing further electrons. Due to a change in the number of electrons a stable atom or molecule changes to highly reactive ionic state. As X-ray produces a train of ions or series of chemical reactions along its track, it is a type of ionizing radiation. Ionising radiations can also be produced by protons, neutrons, as well as by alpha, bets and gamma rays from radioactive sources like cobalt-60, radium, etc., Ionising radiation is measured in terms of rads or roentges units (r). Exposure to one rad produces a slightly larger value (100 ergs) then the energy absorbed by a similar value of material expose to one roentgen (93 ergs.)

Seven years after the poinering work of *Muller, Altenburg* (1934) found that ultraviolet (UV) irradiation too is mutagenic. Ultraviolet rays are not as effective as X-rays because they are not that penetrating and are partly absorbed by the protoplasm. UV has been found to be very useful for inducing mutations in microorganisms because they can be irradiated at a unicellular stage and they have a thin nuclear membrane. In prokaryotes there is no nuclear membrane. Many hereditary changes caused by ionising radiations are irreversible, whereas most of the hereditary changes caused by UV are reversible, indicating thereby that ionising radiations very often bring about chromosomal aberrations whereas UV mostly causes gene or point mutations. UV light is absorbed by DNA and its bases. Enzyme system are also affected as the cells are sensitive to UV because it causes lethal mutations in them. *Xeroderma pigmentosa* is a human hereditary disease. Also causes skin cancer. *Drosophila* fed on UV irradiated food show higher mutation rates. This is the reason why UV sterlization for food preservation and canning is on its way out.

Mutagenic Chemicals

A number of chemicals have been successfully used to produce mutations. Some of these are: organic peroxides and alkaloids like colchicine, caffeine, morphin and camphor. These reagents effect the organisms externally, but there are also some mutagenes which are present within the individuals. Studies of *Ive* (1950) on *Drosophila* revealed that different strains of the same species of fly show different rates of mutation in the same invironment. He studied this difference and found that it was due to presence of a gene, which cause the increase in the rate of mutations in those strains in which it was present. Such a case was found in maize, where the presence of a gene Dt (dotted) caused the gene a, to mutate to A_1 which produces Anthocyanin pigment in the cells of endosperm and other parts of body. Such genes are therefore agents of causing or increasing mutation, and they effect on the mutation of other genes and keep themselves stable.

Heat Shocks

High temperatures have been recorded to increase the rate of mutation by two or three times.

The direction of mutation is always unknown but it is never governed by the requirements of the organisms or by evolution. The number of ways in which gene mutants is determined only by a chance. The structure of gene actually determines the possible ways in which it can mutate. These ways are limited because a gene, like the other chemicals, has a limited ratio of chemical reaction which is restricted by its physical and chemical properties.

Frame Shift Mutations

The mutations caused by the addition or deletion of nitrogenous bases in DNA or mRNA are known as frame shift mutations, because these shift the reading frame of codons from the site of change onward. One deletion may be neutralized by one addition or *vice versa,* provided they occur at the same place or very close to each other. Three deletions or three additions may disturb only a few frames.

Types of Frame Shift Mutations

The frame-shift mutations are two types:

1. *Deletion Mutations*. These mutations are caused due to the loss or deletion of one or more nucleotides.

2. *Insertion mutations*. These mutations are caused by the addition of one or more extra nucleotides in a DNA molecule at one or more places.

Mechanism of Origin of Frame shift Mutation: Acridine dyes have been found to cause deletion or insertion of a single base pair.

Acridines like 5-aminoacridine and proflavin become intecalated between two adjacent purines and thus increase the distance between them from 3.4A to 6.8A. At the time of DNA replication, either a nitrogenous base pair is introduced in the gap or a nitrogenous base pair is lost. Frame-shift mutations caused by the deletions are also introduced by the removal of ethylated bases.

Mutation due to Chromosomal Aberration

The occasional breaks in the chromosomes and fusion of the broken ends causes many changes in the chromosomes called chromosomal aberrations. Likewise chromosomal aberrations are also the kinds of mutations produced as a result of segmental change in the chromosome. These were observed by *C.B Bridges and A.H.Sturtevant* in *Drosophila* as early as 1923. The breaks may occur either in single chromosome, in a pair or in non-homologous chromosomes.

The structural changes in the chromosome are of various types. These changes may involve either the loss of broken fragments of chromosome or addition of same fragments. The hanges which occur in a single chromosome are termed *transpositions*. These transpositions involve *inversions* in which segments containing genes become arranged in reverse order and shifts in which a broken segment moves or shifts from his normal position ot a new point in the same chromosome. In more complicated changes segments get exchanged between two different chromosomes being temed as translocations. Thus, in brief structural changes may be of the following types:

1. Translocations.
2. Deletions.
3. Duplications (repeats).
4. Inversions.

1. Translocation

Translocation is the transfer of a portion of one chromosome to a non-homologous chromosome. At present, a variety of translocations are recognized which fall among the following three categories.

1. Simple translocations: These involve a single break in the chromosome and the transfer of a broken piece of this chromosome to the end of another. However, since the temeres of unbroken chromosomes are 'nonsticky'. such terminal chromosome attachments are rarely, if ever, found.

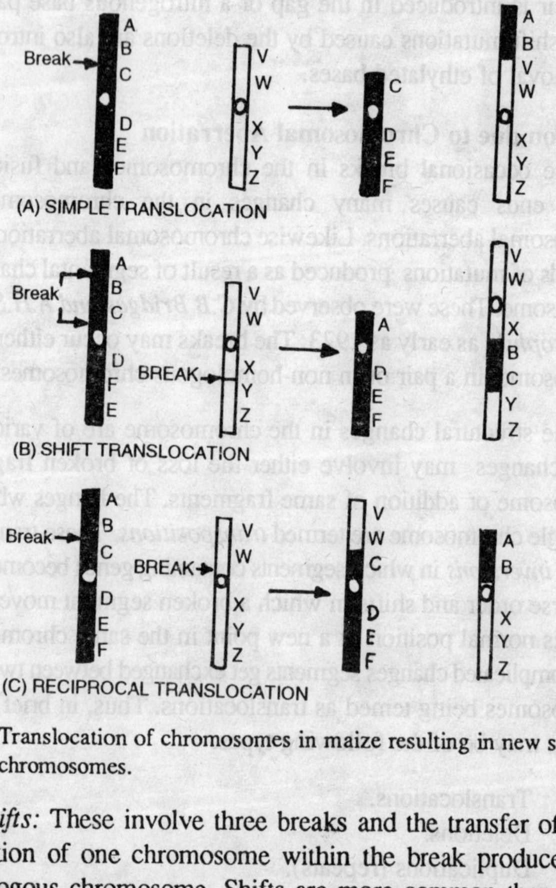

Fig. 22.1. Translocation of chromosomes in maize resulting in new synthetic chromosomes.

2. *Shifts:* These involve three breaks and the transfer of a two break section of one chromosome within the break produced in a nonhomologous chromosome. Shifts are more common than simple translocations.

3. *Reciprocal translocation or interchanges:* These are produced by single breaks in two homologous chromosome and an exchange of chromosome section between them. These are the most frequent and best studied translocations.

Reciprocal translocation is of two types, one resulting in

homozygotes and the other in heterozygotes. The former may have normal meiosis and in fact, be difficult to detect cytologically. However, the altered position of the genes adjacent to 'new neighbours", may produce position effect, about which we shall consider shortly.

Translocation heterozygotes are marked by considerable degree of meiotic irregularity. At synapsis, peculiar and characteristic formations occur because of the difficulty of attaining pairing between homologous parts. Typically a cross-shaped formation occurs in prophase-I, and this often opens out into a ring as the chiasmata terminalise. Translocation heterozygotes are often partially sterile because about 60% of the gametes fail to receive the full complement of genes. Such semisterility is easily observed in corn in which abortive pollen of reduced size, and ears lacking about half the kernels, are seen.

In reciprocal translocation, it is possible for one chromosome to get the centromeres of both the chromosomes and become dicentric while the other is left with none, to become an acentric fragment. The later will be lost, of course, but the former may be pulled apart during anaphase when the centromeres move to opposite poles. This may result in breakage and formation of two chromosomes.

When translocation involves more than two non-homologous pairs of chromosomes, meiotic rings containing six, eight or more chromosomes may be formed as in the case of *Oenothera* species.

2. Deficiency or Deletion

The deficiency or deletion is the loss of a chromosomal segment from any chromosome. Depending upon the length of the lost chromosomal segment, the loss of genes involved may very from a single gene (minimum) to block containing several genes. The loss of chromosomal segment occurs when a portion of chromosome gets detached due to certain reasons and the lost segment does not survive, because it lacks the centromere. The portion of the chromosome carrying the centromere function as a genetically deficient chromosome.

Types of Deletions

Deletion can be two types:-

(i) *Terminal deletion.* It refer to the loss of segment from one or the other end of the chromosome. The terminal acentric part of the chromosome is unable to survive and causes teminal deletion. The teminal deleton is, therefore, caused by a single break in the chromosome.

(ii) Intercalary or intestitial deletion. It denotes the loss of an intercalary segment of the chromosome with the reunion of teminal segments. Therefore, the intercalary deletion is characterised by two points breaks and the reunion of terminal parts.

Deletions was presented by *Bridges* (1917) and *Molar* (1923) in *Drosophila.* A recessive mutant in the X-chromosome causes the tip of the wings to be notched and is, therefore known as *notch character*. It is inherited as a sex linked dominant in female but is lethal in male. Since the males with notch character are not viable, the females help in the inheritance of this character. When such notched females with red eye colour were crossed to white-eyed males having normal wings, the F_1 females with notched wings all exhibited white eye colour indicating as if the white colour is dominant. This pseudodominance has also been shown by certain genes present in the vicinity of white eye locus and is explained to occur by the loss of a piece of the X-chromosome containing these genes.

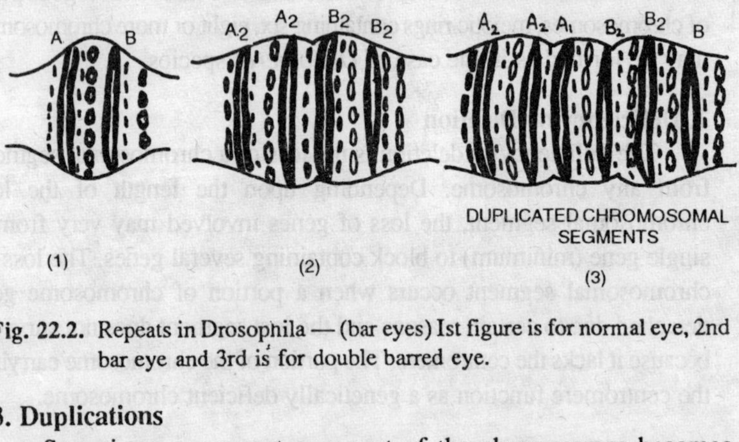

Fig. 22.2. Repeats in Drosophila — (bar eyes) Ist figure is for normal eye, 2nd bar eye and 3rd is for double barred eye.

3. Duplications

Sometimes a segment or a part of the chromosomes becomes repeated in the same chromosome. These additional duplicated segments are called duplications. They arise as a result of unequal crossing over between chromosomal segments. In *Drosophila* mutant bar eyes which are narrow and constricted are due to small duplicated gene. It was observed that each additional duplicate segment made the eye smaller and smaller. Sometimes duplication when act in opposite direction serve as suppressors.

Duplication may sometimes lead to cyclic production of broken and dicentric chromosomes, as the one observed in chromosome 9 of corn with a reverse tandem duplication. Crossovers in the duplicated region may occur either between non-sister chromatids (two homologues) or sister chromatids (one homologue). In both cases, a dicentric chromosome is produced that forms a bridge between the poles during division. Breakage then takes place at any point along the bridge causing each of the separating chromosomes to retain one broken end. The broken ends do not 'heat' and remain 'sticky' i.e., capable of uniting with another broken end. When the chromosomes replicate during subsequent prophase, the two sticky ends produced join, thereby causing the reappearance of the dicentric chromosome and consequently the bridge breakage fusion cycle.

4. Inversion

Inversion is a kind of chromosome aberration in which a chromosomal segment exists in reverse relationship to the rest of its chromosome. Inversions are believed to arise by breakages at the point of intersection of a chromosome loop and reunion with new partners. An organism may be homozygous or heterozygous for an inversion. During meiosis, inversion heterozygotes synapse by forming a looped configuration. In salivary gland chromosomes and other favourable material, an material, an inversion loop makes it possible to determine with precision the extent of the aberration.

The inversion can be of two types:

1. Paracentric Inversion

A single crossing over or an odd number of crossovers in the inverted region will result into the formation of a dicentric chromosome (having two centromeres) and an acentric chromosome (with no centromere.) Of the remaining two chromatids one will be normal and the other will carry the inversion. The dicentric chromatid and acentric chromatid will be observed at anaphase I in the form of a bridge and a fragment. Double crossovers and crossovers within and outside the inversion will give various kinds of configurations.

2. Pericentric inversion

In a pericentric inversion (where centromere is present within the inverted segment), although at pachytene, the configuration observed is similar to those described above for paracentric inversion, the products of crossing over the configurations at subsequent stages of meiosis differ. In the case, two of the four chromatids resulting after meiosis will have deficiencies and duplications. However, unlike paracentic inversion, no dicentric bridge or acentric fragment will be observed. Consequently, at anaphase I, no bridge or fragment will be seen.

However, in pericentric inversion, if the two breaks are not situated equidistant from the centromere, this will result in the change in shape of the chromosome. For instance, a metacentric chromosome (with centromere in the centre) may become submetacentric and *vice versa.*

Detection of Mutations

HJ. Muller devised an easy method for detecting lethal mutations in the sex chromosomes of *Drosophila.* This is called a CIB method in which a special type of female fly is employed which carries a normal X chromosome and an abnormal X chromosome. The abnormal X chromosome contains an inversion mutation C (which prevents the chromosome to do crossing over with the normal X chromosome. Therefore, called *crossovor suppresser)* a recessive lethal mutational gene, I and a dominant gene B for bat-eye. In *Muller's* CIB techniques the CIB female flies are mated with males which are previously treated with some mutagenic agent (such as X rays) to cause mutation in some of their sperms. The resulting zygotes are of four types and one of these, the CIB male, fail to survive because such embryos contain a recessive lethal which expresses itself when hemizygous. Thus, only one class of male (with CIB X chromosome) male remains to fertilize the F_1 females. Each heterozygous CIB female result from the fertilization of a CIB egg and an radiated X-bearing sperm and some of these sperms will contain mutated X-chromosomes. Mated F_1 heterozygous CIB females are distributed individually into culture tubes in which each lays fetile egges and so produces single F_2 cultures. The cultures produced by females bearing an induced lethal mutation contain only females: whereas female bearing irradiated X-chromosome in which no recessive lethal has been induced yield cultures containing some wild type males. Thus, if in a population of 1000 cultures 990 contained some males and 10 contained only females, the induced rate of sex-linked, recessive lethal mutation would be 1 per cent.

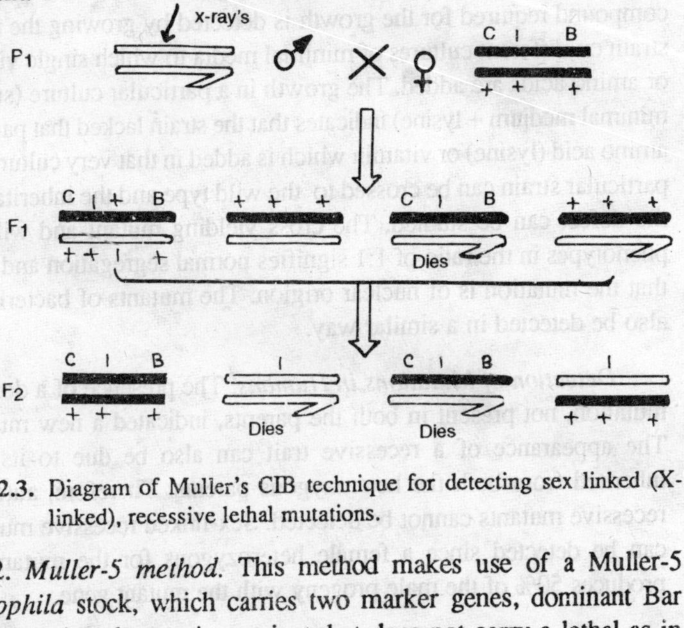

Fig. 22.3. Diagram of Muller's CIB technique for detecting sex linked (X-linked), recessive lethal mutations.

2. *Muller-5 method.* This method makes use of a Muller-5 *Drosophila* stock, which carries two marker genes, dominant Bar (barred eye) and recessive apricot, but does not carry a lethal as in CIB. Moreover, Muller-5 contains a more complex in version and, therefore, has a better crossover suppressor than in CIB. In the F_2 generation, 50% males are Muller-5 in phenotype and the remaining 50% are wild type. If a lethal mutation is induced in X-chromosome of irradiated male, no wild type males would appear in the F_2 generation. Therefore, the absence of wild type males of F_2 is an indication of an induced lethal mutation.

Detection of Mutation in Neurospora. The normal wild type of *Neurospora* and the mutant of *Neurospora* can grow on a complete medium. The complete medium consists of inorganic salts, vitamins and amino acids. The wild type of *Neurospora* can also grow on a *minimal medium* which consists of glucose (as the carbon source) and certain inorganic salts and biotin vitamin. It can synthesis all vitamins and amino acids from this source. The nutritional mutants cannot grow on the minimal medium since they cannot synthesise from it all the vitamins and amino acids needed for their growth. If certain amino acids or vitamins are added to the minimal medium, the mutant which lacks the capacity to synthesise these, grows on it. The particular compound required for the growth is detected by growing the mutant strain on different cultures of minimal media to which single vitamins or amino acids are added. The growth in a particular culture (suppose minimal medium + lysine) indicates that the strain lacked that particular amino acid (lysine) or vitamin which is added in that very culture. This particular strain can be crossed to the wild type and the inheritance of the defect can be studied. The cross yielding mutant and wild-type phenotypes in the ratio of 1:1 signifies normal segregation and shows that the mutation is of nuclear origion. The mutants of bacterian can also be detected in a similar way.

Detection of Mutations in Humans. The presence of a dominant mutation, not present in both the parents, indicated a new mutation. The appearance of a recessive trait can also be due to its being inherited from both the heterozygous parents. Therefore, autosomal recessive mutants cannot be detected. Sex-linked recessive mutations can be detected since a female heterozygous for the mutant gene produces 50% of the male progeny with the mutant gene.

Position Effects. The effects of a gene of the development of an organism depend not only the structure of a gene, but also on its position on the chromosome. Such effects, which appear in addition to the other factor, due to the specific position of gene, are termed as *position effects*. The phenomenon of position effects is important for understanding the general theory of the gene and for appreciation of the gene as functional unit of inheritance.

Since the discovery of this phenomenon by *Sturtevant* in 1925, studies have been made in *Drosophila* (Lewis, 1950, 51, 52, 55; *Green* and *Green,* 1949; Green, 1954, 53), *Oenothera (catchside* 1947), maize (*Mc Clintock,* 1951, 58) and in other plants and animals. But they are not universal.

The position effects have been found to be related to the loss of the dominance of the genes. *Dubinin* and *Sidirov* observed that white type allele of gene in *Drosophila melanogaster* lost its dominance over the recessive allele, when the former in position changed in respect to heterochromatin. The change in action is due to change in position and not to any alternation of its properties.

The occurrence of position effects, change in position of genes in relation to other genes on a chromosome, lends support to the controversial view that chromosome, as a whole is one integrated, basic, functional unit and not the genes. In another way, it could be said that a gene operates in conjunction with other genes and not as an isolated unit. *Lewis* (1950) has recognized two types of position effects in *Drosophila melanogaster*. The first is the stable or S-type which is confined to the euchromatic region of the chromosome and the second is V-type (variegated). This is associated with the heterochromatin regions of the chromosome. The total effect of mutation appears only when the genes of both these regions come together come together. They can be brought together by chromosomal aberrations. The chromosomal translocations and end to end invesions are two important phenomenon which could bring a change in the position of the genes and consequently the position effects. In general, these changes which alter the order of the genes, may also bring variations in the linkage relationships.

Two hypotheses have been put forth to explain position effect. According to *structural hypothesis (Ephrussi* and *Sutton,* 1944), there is some alternation in the gene, so that its functional activity and consequently it phenotype expression is changed. The change is probably brought about geometrically by the forces of somatic pairing. This type of alteration is revesible in nature.

The second hypothesis, the *kinetic hypothesis (Lewis,* 1955) envisages that the position effects are produced by and interaction of adjacent genes at the level of gene products.

REVIEW QUESTIONS

1. Persons with mongolism are feeble minded and usually do not reproduce. In the event that a mongoloid does have children, what are their chances of mongolism?

2. Fewer recessive phenotypes appear among the offspring of self-pollinated tetraploids than among those of self-pollinated diploids. Why?

3. A male *Drosophila* heterozygous for the translocation mentioned in problem 1 and with the recessive genes brown eye (bw) and ebony body (e) on the non-translocated second and third chromosomes respectively, is crossed to a female with normal chromosomes and homozygous for the genes bw and e. What offspring would be expected? (Any zygotes with an extra chromosomes arm or deficient for one do not survive.)

4. A *Drosophila* is heterozygous for a reciprocal translocation between the second and third chromosomes, each break having occurred near the spindle attachment (which for each of these chromosomes is near the center). Draw a diagram showing how these chromosomes would synapse at meiosis.

5. Sturtevant found a gynandromorph *Drosophila* that began as a zygote with the following X chromosomes.

$$\frac{s + e + c\ v\ g\ f}{+ w + r + + + f}$$

The male parts of the fly showed the mutants s, e, c, g and f. The female parts showed only f.

a. How was this gynandromorph formed?

b. Which mutants are autonomous, which non-autonomous, and for which does this fly offer no evidence?

ANSWERS

1. Approximately half the children should be mongoloid.

2. Because the trait, if completely recessive, must be present four times to appear in a tetraploid, while only two times in a diploid.

3. 1/2 brown ebony, 1/2 normal (among surviving progeny).

4.

5. By loss of lower chromosome in the male parts.
s, e, c, g autonomous; v non-autonomous; no evidence on w, r and f.

23

Biochemical Genetics

The cell is a very complex and its function depends upon the integration of the activities of many genes. The genes may interact within the organism. The genes control various traits in an organism through a control exercised on the developmental processes. Such a control is due to the synthesis of enzymes or proteins. It is a well-known fact that the synthesis of protein is under the direct control of gene or genes. Thus, we see that the genes express themselves through the synthesis of enzymes and it was demonstrated for the first time 1941 by *G.W. Beadle* and *E.L. Tatum* while working on the biochemical mutations is *Neurospora.* Based on their this work, *Beadle* and *Tatum* proposed a concept called "one gene-one enzyme" hypothesis. This hypothesis means that all the steps which transform a precursor substance to its end product that is ultimately expressed into a structural and functional phenotypic trait, constitutes a *biosnthetic or metabolic pathway,* and each step of a biosynthetic pathway is catalysed by a specific enzyme which in turn is synthesized under the control of a specific gene.

Biochemical genetics has been widely studied in haploid organisms whose growth requirements are known. Any error in the metabolic pathway is expressed in the form of a disease as in human beings.

Genetic Control of Metabolism in Humans

Early evidence for the relationship between genes and enzymes came from the work of the physician *Archibald Garrod* in 1909, the results of which were published as a book called *Inborn Errors of Metabolism. Garrad* was interested in human diseases that had an apparent genetic basis. He postulated and rightly so, that biosynthesis and degradation of a product takes place in several sequential steps and these steps are controlled by genes, one gene controlling one step. He suggested further that if any of these genes mutates in the metabolic path, all the other reactions ahead of it, stop and there is accumulation of the substance which the intact gene preceding the mutant one, synthesizes as shown under the following sub-heads:

1. Phenylketonuria or Folling's disease (—PKU)—This disease is caused due to the absence of enzyme which catalyses the *conversion of phenylpyruvic acid into hydroxy-phenylpyruvic acid.* As a result phenylpyruvic acid accumulates in the blood and causes impairment of brain tissue, pale skin and a tendency to epileptic seizure. Thus PKU children are mentally retarded and are known as *phenylpruvic idiots.* Such persons have recessive genotype *pp* that fails to produce enzyme phenylainine hydroxylase.

Persons suffering from phenylketonuria also show mild symptoms of albination; this is because melanin pigment is not produced in such individuals in as much abundance as is necessary for normal colour development.

2. *Alcaptonuria.* The urine of persons suffering from hereditary disease alcaptonuria turns black on exposure to air. This is due to presence of a substance alcaption (homogentisic acid) in their urine, because in them the enzyme-*homogentisic acid oxidase* required in the *conversion of homogentisic acid* into *maleylacetoacetic acid* is absent. The homogentisic acid is passed out in the urine, and latter turns black on exposure to air. The recessive mutant gene fails to synthesize the enzyme and blocks the step in the normal degradation of alcapton or homogentisic acid.

3. *Tyrosionsis.* The recessive gene, t in its homozygous condition, blocks the conversion of p-hydroxy pyruvate into 2,5 dihy droxyphenyl pyruvic acid. This leads to the accumulation of tyrosine. Its excess is excreted with the urine.

4. *Goitrous cretinism.* Sometimes persons lack enzyme which is required for the conversion of tyrosine into thyroxine and trioido thyronine hormone. Deficiency of thyroid hormone causes goitrous cretinism characterised by physical and mental retardation and hypertrophy of thyroid gland.

Sickel cell anemia: It is a type of abnormality in the red blood cells whose genetical basis has been determined fairly recently. It is due to a recessive genes. The immediate effect of this gene is seen on the shape of RBC affecting the transportation of oxygen in lesser proportion. The homozygous stage is more susceptable to reduction of oxygen than heterozygous. In fact, in the former, the blood sample may show a few sickel-shaped cells before any substantial reduction of oxygen.

The homozygous recessive individuals rarely survive to leave any progeny because of secondary (pleiotropic) effect of the gene (the arms and legs are elongated and there are other physical deformities). The heterozygous individuals are indistinguishable from the normal except that the shape of their RBC changes consequently on lowered oxygen supply. These heterozygotes have been found to be superior over normal in respect of their immunity to the attack of malarial parasites.

When the haemoglobin from normal and sickle cell anemic cells were analysed, they were seen identical in all the fragments excepting for the number four. From the determination of amino acid sequence, it was seen that the fragment four of the abnormal haemoglobin contained *valine* in place of *glutamic acid,* other amino acids being identical. In other words, *valine* was seen to replace *glutamic acid* and the rest of the 285 amino acids remained undisturbed.

Genetic Control of *Drosophila* eye Pigments

Garrod's pioneering work provided the first clue to the link between Mendelian factors and proteins. In 1935 *G. Beadle* and *B. Ephrussi* obtained more evidence for gene enzyme relationships in a biochemical pathway in their studies of the synthesis of the eye pigments in the fly, *Drosophila melangaster.*

There are two pigments in the eye of *Drosophila* the bright red pterins and the brown ommochromoes. As we now know, these two pigments are made in two multistep biochemical pathways, each reaction of which is catalyzed by an enzyme. The bright red and brown pigments that are the end products of these pathways become attached to protein granules and are deposited in the eye cells. The combination of the two pigments results in the dull red eye colour characteristic of the wild type. Further, in *Drosophila,* the larval stages contain groups of cells called *discs,* each of which develops into a particular adult structure during metamorphosis. Two discs can be identified as the progenitors of eyes. In their initial experiments, *Beadle* and *Ephrussi* showed that an eye disc from a larva can be transplanted into the abdomen of a second larva and the disc will develop into an identifiable eye structure that can be found in the abdomen of the adult following metamorphosis. This paved the way for a series of elegant experiments in which eye discs were transplanted between larvae of different genetic constitutions.

At the time of *Beadle* and *Ephrussi's* experiments, three gene lock were known that were implicated in the production of brown pigment: *scarlet, cinnabar* and *vermilion.* Mutations in any one of these genes resulted in a bright orange eye (the shade of orange differs in each mutant type). *Beadle* and *Ephurssi* first addressed the question of whether embryonic eye tissue from the three mutant types *st* (scarlet), *cn* (cinnabar), or *v* (vermilion) transplanted into a wild type larval host would develop a wild type eye colour. In the case of the *st* mutant the transplanted disc developed into a scarlet-coloured eye and is an example of *autonomous development* in which the wild type host was unable to provide substances to the disc to enable the brown pigment to be produced. By contrast, transplanted discs from either *v* or *cn* developed into normal-coloured eyes in the wild type host and thus showed *non-autonomous development.* The conclusion in these cases was that the wild type provided a diffusible substance that the discs used to bypass the genetic block and to make brown pigment.

Of more importance historically were the results of reciprocal transplant experiments between *v* and *cn* larvae. The experiments showed that *cn* discs transplanted into *v* hosts produced cinnabar-coloured eyes, whereas *v* discs transplanted into *cn* hosts developed into wild type coloured type. They concluded from these results that the production of the brown pigment involved a biochemical sequence with at least two precursors, the wild type having both, *cn* having one, and *v* having neither; the gene-reaction step relationships are shown in

Figure 23.1. Thus, the production of a wild type coloured eye when a *v* diffusible substance so that it becomes wild type. In the reciprocal experiment, the *cn* disc cannot convert the v+ substance to the cn+ substance, and the *v* host is deficient in the production of v+ substance so no brown pigment can be produced.

TABLE 23.1

Results of Eye Disc Transplantation Experiments Involving Eye Colour Mutants

Source of eye disc	*Host fly*	*Colour of transplanted eye after metamorphosis*
+	v	+
v	+	+
+	cn	+
cn	+	+
cn	v	cinnabar
v	cn	+

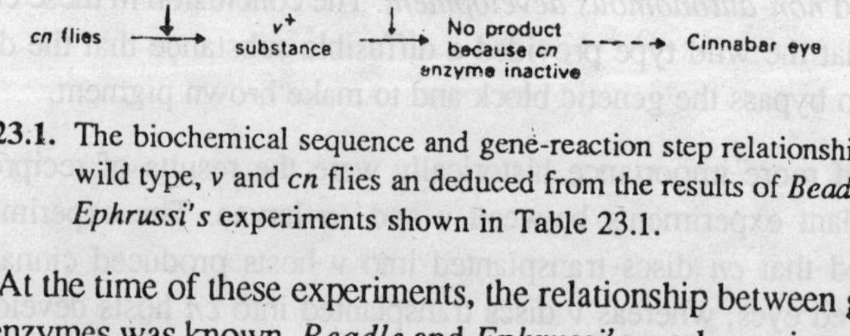

Fig. 23.1. The biochemical sequence and genė-reaction step relationships for wild type, *v* and *cn* flies an deduced from the results of *Beadle* and *Ephrussi's* experiments shown in Table 23.1.

At the time of these experiments, the relationship between genes and enzymes was known. *Beadle* and *Ephrussi's* work was important historically since it indicated a strong link between a phenotype (in this case eye colour) and genotype (*v* and *cn*) *E. Tatum* further investigated this system by making extracts of *v* and *cn* flies by homogenization and by injecting the extracts into host larvae. He

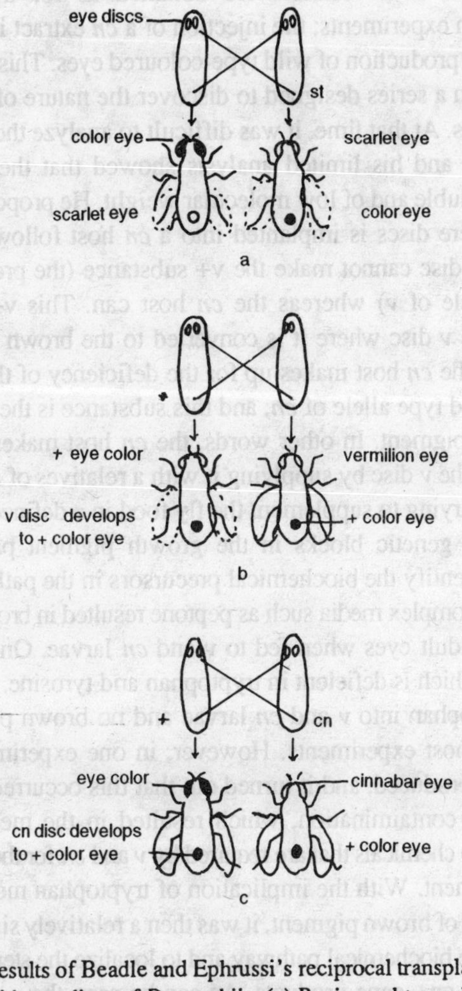

Fig. 23.2. Results of Beadle and Ephrussi's reciprocal transplant experiments with eye discs of *Drosophila*. (a) Reciprocal transplants between + and *st* (scarlet)., The *st* disc develops into a scarlet-coloured eye in the wild type host, indicating autonomous development for *st*. (b) Reciprocal transplants between + and *v* (vermilion). The v disc develops into a wild type-coloured eye in the wild type host, indicating non-autonomous development of the *v* disc. (c) Reciprocal transplant between + and *cn* (cinnabar). Like *v*, the *cn* disc shows non-autonomous development in that it produces a wild type-coloured eye in the wild type host.

found that the same results were obtained as for the eye disc transplantation experiments; the injection of a *cn* extract into *v* larvae resulted in the production of wild type-coloured eyes. This experiment was the first in a series designed to discover the nature of the v+ and cn+ substances. At that time, it was difficult to analyze the extracts he had prepared, and his limited analysis showed that the substances were water soluble and of low molecular weight. He proposed that the substances were discs is implanted into a *cn* host follows logically. That is, the *v* disc cannot make the v+ substance (the product of the wild type allele of *v*) whereas the *cn* host can. This v+ substance diffuses to the *v* disc where it is converted to the brown pigment. In other words, the *cn* host makes up for the deficiency of the *v* disc by carries the wild type allele of *cn*, and this substance is then converted to the brown pigment. In other words, the *cn* host makes up for the deficiency of the *v* disc by supplying it with a relatives of amino acids and set about trying to supplement the fly food in a defined way to try to bypass the genetic blocks in the growth pigment pathway and therefore to identify the biochemical precursors in the pathway in this way. Several complex media such as peptone resulted in brown pigment formation in adult eyes when fed to *v* and *cn* larvae. One exception was gelatin, which is deficient in tryptophan and tyrosine. *Tatum* then injected tryptophan into *v* and *cn* larvae and no brown pigment was produced in most experiments. However, in one experiment, brown pigment was produced, and it turned out that this occurred as a result of a bacterial contaimination, which resulted in the metabolism of tryptophan into chemicals that are required by *v* and *cn* for the production of brown pigment. With the implication of tryptophan metabolism in the production of brown pigment, it was then a relatively simple matter to work out the biochemical pathway and to localize the steps catalyzed by the v+ and cn+ gene products. As can be seen the *v* mutants are unable to convert tryptophan to formylkynurenine and the *cn* mutants cannot convert formlyk ynurenine to 3-hydrozykynurenine because the enzymes they have for these respective steps are non-functional.

The pathway shown in Fig. 23.4 can serve to illustrate a general point about the sequence of reactions and the consequences of genetic mutations. In the v mutations, for example, the flies accumulate high amounts of tryptophan since it is not converted to the brown pigment in this pathway. This is analogous to the accumulation of homogentistic acid in alkaptonuriacs. Here it provided further evidence that tryptophan metabolism is involved with eye pigment formation. Indeed, if one feeds wild-type larvae with radioactive tryptophan, the brown ommochrome pigment of the adult eye is found to be radioactive.

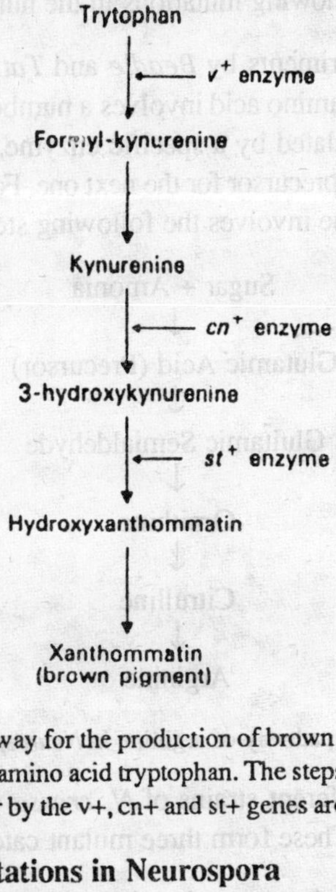

Fig. 23.3. The pathway for the production of brown eye pigment of *Drisophila* from the amino acid tryptophan. The steps catalyzed by the enzymes coded for by the v+, cn+ and st+ genes are indicated on the diagram.

Biochemical Mutations in Neurospora

The unicellular fungus *Neurospora* has been most widely utilised in the study of biochemical mutations influencing or amino metabolic pathways, concerned with the synthesis of acids, molecules of nucleic acids and some of the vitamins.

The normal strain of *Neurospora* is able to grow in the *minimal medium* containing sucrose, inorganic salts and one of the vitamins—*biotin.* It can synthesize all other complex substances, like vitamins, proteins, nitrogenous bases and the nucleic acid from the simple substances of culture medium. *Beadle* and *Tatum* exposed some of the asexual spores (conidia) of *Neurospora crassa* to mutagen (X-rays or ultraviolet rays). These mutated ascospores were found unable to grow in minimal medium but required the addition of some supplement for the normal growth. The mutated strains of *Neurospora* are called *nitritional mutants* showing mutations in the nutritional genes.

Controlled experiments by *Beadle* and *Tatum* have shown that biosynthesis of each amino acid involves a number of sequential steps and each step is regulated by a specific enzyme. The end product of one reaction acts as a precursor for the next one. For example, synthesis of amino acid arginine involves the following steps :

Sugar + Amonia
↓
Glutamic Acid (Precursor)
↓
Gluitamic Semialdehyde
↓
Ornithine
↓
Citrulline
↓
Arginine

Fig. 23.4. Biosynthetic pathway of arginine in *Neurospora.*

There are 12 different strains of *N. crassa* which are unable to synthesize arginine. These form three mutant categories.

(a) Mutants which grow only when either citrulline or arginine is supplied but not grow if ornithine or cutrulline.

(b) Mutants which grow when either citrulline or arginine is supplied but not grow if ornithine is added to the medium.

(c) Mutants which can grow in the medium containing either ornithine, citrulline or arginine.

These mutants indicate that in first case (a) the mutants are incapable of utilizing intermediate products ornithine or citrulline. In them the gene responsible for the conversion of citrulline into arginine has undergone mutation and fails to synthesize the specific enzyme. (b) The mutants are incapable of utilizing ornithine. It means the enzyme responsible for transformation of ornithine into citrulline is lacking and the gene concerned with the synthesis of this enzyme has undergone mutation. In third case (c) the mutant can utilize intermediate products but not the precursor (i.e. glutamic acid).

Biochemical Mutants in Flowering Plants

Arabidopsis thaliana is a small rapidly growing crucifer with about a 28 day life cycle. It has a low chromosome number (2n = 10). The number of seeds per fruit is quite large (150-300 seeds) and 25000 per plant. It cannot however tolerate a temperature higher than 25. Requirement of such low temperature therefore makes it extremely difficult to work with this material in the tropics except in winter. When grown under controlled condition, it is supplied with 20 hours photoperiod and high intensities of 10-12 thousand lux. Because of their small size and fast growth rate they can be grown continuously through out their life cycle in the test tube on sterile nutrient medium +2 per cent sucrose solidified in suitable concentration of agar. The trace elements used in the culture solution is chelated.

By means of various mutagenic agents e.g. X-rays, EMS (1 per cent for two hours) etc. auxotrophs are easily obtained. Different types of lethal embryos (vanda, fusca, parwa, diffusa) can be picked up under the dissection microscope. The unripe fruits are opened and the undeveloped embryos can be seen without dissecting the seeds. Chlorophyll mutants appear in the form of albina, X antha, chlorine (yellow green). As in the case of biochemical mutants in *N. crassa,* supply of different growth requiring substances (amino-acids, vitamins) either individually as in combination in the culture medium enables the mutants grow to different lends of development. Some chlorophyll mutants die at the cotyledon stage some at the 3 rosettle stage without turning green, still others at the bud stage. Some of the mutants which deserve special mention are *lencine, tryptophan, thiamine anxotrophs.* These, on supply of their specific requirements, turn green and grow to full length and produce seeds. One of the interesting observation in these studies 15 that in leucine auxotroph, addition of this amino acid by itself restores the normal growth but in combination with valine in different proportions inhibits growth.

Gene Control of Protein Structure

So far we have learned that there is a direct relationship between genes and enzymes. Since an enzyme catalyzes one reaction in a biochemical pathway, a mutation in the gene for the organism depends on the nature of the pathway, the effects of the compound accumulated prior to the block, and the effects of a deficiency of the end product.

However, while all enzymes are proteins, not all proteins are enzymes. Nevertheless, the functions of proteins are also affected by changes in amino acid sequence and tertiary structure. For example, the human disease sickle cell anemia is characterized by a sickling of the red blood cells. This leads to problems with the red blood cells passing through the capiliaries, hereby causing tissue and organ damage, anemia, and possibly death. The life of the red cells is also greatly shortened. Sickle cell anemia is caused by homozygosity for a mutation in the gene for the β-polypeptide of haemoglobin. (Haemoglobin consists of two copies each of ∝ and β polypeptides). The mutation results in the substitution of the neutral amino acid valine for the normal acidic amino acid glutamic acid at the sixth position form the N-terminal end. This change is at an important part of the molecule, and the presence of a neutral amino acid instead of an acidic amino acid makes the region hydrophobic (water-hating) instead of hydrophilic (water-loving). As a result, the β-polypetide folds to place the altered region away from the aqueous environment, and this causes an unusual stacking of haemoglobin molecules and sickling of the red blood cells.

Colinearity

A more precise understanding of the relationships between a gene (cistron) and the sequence of amino acids in the polypetide for which it codes came from the work of *C. Yanofsky* and his group in 1967. Their studies centered on the enzyme *tryptophan synthetase* from *E. coli.* This enzyme consists of two copies each of two distinct polypeptides A and B, which are coded for by two adjacent genes. Here is an example of where two genes code for one enzyme—a clear exception to *Beadle* and *Tatum's hypothesis*. The reactions that the enzyme carries out are in the biosynthetic pathway for the amino acid tryptophan.

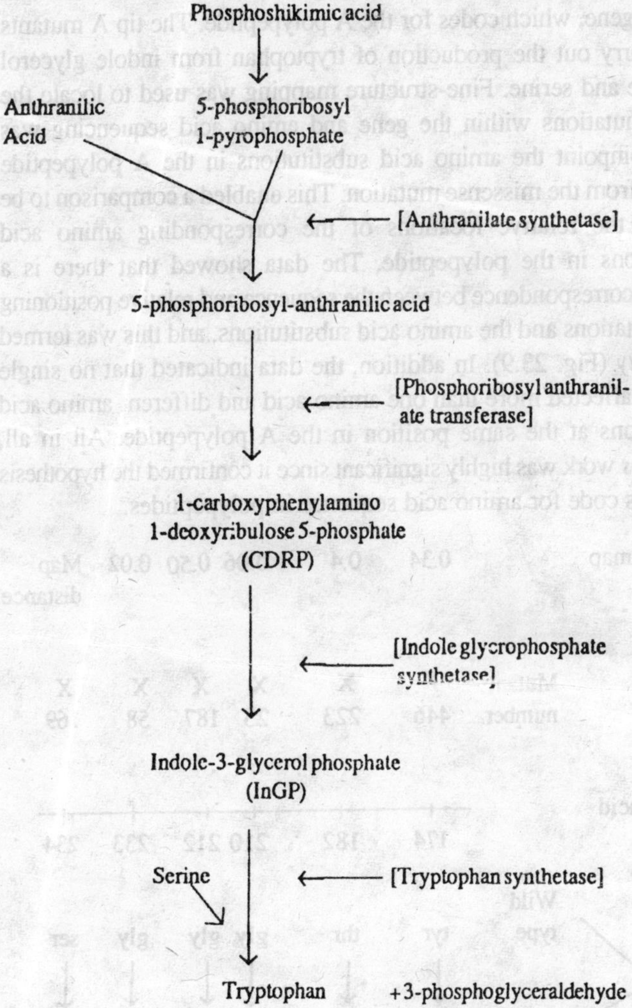

Fig. 23.5. The tryptophan biosynthetic pathway. The enzymes that catalyze the reaction steps are shown in square brackets Tryptophen synthetase catalyzes the last step in the pathway.

Tryptophan synthetase is easily isolated and the two polypeptides A and B can be purified. Further, the amino acid sequence of the 267-

amino acids-long A polypetide was determined at the ouset of their experiments. They then isolated a series of tryptophan auxotrophs and, with further tests, identified those that carried missense mutations in the tip A gene, which codes for the A polypeptide. The tip A mutants cannot carry out the production of tryptophan from indole glycerol phosphate and serine. Fine-structure mapping was used to locate the various mutations within the gene and amino acid sequencing was used to pinpoint the amino acid substitutions in the A polypeptide resulting from the missense mutation. This enabled a comparison to be made of the relative locations of the corresponding amino acid substitutions in the polypeptide. The data showed that there is a complete correspondence between the sequence and relative positioning of the mutations and the amino acid substitutions,.and this was termed *colinearity* (Fig. 23.9). In addition, the data indicated that no single mutation affected more than one amino acid and different amino acid substitutions at the same position in the A polypeptide. All in all, *Yanofsky's* work was highly significant since it confirmed the hypothesis that genes code for amino acid sequence in polypeptides.

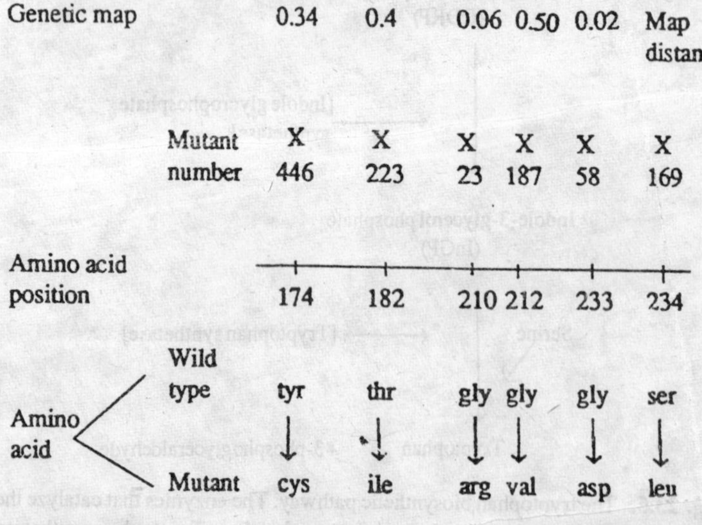

Fig. 23.6. Part of the tip A gene of *E. coli* and its product, the tryptophan synthetase A polypetide, showing the colinearity of mutation positions and amino acid substitutions.

24

Population Genetics

The concept of a population is generally used by demographers simply to denote an assembly of individuals that have a common definition. For instance, the human species can be subdivided into population of nations, populations of cigarette smokers versus non-smokers, a populations of frovers versus man fishers. Each scheme of subdivided could some as a valid cretenion for defining populations in answer to certain types of questions about humans.

Genetics in general concerns the genetic constitution of organisms and the laws governing the transmission of this hereditary information from one generation to the next. *Population genetics* is that branch of genetics concerned with heredity in groups of individuals, that is, in populations. Population geneticists study the genetic constitution of populations and how this genetic constitution changes from generation to generation.

Hereditary changes through the generations underlie the evolutionary process. Hence population genetics may also be considered as *evolutionary genetics*. However, these two concepts can be distinguished. It is often understood that population genetics deals with populations of a given species, while evolutionary genetics deals with heredity in any populations, whether of the same or of different species. By these definitions, evolutionary genetics is a broader subject than population genetics—it includes population genetics as one of its parts. Evolutionary genetics is the subject of the rest of this book.

Populations and Gene Pools

The most obvious unit of living matter is the individual organism.

In unicellular organisms, each cell is an individual; multicellular organisms consist of many interdependent cells, many of which die and are replaced by other cells throughout the life of an individual. In evolution, the relevant unit is not the individual but the population. A *population* is a community of individuals linked by bonds of mating and parenthood; in other words, a population is a community of individuals of the same species. The bonds of parenthood that link members of the same population are always present, but mating is absent in organisms that reproduce asexually. A *Mendelian population* is a community of interbreeding, sexually reproducing individuals, that is, Mendelian populations are those in which reproduction involves mating.

The reason why the individual is not the relevant unit in evolution is that the genotype of an individual remains unchanged throughout of its life; moreover, the individual is ephermeral (even though some organisms, such as conifer trees, may live up to several thousand years). A population, on the other hand, has continuity from generation to generation; moreover, the genetic constitution of a population may change—evolve—over the generations. The continuity of a population through time is provided by the mechanism of biological heredity.

The most inclusive Mendelian population is the *species*. As a rule, the genetic discontinuities between species are absolute; sexually reproducing organisms of different species are kept from interbreeding by reproductive isolating mechanisms. Species are independent evolutionary units: genetic changes taking place in a local population can be extended to all members of the species, but are not ordinarily transmitted to members of a different species.

The individuals of a species are not usually homogeneously distributed in space; rather, they exist in more or less well-defined clusters, or local populations. A *local population* is a group of individuals of the same species living together in the same territory. The concept of local population may seem clear, but its application in practice entails difficulties because the boundaries between local populations are often fuzzy. Moreover, the organisms are not homogeneously distributed within a cluster, even when the clusters are quite discrete, as in true of organisms living in lakes or on islands; the lakes or islands may be sharply distinct, but individuals are not evenly distributed within a lake or on an island. Animals often migrate from one local population to another, and the pollen or seeds of plants may also move from population to population, all of which makes local populations far from completely independent of each other.

The concept of a *gene pool* is useful in the study of evolution. The gene pool is the aggregate of the genotypes of all the individuals in a population. For diploid organisms, the gene pool of a population with N individuals consists of 2N haploid genomes. Each *genome* consists of all the genetic information received from one parent. Thus, in the gene pool of a population of N individuals, there are 2N genes for each gene locus, and N pairs of a homologous chromosomes. The main exceptions are the sex chromosomes and sex-linked genes that exist in a single dose in heterogametic individuals.

Genetic Variations and Evolution

The existence of genetic variation is a necessary condition for evolution. Assume that at a certain gene locus all individuals of a given population are homozygous for exactly the same ***. Evolution cannot take place at the locus, because the allelic frequencies cannot change from generation to generation. Assume now that in a different population there are two alleles at that particular locus. Evolutionary changes can take place in this population: one allele may increase in frequency at the expense of the other allele.

The modern theory of evolution derives from *Charles Darwin* (1809-1882) and his classic, *On the origin of Species*, published in 1859. The occurrence of hereditary variation in natural populations was the starting point of Darwin's argument for evolution by a process of natural selection. *Darwin* argued that some natural hereditary variations may be more advantageous than others for the survival and reproduction of their carriers. Organisms having advantageous variations are more likely to survive and reproduce than organisms lacking them. As a consequence, useful variations will become more prevalent through the generations, while harmful or less useful ones will be eliminated. This is the process of *natural selection,* which plays a leading role in evolution.

A direct correlation between the amount of genetic variation in a population and the rate of evolutionary change by natural selection was demonstrated mathematically with respect of *fitness* by *Sir Ronald A. Fisher* in his *Fundamental Theorem of Natural Selection* (1930): *The rate of increase in fitness of a population at any time is equal to its genetic variance in fitness at that time.* (Fitness, in the technical sense used in the theorem, is a measure of relative reproductive rate).

The *Fundamental Theorem* applies strictly to allelic variation at a single gene locus, and only under particular environmental conditions. But the correlation between genetic variation and the opportunity for evolution is inturitively obvious. The greater the number of variable loci and the more alleles there are at each variable locus, the greater the possibility for change in the frequency of some alleles at the expense of others. The requires, of course, that there be selection favouring the change of some trait(s) and that the variation be relevant for the trait(s) being selected.

Two Models of Population Structure

Two conflicting hypotheses were advanced during the 1940s and 1950s concerning the genetic structure of population. The classical model argues that there is many little genetic variations; the balance model, that there is a great deal.

Classical Model

According to the *classical model*, the gene pool of a population consists, at the great majority of loci, of a wild-type allele with a frequency very close to 1, plus a few deleterious alleles arisen by mutation but kept at very low frequency by natural selection. A typical individual would be homozygous for the wild-type allele at nearly every locus, but at a few loci it would be heterozygous for the wild allele and a mutant. The "normal", ideal genotype would be an individual homozygous for the wild-type allele at every locus. Evolution would occur because occasionally a beneficial allele arises by mutation. The beneficial mutant would gradually increase in frequency by natural selection and become of the new wild-type allele, with the former wild-type allele being eliminated or reduced to a very low frequency.

Balance Model

According to the *balance model,* there is often no single wild-type allele. Rather, at many—perhaps most—loci, the gene pool consists of an array of alleles with various frequencies. Hence individuals are heterozygous at a large proportion of these loci. There is no single "normal" or ideal genotype; instead, populations consist of arrays of genotypes that differ from one another at many loci but are satisfactorily adapted to most environments encountered by the populations.

The balance model sees evolution as a process of gradual change in the frequencies and kinds of alleles at many loci. Alleles do not act in isolation; rather, the fitness conferred by one allele depends on which other alleles exist in the genotype. The set of alleles present at any one locus is coadapted with the sets of alleles at other loci; hence allelic changes at one locus are accompanied by allelic changes at other loci. However, like the classical model, the balance model accepts that many mutants are unconditionally harmful to their carriers; these deleterious alleles are eliminated or kept at low frequencies by natural selection, but play only a secondary, negative role in evolution.

The Hardy-Weinberg Law

It is indeed rare in any science for the fundamental model on which it is based to be the first to be developed. But such is the case with the formulations set forth independently in 1908 by *Hardy* and by *Weinberg*. They were faced with the question of how particulate Mendelian inheritance alone, in the absence of all other forces except segregation, could maintain genetic variability at a locus. At that time, the competing hypothesis to particulate inheritance was that an offspring was a blend of the mother's and father's genetic constitution. This *blending hypothesis* was alternative to the *Mendelian particulate hypothesis* that had been recently rediscovered (in 1900). The observations of certain experimental mating that supported particulate inheritance suggested that one half of the genetic variability would be lost in each generation. Such an outcome was also predicted by the blending theory. For example, the cross *Aa X Aa* on the average yields one half homozygous progeny. This homozygosity was thought to cause a "decay" in the ability of the progeny to generate genetic variability through segregation in the next generation. The key fallacy of this argument, which was exposed in the model developed by these two men, was that the Mendelian population is a mixture of mating types, so that the matings that contribute to a loss of heterozygosity are compensated for by other matings that generate heterozygotes. *Hardy* and *Weinberg* realized that when all possible mating types involved in the reproduction of the population were considered, the overall ability of the population to maintain segregating alleles (genetic variability) would not "decay" in each generation, as predicted by the blending theory. The H.W. law simply states that the genotype frequencies in the parental generation. The law assumes random mating of parents and Mendelian segregation involved in reproduction is very large. If p is the frequency of one allele (A) and q = (1 - p) is the frequency of a second allele (a) in the parental generation, the H-W law states that the genotype frequencies expected in the offspring are given by the binomial expansion $(p + q)^2$. The result is

$$\text{relative frequency of } AA = p^2$$

$$\text{relative frequency of } Aa = 2pq$$

$$\text{relative frequency of } aa = q^2$$

By definition, $p^2 + 2pq + q^2$ equals one. If the H-W law is an appropriate model for the locus in the population, the progeny will always have *H-W equilibrium* frequencies. The law assumes a static equilibrium. It must be emphasized that factors that could lead to a change in allele or genotype frequencies throughout the life cycle are not represented in the H-W law.

For convenience, we will use *S*, *Y* and *Z* as the observed frequencies of the genotypes *AA*, *Aa*, and *aa*, respectively. A population is said to be in H-W equilibrium if these observed frequencies are equal to the expected H-W frequencies of p^2, $2pq$ and q^2, respectively. In the interpretation of genotype frequencies, it is important to realize that the H-W equilibrium will result if the H-W law is a valid model for the population. But the observation of H-W equilibrium frequencies in a particular population does not prove that the H-W law is the only possible explanation for the reproduction of the population. For example, balancing forces throughout the cycle of reproduction could lead to H-W equilibrium frequencies resulting from a dynamic equilibrium, whereas the H-W law assumes a static equilibrium. In practice, the H-W equilibrium property is used to relate genotype frequencies at some point in the life cycle of a population to allele frequencies at that point. We will next consider the derivation of the static H-W equilibrium caused by the operation of the H-W law.

Consequences of the Hardy-Weinberg Equilibrium

A population that reproduces by random mating, in which no forces are disturbing the allele or genotype frequencies during the life cycle, has a number of properties that are a consequence of the H-W equilibrium. If we can assume that the H-W law is an adequate description of genetic variation, the properties of the H-W equilibrium can also be used to understand allele and genotype frequencies in that population. Before considering the mathematics of how populations might deviate from the H-W equilibrium, we will review the consequences and uses that can be expected for the H-W equilibrium, as we have defined it here.

Prediction of Allele and Genotype Frequencies

The key consequence of the H-W law is the fixed relationship between allele frequencies and genotype frequencies for autosomal loci that define the gene pool of a population in H-W equilibrium; that is,

$$p^2 = X\ (AA)$$
$$2pq = Y\ (Aa)$$
$$q^2 = Z\ (aa)$$

By knowing the allele frequencies we can determine the genotype frequencies, we can determine the allele frequencies in the population. This fundamental binomial property of the frequencies of alleles and genotypes in diploid species may also be used to relate the allele frequencies in one generation to the genotype frequencies in the next generation.

Conservation of Genetic Variability

The static H-W equilibrium will continue indefinitely as long as there are no disturbing influences that alter random mating and as long as the normal Mendelian segregation ratios occur in the offspring of all mating types. Consequently, any genetic variability that exists in the population will be maintained as long as the H-W equilibrium persists.

Restoration of H-W Equilibrium

If the allele frequency is disturbed at any point in the life cycle by the intervention of mutation, drift, selection, or migration, the parental genotype frequencies may not be in H-W equilibrium. But the genotype frequencies in the next generation of parents will return to the H-W equilibrium values after just one generation of random mating if the disturbing forces have been removed from influencing the population.

The Dynamic H-W Equilibrium

When there is a static equilibrium, both the allele and genotype frequencies are predicted by the H-W law. When the allele frequencies are disturbed, the population does not return to the former equilibrium, but seeks a new equilibriun in one generation. But an alternative kind of equilibrium may be more typical in natural populations, in which the allele frequencies remain stable from one generation to next even though the genotype frequencies are disturbed during the life cycle. Most natural populations are likely to experience this kind of dynamic equilibrium because a balance of the forces of mutation, genetic drift, selection, and migration is usually operating in a real population. An idealized population in which these factors do not play a role in reproduction rarely, if ever, exists in nature. Consequently, the consideration of the dynamic equilibrium is important. We will discuss the dynamic equilibrium in more detail later in the chapter, when we consider deviations from the H-W law.

Prediction of Heterozygote Frequencies

The assumption that a static H-W equilibrium holds in a population allows us to make a number of inferences about genotype frequencies from allele frequencies. Figure 24.1 shows a plot of genotype frequencies against allele frequencies for a locus with two alleles. We may note two properties of genotype frequencies as a function of allele frequencies. First, those populations with an H-W equilibrium allele frequency near 0.5 will have higher frequencies of heterozygotes than those populations with allele frequencies of heterozygotes than those populations with allele frequencies nearer zero or one. Maximum heterozygosity is achieved when $p = q = 2pq = 0.5$. as the allele frequency moves away from 0.5, the frequencies of the homozygotes increase. If a population is at an H-W equilibrium near zero or one, we know that it has fewer heterozygotes and less genetic variability than a population with an equilibrium frequency near 0.5. In the range of p from 0.33 to 0.66, heterozygotes will have a higher frequency than either type of homozygote, but when p is less than 0.33 or greater than 0.66, one or the other of the homozygote genotypes will have a higher frequency than heterozygotes.

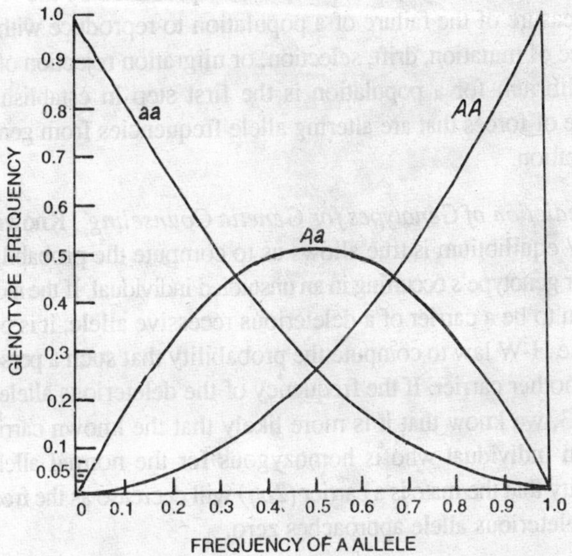

Fig. 24.1. Two-coordinate graph of the three genotypes for a single pair of alleles at one locus as a function of allele frequency, assuming random mating.

A second consequence of the differences in allele frequency between populations is also illustrated in Figure 24.1. We note that as the recessive allele becomes rarer, an increasing proportion of the recessive allele is carried by heterozygotes (Aa) as opposed to recessive homozygotes (aa). This property helps to explain the persistence of recessively inherited diseases in a population. The consequences of this relationship between allele frequencies and genotype frequencies for removing the recessive allele by section will be discussed in the final section of this chapter. We will then demonstrate what is intuitively apparent from namely, that it becomes increasingly difficult to remove a recessive allele from the population as its frequency approaches zero, because it is being carried more frequency by heterozygotes, who are not affected by the selective forces that operate on the recessive homozygotes.

Uses of the Hardy—Weinberg Equilibrium

Testing the H-W Equilibrium : Deviations of the observed genotype frequencies from the expected frequencies predicted by the H-W law are a measure of the failure of a population to reproduce without the influence of mutation, drift, selection, or migration rejection of the H-W equilibrium for a population is the first step in establishing the presence of forces that are altering allele frequencies from generation to generation.

Prediction of Genotypes for Genetic Counseling : Knowing that the H-W equilibtium is true allows us to compute the probability of a particular genotype's occurring in an unstudied individual. If the individual is known to be a carrier of a deleterious recessive allele, it is possible to use the H-W law to compute the probability that such a person will marry another carrier. If the frequency of the deleterious allele is less than 0.33, we know that it is more likely that the known carrier will marry an individual who is homozygous for the normal allele. The probability that the mate is a carrier *(2pq)* will decrease as the frequency of the deleterious allele approaches zero.

Mathematics of Allele Frequency Changes Owing to Deviations from the Hardy-Weinberg Law

Biological evolution depends on the differences in allele frequencies that occur (1) among the breeding units of a species that are isolated in space and (2) over the generations within a unit. The Hardy-Weinberg law describes an ideal population of infinite size in which the allele frequency is constant over time. Under these steady-state conditions, evolution could not proceed. In virtually all species, however, the allele frequency is not constant from generation to generation and from one population to the next. Biological evolution is measured in terms of the change in the hereditary composition of pouplations over time. Genetic change within population (called *anagenesis)* or between populations (called *cladogenesis)* over time is a direct outcome of the deviations of a population from the static expectation of the H-W equilibrium. Differential adaptation of groups of organisms to new environments cannot take place unless the allele frequencies change. Each of the disturbing forces may operate to influence the allele and genotype frequencies for a given locus in a way that is not predicted by the H-W law, to produce patterns or distributions of allele frequencies in time within populations and in space among populations of a species. Consequently, for most loci the H-W law alone is an inadequate mathematical model for predicting the behaviour of the allele frequencies of a population from generation to generation.

Sources of Genetic Variations and Their Impact

Genetic variation is fundamental to the survival of a species. Without inherited variation there is not potential for adapation to a new of changing environments. The evolution of a species depends on the presence of inherited allelic variations that will be adaptive, will result in a higher reproductive fitness, and will subsequently replace the less fit alleles.

As evolution proceeds, the forces of selection tend to reduce the amount of genetic variability. Yet continued evolution depends on a supply of new genetic types. Without a constant renewal of genetic variation, evolution would ultimately grind to a halt, and the species would be at risk for extinction because it could not adapt. It should come as no surprise that the successful populations are those that have an ample storehouse of heritable variations from which to draw as the environment changes.

There are these fundamental sources of genetic variations that are available to a population. They are mutation, recombination, and migration into the population from another population of the species. Because recombination and migration are simply a reorganization of the genotype and a movement of variants, respectively, the ultimate source of all heritable variability is mutation.

Mutation

Although it is the ultimate source of variation, *mutation* is not strong force in changing the geno pool. A mutation rate of a wild type allele to a mutant from on the order of 10^{-5} to 10^{-7} pet generation would require hundreds of thousands of generations for the mutant to replace the wild type.

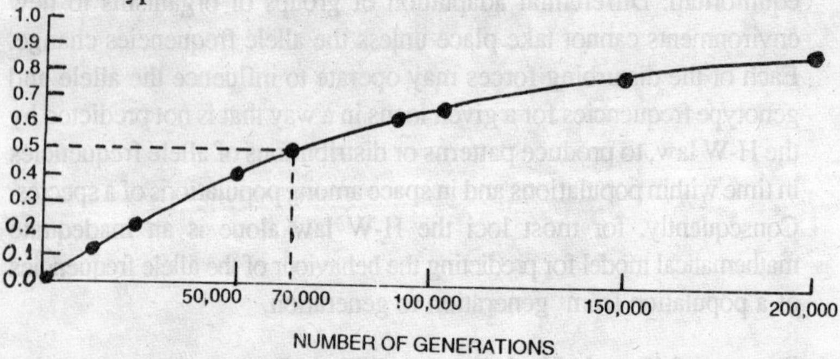

Fig. 24.2. The change with time in the relative of allele, a, caused by recurrent mutation of A to a at the rate of u = 10^{-5} per generation.

Figure 24.2 gives the increase in the relative frequency of mutant *a* as a function of time, when the mution rate (u) is 10^{-5} per generation.

Migration

Mutations in each generation result in changes in a small fraction of the genetic material of a population. Another source of variation is *migration* into the population of an individual who carries an inherited form of a gene, chromosome, or genome that is not already present. This process introduces, rather than creates, a new variation in the population. The initial success of this new variation depends on the success of the individual to interbreed with members of the population. A migrant who enters a population but does not contribute to the gene pool of that population would be ineffective in introducing new genetic variants. The contribution of migration to the introduction of new alleles into a population may vary from about the mutation rate to a substantial contribution, depending on the degree of isolation from other populations.

Selection

The term *selection* is used to include all of the effects occurring during the life cycle that contribute to differential survival of an allele in the reproduction of the population. Selection may occur as differential survival of different alleles during gametogenesis or of genotypes during embryognesis and development. It may also occur as differential *fecundity* (capability to reproduce) of genotypes during the reproductive span of the organism. The sum of all the selective differences among alleles during the reproduction of the population is the fitness value of an allele. The fitness value summarizes the differential contribution of each allele in the parental gene pool to the gene pool of the next generation.

One of the most convincing examples of the power of selection to change the genetic make up of a population was established by *Kettlewell.* He investigated the phenomenon of industrial melanism in the English peppered moth, *Biston betularia.* All samples of this species collected before the mid-1800s were grayish- white with black spotting. These moths typically perch for long periods on light-grayish lichen-covered bark that closely resembles their own colouration. But in some populations of moths vary a different, totally black from occurs that was first identified in Manchester, England, about 1850. Studies have established that the difference in colouration between the two forms is controlled by a single locus with two alleles. The allele that determines the black (or melanic form is dominant to the allele that determines the typical grayish colouration that resembles the lichen-covered bark. It was found that in the smoke-polluted woods around certain English factories, the lichen was absent and the trees had become darkened by soot. In this environment, the dark melanic form was considerably more prevalent than the lighter form. The two types of moths. The moths with the typical lighter colouration were found to be most prevalent in the unpolluted countryside, where lichen covered bark was common. But in the wooded areas where the liches were killed by the industrial soot, the black form was found in highest frequency. Further investigation showed that the selection pressure in this case was prediation by birds. Birds that feed on the moth use vision to locate their prey. As a consequence, melanic forms have a higher survival rate in the industrialized environments and a lower survival rate in the non-polluted areas where lichens create a lighter background. It was hypothesized from this work that it took less than 40 years for the moth to adapt genetically to the industrial revolution by developing the dark colouration.

Genetic drift

The theoretical population consideral at the outset of this chapter to be in *Hardy-Weinberg* equilibrium was of infinite size. Even though natural populations are smaller than that, many of them are large enough for random mating to occur in effect, and genetic equilibrium is maintained through the generations. On the other hand, if the sample of alleles contributing to the zygotes for the next generation is not representative of the overall allelic composition of the population's gene pool, deviations from genetic equilibrium can occur. This will be observed either as chance variations of allele frequencies in the population or possibly as chance fixation of an allele (i.e. P or q = 1) in the population. Such a random change in allele frequencies is called *genetic drfit*. The effect of genetic drift is very small in large population but can be large in a small population. Fixation of one allele or the other becomes more likely the smaller the population is.

In natural population, the risk of loss of an allele from a population as a result of genetic drift is very high for alleles that are present at very low frequencies. This results in fixation of the frequent allele. Clearly, if the rare allele is lost in the sampling process in one generation, it cannot be replaced by a compensating variation of sampling in another generation. Instead, it can only be replaced by mutation. When the occurs, the new allele is again very rare in the population and is again subject to loss through genetic drift. Those alleles that remain in the population may spread through a population as a result of genetic drift, regardless of their selective advantage or disadvantage. Further, in small populations, genetic drift, will action all loci represented by two or more alleles, but the direction and magnitude of the effect is likely to be different at each locus.

There can be other effects of population size on allele frequencies. For example, in human populations, rare alleles are often found in particular areas at relatively high frequencies. The reason for this is that small isolates (self-contained breeding units) within a larger population are particularly susceptible genetic drift. A good illustration of this is found in the Dunkerys' population of eastern Pennyslvania, USA. The Dunkers are a small religious sect who are descended from 28 West German immigrants who arrived in the United States over 250 years ago. The current population consists of about 300 individuals, of whom about 90 are parents in each generation. This is undoubtedly a very small breeding population and one which is isolated from other populations because of their religious beliefs. From studies of a variety of genetic traits, it has been concluded that the frequencies of some genes are very different in the Dunker population compared with either the United States or West German population. This illustrates what is called the *isolate effect or founder effect;* that is, the present allele frequencies reflect a chance sampling of alleles in the original immigrants and pairings occurring only within the population.

Conclusions

We have seen in this chapter a little of how genes in populations are distributed and the effects of various factors, such as mutation, selection, and migration, on the gene pool of a population. All of the genes of an organism, in association with the environment, are responsible for the phenotype of that organism. Therefore, as the environment changes, different combinations of gene frequencies in the population will result over many generations by the forces described. This is the process of evolution, at least in simple terms. However, evolution is an extremely complex process in which numerous factors are intertwined, and a lot remains to be learned about it. This chapter has presented a simplified view of the genetics of population and it is hoped that the reader will extrapolate the basic concepts that have been discussed to natural populations and to the evolutionary process.

REVIEW QUESTIONS

1. In man the two genotypes D/D and D/d are Rh⁺ and d/d is Rh⁻. The frequency of Rh⁻ people in the population is about 16/100. What proportion of the children in a series of families in which one parent is Rh+ and the other Rh⁻ would be expected to be Rh⁻? What proportion of the children from

a series where both parents are Rh^+ would be expected to be Rh-?

2. Island A in the South Pacific has one per cent albinos in the population. Island B has 4 per cent. Recall that albinos result from a homozygous recessive gene, and assume that both populations are of equal size and have married at random with respect to the albino gene. Due to a geological upheaval these two islands become connected by an isthmus, and the two populations become a single randomly mating unit. What will be the frequency of the albino gene in the combined population? What proportion of the combined population will be albino next generation? How does this compare with the average frequency of albinos in the two populations before the merger? (If you like algebra, prove that it will always be true that combining two populations will give a lower incidence of homozygous recessives than the average of the two original populations.)
3. In purebred Holstein cattle about one calf in 100 is spotted with red rather than black, red being due to a recessive gene. What is the proportion of red genes in the population? What proportion of the black population is homozygous and what heterozygous?
4. In humans about one man in 20 is colour blind, colour blindness being due to sex-linked recessive factor. What proportion of women would be expected to be colour blind?
5. Among 6000 shorthorns chosen at random from the herdbook, 516 were white, 2856 were red, and 2628 roan. Are these roughly consistent with the assumption that roan is due to the heterozygous condition and that mating has been at random for this pair of genes?
6. Using the data in problems 4 and 5, and knowing that the frequency of the M factor is .56, compute the frequency of persons in the population that are O, MN, and Rh^+.
7. In humans baldness is due to a dominant gene that is expressed only in males; females are non-bald, irrespective of genotype. Assume that the frequency of the gene for

baldness is P, and that marriage is random with respect to this gene.

a. What proportion of the sons would be bald from a large series of families where the father was non-bald?

b. What proportion of the sons would be non-bald from a large series of families where the father was bald?

8. The R locus controls the production of one system of antigens on the red blood cells of humans. The dominant allele results in Rh⁻ positive individuals, whereas the homozygous recessive condition results in Rh-negative individuals. Consider a population in which 85% of the people are Rh-positive. Assuming the population to be at equilibrium, what is the gene frequency of alleles at this locus?

9. An isolated field of corn was found to be segregating for yellow and white endosperm. Yellow is governed by a dominant allele and white by its recessive allele. A random sample of 1000 kernels revealed that 910 were yellow. Find the allelic frequency estimates for this population.

10. Gene A governs purple stem and its recessive allele a produces green stem in tomatoes; C governs cut-leaf and c produces potato-leaf. If th;e observations of phenotypes in a sample from a tomato population were 204 purple, cut : 194 purple, potato : 102 green, cut : 100 green, potato, determine the frequency of (a) the cut allele, (b) the allele for green stem.

11. The ability of certain people to taste a chemical called PTC is governed by a dominant allele T, and the inability to taste PTC by its recessive allele t. If 24% of a population is homozygous taster and 40% is heterozygous taster, what is the frequency of t? Hint. Use the same methods as that employed for codominant alleles for greatest accuracy.

12. The M.N blood group system in man is governed by a pair of codominant alleles (L^M and L^N). A sample of 208 Bedouins in the Syrian Desert was tested for the presence of the M and N antigens and found to contain 119 group M ($L^M L^M$),

76 group MN (L^ML^N) and 13 group N (L^NL^N). (a) Calculate the gene frequencies of L^M and L^N. (b) If the frequency of L^M = 0.3, how many individuals in a sample of size 500 would be expected to belong to group MN?

13. A population of soybeans is segregating for the colours golden, light green, and dark green produced by the codominant genotypes C^GC^G,C^GC^D,C^DC^D respectively. A sample from this population contained 2 golden, 36 light green and 162 dark green. Determine the frequencies of the alleles C^G and C^D.

ANSWERS

1. 5/7. 4/49.
2. .15. .0225. Before merger, .025.
3. 1/10. 81/99 homozygous, 18/99 heterozygous.
4. 1/400.
5. Roughly consistent. (However, if you do a chi- square test you will see that there is small, but statistically significant excess of roans).
6. 18.3%.
7. a. p.
   b. $q^2/(1+q)$.
8. R = 0.613, r = 0.387.
9. Y = 0.7, y = 0.3.
10. (a) C = 0.30, (b) a = 0.58.
11. t = 0.56.
12. (a) L^M = 75.5%, L^N = 24.5% (b) 210.
13. C^D = 0.9, C^G = 0.1.